Aural Architecture in Byzantium

Emerging from the challenge to reconstruct sonic and spatial experiences of the deep past, this multidisciplinary collection of ten essays explores the intersection of liturgy, acoustics, and art in the churches of Constantinople, Jerusalem, Rome, and Armenia, and reflects on the role digital technology can play in re-creating aspects of the sensually rich performance of the divine word. Engaging the material fabric of the buildings in relationship to the liturgical ritual, the book studies the structure of the rite, revealing the important role chant plays in it, and confronts both the acoustics of the physical spaces and the hermeneutic system of reception of the religious services. By then drawing on audio software modelling tools in order to reproduce some of the visual and aural aspects of these multi-sensory public rituals, it inaugurates a synthetic approach to the study of the pre-modern sacred space, which bridges humanities with exact sciences. The result is a rich contribution to the growing discipline of sound studies and an innovative convergence of the medieval and the digital.

Bissera V. Pentcheva is professor of medieval art at Stanford University, USA. She has published three books to date: *Icons and Power: The Mother of God in Byzantium*, *The Sensual Icon: Space, Ritual, and the Senses in Byzantium*, and *Hagia Sophia: Sound, Space, and Spirit in Byzantium*. Her articles on phenomenology and aesthetics of medieval art have appeared in *The Art Bulletin*, *Gesta*, *Speculum*, *RES Journal of Anthropology and Aesthetics*, *Performance Research International*, and *Dumbarton Oaks Papers*.

Aural Architecture in Byzantium

Music, Acoustics, and Ritual

Edited by
Bissera V. Pentcheva

Routledge
Taylor & Francis Group
LONDON AND NEW YORK

First published 2018 by Routledge

2 Park Square, Milton Park, Abingdon, Oxfordshire OX14 4RN
52 Vanderbilt Avenue, New York, NY 10017

Routledge is an imprint of the Taylor & Francis Group, an informa business

First issued in paperback 2019

British Library Cataloguing in Publication Data
A catalogue record for this book is available from the British Library

Library of Congress Cataloging in Publication Data
Names: Pentcheva, Bissera V., editor.
Title: Aural architecture in Byzantium : music, acoustics, and ritual / edited by Bissera V. Pentcheva.
Description: Abingdon, Oxon ; New York, NY : Routledge, 2017. | Includes bibliographical references and index.
Identifiers: LCCN 2016053632| ISBN 9781472485151 (hardback) | ISBN 9781315203058 (ebook)
Subjects: LCSH: Architectural acoustics. | Architecture, Byzantine. | Music, Byzantine. | Orthodox Eastern Church—Liturgy.
Classification: LCC NA2800 .A873 2017 | DDC 723/.2—dc23
LC record available at https://lccn.loc.gov/2016053632

ISBN: 978-1-4724-8515-1 (hbk)
ISBN: 978-0-367-23184-2 (pbk)

Typeset in Times New Roman
by Florence Production Ltd, Stoodleigh, Devon, UK

Contents

Figures

Contributors

Jonathan S. Abel is consulting professor at the Center for Computer Research in Music and Acoustics in the Department of Music at Stanford University, where he explores audio and music applications of signal and array processing.

Ravinder S. Binning is a PhD candidate in medieval art history at Stanford University as well as a Mellon fellow in the Cathedral Program, Santiago de Compostela, Spain and the 2016–2019 Paul Mellon predoctoral fellow at the Center for Advanced Study in the Visual Arts at the National Gallery, Washington, DC. His current research focuses on the relationship between fear, art, and architecture.

Peter Jeffery is the Michael P. Grace II Chair of Medieval Studies at the University of Notre Dame, where he is also professor of music, concurrent professor of theology, and concurrent professor of anthropology. He is simultaneously the Scheide Professor of Music History at Princeton University. He has published many books and articles on the medieval eastern and western traditions of Christian chant and liturgy. He is currently working on a book tentatively entitled "Religious and Civic Ritual in Eighth-Century Rome: Ordo Romanus Primus."

Christina Maranci is the Arthur H. Dadian and Ara T. Oztemel Chair of Armenian Art and Architectural History at Tufts University. Her books include *Medieval Armenian Architecture: Constructions of Race and Nation* (Peeters, 2001), and *Vigilant Powers: Three Churches of Early Medieval Armenia* (Brepols, 2015).

Bissera V. Pentcheva is professor of medieval art at Stanford University. Her books include *Icons and Power: The Mother of God in Byzantium* (Pennsylvania State University Press, 2006); *The Sensual Icon: Space, Ritual and the Senses in Byzantium* (Pennsylvania State University Press, 2010) and *Hagia Sophia: Sound, Space, and Spirit in Byzantium* (Pennsylvania State University Press, 2017).

Walter D. Ray is associate professor in library affairs at Southern Illinois University Carbondale. His primary area of research is early and Eastern Christian liturgy. He has published studies on early Egyptian Eucharistic prayers and on the early Jerusalem liturgical calendar, and a recent book on Byzantine liturgy, *Tasting Heaven on Earth: Worship in Sixth-Century Constantinople* (Eerdmans, 2012).

Laura Steenberge is a performer–composer who recently completed a DMA in music composition at Stanford University. As a researcher she studies language, the voice, harmony, and acoustics. Her creative work engages with questions about the origins and transformations of instruments, songs, myths, and science through the ages.

Lora Webb is a PhD candidate in medieval art history at Stanford University. In addition to architecture, her current research centers on treasury arts: understanding how they were made, used, traveled, and were altered over time.

Ruth Webb is professor of Greek language and literature at the Université Lille 3 (France). She has published widely on the use of appeals to the visual imagination in Ancient Greek and Byzantine literature and rhetoric as well as on theatre and dance in Late Antiquity and Byzantium.

Christian Troelsgård is associate professor of medieval Greek and Latin at University of Copenhagen. He is executive secretary of the international editorial project Monumenta Musicae Byzantinae. His books include *Byzantine Neumes, A New Introduction to the Middle Byzantine Musical Notation* (Museum Tusculanum Press, 2011) and (ed.) *Chants of the Byzantine Rite: The Italo-Albanian Tradition in Sicily* (Museum Tusculanum Press, 2016).

Kurt James Werner is a lecturer in audio at the Sonic Arts Research Center (SARC), Queen's University, Belfast. His work focuses on theoretical aspects of wave digital filters, computer modeling of circuit-bent instruments, virtual analog and physical modeling more broadly, and the history of music technology. He obtained a PhD in 2016 in computer-based music theory and acoustics at Stanford University's Center for Computer Research in Music and Acoustics (CCRMA).

Wieslaw Woszczyk is James McGill Professor in the Department of Music Research at the Schulich School of Music of McGill University. He is the founding director of the Graduate Program in Sound Recording and of the Centre for Interdisciplinary Research in Music Media and Technology (CIRMMT), and directs the Virtual Acoustics Technology Laboratory at McGill. He served as president of the Audio Engineering Society and chair of its technical council.

Acknowledgments

I owe a debt of gratitude to Dr. Maria Sereti and the Onassis Foundation (USA), who generously sponsored this seminar; to all the authors and participants of this project; and to Lory Frankel and Fronia Simpson for their amazing copyediting work. I am also beholden to the graduate students who helped me in the last stage of turning this material into a book: Ravinder Binning, Daniel Smith, Laura Steenberge, Lora Webb, and Erik Yingling. A debt of gratitude is owed to the Stanford Global Studies Program, which offered the infrastructure for the Onassis seminar, and, in particular, the director, Prof. Norman Naimark, and staff members Jane Stahl and Mark Rapacz, who created the Website. I also thank Mike Seely, who filmed and edited the videos of the sessions.

Introduction

Bissera V. Pentcheva

This book emerged from the challenge to reconstruct sonic and spatial experiences of the deep past and was realized through cross-disciplinary collaboration and new research and experimentation. It invites readers into the complex world of the Byzantine liturgy, experienced in its chanted form in interiors covered with monumental mosaics and frescoes. The essays that compose it draw on diverse approaches informed by sound studies, phenomenology, art history, musicology, liturgy, affect studies, digital technology, and computer modeling.

Aural Architecture in Byzantium has two main goals: first, to explore the intersection of Byzantine liturgy, music, acoustics, and architecture, and second, to reflect on the role digital technology can play in re-creating aspects of the sensually rich performance of the divine word. It also intends to open rather specialized discourses to a larger intellectual community interested in the intersection of art, ritual, and sound studies.

By reconstructing aspects of the visual and acoustic conditions of the Byzantine liturgy, this array of studies pushes the digital and the medieval to converge in the meeting between imagistic evidence and computer synthesis. Digital technology has enabled the re-creation of some of these aural environments and transformed them into instruments for modern performers to inhabit and play. Further, the use of live **auralizations** has restored the possibility for some of the historical buildings to recuperate their lost voice. This is particularly the case with Hagia Sophia, the cathedral of Constantinople, which as a museum today lives under a strict interdiction banning any vocal or instrumental performance in its interior. Yet by means of auralizations, modern choristers and audiences can interact live with this extremely reverberant chamber.

Sound studies and Byzantium

Aural Architecture in Byzantium responds to new trends in humanities research particularly informed by the emergence of sound studies. Sound studies have set their research focus outside the strict domain of musicology, defending the premise that there is no set epistemology of what constitutes sound and that there are many competing systems of knowledge to grapple with the sonic.[1] The concept of soundscape, introduced by Murray Schafer, plays an important role in this new field. He defined it as an acoustic environment and sonic ecology.[2]

More to the point of this research on the aural architecture of Byzantium is Emily Thompson's redefinition of soundscape as a complex construct of culture, ideology, and technology. Thompson has argued that soundscape, like landscape, is shaped by the interaction of multiple societal forces. Her study traces the formation and development of the science of acoustics in the early twentieth century. She then considers how sound technology in that period changed the culture of listening, introducing the expectation of clear, intelligible sound (known in the audio field as "dry" sound), as well as control over and ultimate suppression of reverberation (which produces a "wet" sound).[3] Against what Thompson defines as the early twentieth-century negative attitude to wet sound, Byzantium presents a strong contrast. The acoustics of the Great Church, Hagia Sophia, are very reverberant; as a result, they are not suited for the delivery of intelligible speech, but support well the singing of monody.[4]

Research in the architecture and acoustics of the distant past has also given rise to the new field of archaeoacoustics. Chris Scarre and Graeme Lawson's edited volume contains a collection of essays on multiple case studies, all exploring the role of sound in shaping human behavior and in determining the human use of places and the construction of buildings. The editors and some of the contributors to this volume aim to detect intention between acoustics and the design of buildings. Scarre proposes the study of two specific principles—recurrent patterning and closeness of fit—that would help a researcher determine a causal relationship between acoustics and design.[5] Further support is sought in the way data drawn from acoustic measurements of ancient monuments can be converted into what the authors call an "admissible" evidence of behavioral connection.[6]

Eschewing this more deterministic approach, Barry Blesser and Linda-Ruth Salter's *Spaces Speak, Are You Listening? Experiencing Aural Architecture* presents a larger historical survey on the interconnections among acoustics, architecture, and technology.[7] As the electrical engineer who invented the first commercial digital reverberation system, Blesser offers an important perspective on the development of sound processing technology that led to the current widespread adoption of computational methods over analog systems in the production of virtual aural spaces. For instance, Stanford's "Icons of Sound" live auralization of Byzantine chant addressed in this book stands as an example of the use of computational artificial reverberation.[8]

Blesser and Salter introduced two very useful terms—"aural architecture" and "earcon". Aural architecture identifies those features of a building that can be perceived by the act of listening in them.[9] Since so much in the field of acoustics, especially of the premodern period, depends on the ability of the researcher to address the question of intentionality—that is, was the building designed to have the reverberation it exhibits—and since the textual evidence is so reticent in providing any hard facts, Blesser and Salter have given us a chance to circumvent the direct answer of this question and pursue instead a far more productive line of investigation. To do that, the two authors coined the neologism "earcon." Conceptual rather than material, "earcon" identifies the imprint that the prolonged exposure to the acoustics of a particular space leaves in the memory and consciousness of the participant. This aural matrix is further molded by the societal forces of culture, religion, and politics. Thus, rather than asking whether a building such as Hagia Sophia was designed to have reverberant acoustics, earcon allows us to ask how the prolonged exposure to its resonant acoustics inflected the Byzantine perception of the divine and how this perception was reflected in the liturgy and informed the content of what was being recited and sung in its cathedral service.[10]

Drawing on these diverse recent investigations but centering on a specific urban site, Deborah Howard and Laura Moretti's *Sound & Space in Renaissance Venice* has pursued an interdisciplinary approach to architecture, music, and acoustics.[11] Their book explores the interrelations among three entities: the sixteenth-century ecclesiastical architecture of Jacopo Sansovino and Andrea Palladio; acoustic theory derived from Vitruvius and Leon Battista Alberti; and choral polyphony introduced in Venice by the doge Andrea Gritti (r. 1523–38). *Sound & Space* ties in the humanities research with the evidence of the acoustic measurements of twelve Venetian churches gathered in 2006, the in situ recording of St. John's College Choir, Cambridge, conducted in April 2007, and the subjective response of the audience and choristers to the acoustics of these interiors. The project itself called for multifaceted research involving specialists in the humanities and the sciences, choir directors, and sizable audiences. The swiftness and efficiency in the execution of the project and its timely publication attest above all to the organization and team building efforts of Howard and Moretti. *Sound & Space in Renaissance Venice* has provided a model for the variety of research venues that sound studies can pursue. This has informed the selection of fields and contributors to the Onassis Seminar on Aural Architecture at Stanford and resulted in the final lineup of essays in this volume, ranging from architectural history, musicology, and mystagogy (liturgical exegesis) to an analysis of acoustics and auralizations.

In addition to the rich contribution of sound studies, *Aural Architecture in Byzantium* draws on recent titles in the field of premodern architecture. Bonna Wescott and Robert Ousterhout's *Architecture of the Sacred: Space, Ritual, and Experience from Classical Greece to Byzantium* offers an informative array of essays exploring the architectural staging of pagan and Christian rituals.[12] Vasileios Marinis's *Architecture and Ritual in the Churches of Constantinople: Ninth to the Fifteenth Century* sharpens the focus on the monastic rite and how elements of it fused with the cathedral rite as a new hybrid service evolved in some of Constantinople's aristocratic monastic foundations.[13]

The Onassis Seminar on aural architecture

This investigation into Byzantium's soundscapes began with the interdisciplinary project "Icons of Sound," which I codirected with the electrical engineer Jonathan Abel at Stanford University. "Icons of Sound" explores the aesthetics and acoustics of Byzantium's Great Church, Hagia Sophia.[14] The medieval ecclesiastical space—a large interior, covered in reflective and polished surfaces of stone—was reverberant. Reverberation supported and enhanced chanted vocalizations. Yet in the early twentieth century, reverberation acquired a negative connotation, defined as noise. Acoustically engineered building materials were used in the interiors in order to dampen the reverberation and increase the intelligibility of speech vocalizations.[15] More recent technological developments have reintroduced the aesthetic importance of "wet" acoustics and made it interactive and mobile; this is the direction that "Icons of Sound" has pursued through a concert at Stanford's Bing Hall in 2013 and the aesthetic act of performance.

Using acoustic data of recordings of popping balloons that I gathered in Hagia Sophia in 2010, Abel worked out a method for live auralization. In this process, he imprinted live the sonic signature of Hagia Sophia on the performance given by Cappella Romana, a vocal ensemble, of Byzantine chant at Bing Hall in 2013.[16] The contemporary audience was thus enveloped in the resonant sound of the Great Church.[17]

Many questions arose from this live auralization: Were medieval buildings designed intentionally to produce long **reverberation times**? Did the semantics of the cathedral liturgy engage the wet sound of the reverberant interiors? Was the decorative program designed to interact with the resonant acoustics? How close can modern performance and the use of digital technology bring us to the medieval sensual experience?

The Bing concert inspired the creation of a year-long seminar, "Aural Architecture: Music, Acoustics and Ritual in Byzantium," at Stanford University, generously sponsored by the Onassis Foundation (USA) in 2013 and 2014. It was intended to address some of these questions by developing a methodological framework for an interdisciplinary inquiry into music, acoustics, and ritual in the medieval sacred space.[18] The individual papers featured in this volume began as presentations and discussions with experts and students from a variety of different fields. The sessions were recorded, and edited short videos were posted on our Website.[19]

Alexander Lingas, a Byzantine musicologist and the artistic director of Cappella Romana, opened the Onassis series with a presentation on the Byzantine cathedral office and its interaction with the rite of Jerusalem. Lingas also detailed the challenges for the modern performance of Byzantine chant. Peter Jeffery examined the formation of the cathedral liturgy in Jerusalem, Rome, Constantinople, and Alexandria. Vasileios Marinis focused on the relation of liturgy and architectural design in the Middle Byzantine period. Expanding beyond Constantinople, Christina Maranci introduced Armenian architecture and its investment in the liturgy of Jerusalem. Christian Troelsgård addressed the evolution of Byzantine musical notation and the problems of transcribing and interpreting it. Steven Hawkes-Teeples introduced the mystagogical treatises of Saint Symeon of Thessaloniki (d. 1429) as a synthesis of the Byzantine exegesis of the liturgy. Ruth Webb enlarged on the Byzantine ekphrasis of sacred space, foregrounding the role of agency and embodiment. To this sphere of historical and humanistic research, the Onassis seminar added two studies centered on the acoustics of Hagia Sophia and the digital technology used to model it: Wieslaw Woszczyk shared the extensive survey and measurements of the acoustics of Hagia Sophia he carried out in 2013, while Jonathan Abel put into play his innovative method of live auralization employed for the Cappella Romana's concert in Bing Hall in 2013.

Synthesis of the ten essays

Collectively, the ten essays featured in this book aim to reveal connections among the different lines of research pursued by the seminar. The insights emerging from this assembly of diverse approaches invites us to consider the study of architecture, music, and liturgy as an integrated phenomenon. Musicologist Peter Jeffery opens the collection of essays with a discussion of the rise of the cathedral rite in Jerusalem, Rome, Constantinople, and Alexandria. His analysis draws on the evidence of surviving service books and the scant archaeological evidence of ecclesiastical buildings and their movable furniture. While occurring at different periods, a shift from the cathedral **stational liturgy** (whose characteristics include ostentatious sung office and urban processions) to the monastic rite (which does not originally put an emphasis on chant but requires the daily recitation of all the hundred and fifty psalms) takes place in all four centers. The discussion of Rome is the richest. Here Jeffery furnishes evidence about the eighth-century vespers service for Easter week, which preserves traces of the earlier nonmonastic rite. The melodies for the service include melismatic alleluias with verses in Latin and Greek and singing that alternates between a boys' and men's choirs. It is these elaborate alleluias that bear resemblance to the cathedral chant of Constantinople.

Yet, could there be traces here of the cathedral rite of Jerusalem? In this period, when immigrants from the East streamed into Rome after the fall of Jerusalem in 638, the connection with the Holy Land was very strong. The music of the papal chapel of the eighth and ninth centuries could thus be viewed as one more artistic **mode** expressive of a new synthesis of Eastern and Western traditions that Rome produced in this period. Some of the visual equivalents include the introduction of the iconographies of the Anastasis, echoing developments in the liturgy of Jerusalem (S. Maria Antica), or the invention of cloisonné enamel (the cross of Pope Paschal, r. 817–24), or the rise of illustrations that function as visual commentaries placed in the margins of Greek manuscripts of the Psalter.[20] The visual and sonic are two modes through which the divine is imagined and presented in these works of art. It will be important to investigate in the future how the shaping of the melodic contours of the services interacted with and inflected the form and emotional expressivity of the visual programs.

Next, art historian Christina Maranci formulates and considers a compelling question: Can we gain insight into the medieval experience of ecclesiastical spaces and their symbolic meanings by mobilizing the evidence of both surviving liturgical texts and church facades? This research sharpens our understanding of the extraordinary role Jerusalem played in shaping the spiritual imaginary of Armenia. Taking three seventh-century churches at Mastara, Zuart'noc', and Mren, Maranci carefully explores their physical setting, epigraphy, and relief sculpture. The strong "exteriority" of Armenian monuments dictated this emphasis on the facade, which became the visual focus of the churches' consecration ceremonies. Maranci correlates the material evidence of the facades with the ceremonies performed outside and inside these buildings; recuperated liturgical texts of the consecration rite; hymns (or *šarakans*) of the Holy Cross; and the fifth-century Armenian Lectionary. The combination of liturgical texts and archaeological evidence uncovers the complex encounter with the churches in these ceremonies, which are inherently kinetic and multisensory, requiring climbing, carrying, singing, circumambulating, and smelling. (Aspects of this liturgically driven circumambulatory kinesthesis is addressed in Ruth Webb's study of *periēgēsis* [tour around] structuring the Byzantine texts that describe church buildings.) In the process, the participants could transcend sensorially spatial distances, partaking in the nearness of both a terrestrial hagiopolite (Jerusalem) site, which was sometimes ascertained through pilgrimage, and an imagined celestial realm. Maranci interrogates how the language of the liturgical texts and their performance in procession inflected, if for a few moments, the visual imagery of the church and offered a powerful, but heretofore neglected, aural dimension to the church encounter.

Christian Troelsgård, a Byzantinist philologist, musicologist, and the author of *Byzantine Neumes: Byzantine Musical Notation*, concentrates on the transmission of Byzantine chant, exploring how it was shaped by three factors: oral (performance practices), written (musical notation), and aural (architectural setting).[21] He argues that the written notation was just a technological vehicle for transmission and not the leading factor in determining the contours of Byzantine chant. In the early period, only a small portion of the hymnography of the daily offices was recorded in a notation known as "partial" or "Paleo-Byzantine," which is not intervallic, and most of the evidence suggests that this written record served chiefly a didactic rather than performance function. By the late twelfth and thirteenth centuries a diastematic system developed, today identified as "the Middle Byzantine notation." It was used to record the melodies of some of the elaborate settings for the cathedral rite of Hagia Sophia. This repertoire is featured in the ***asmatikon*** (music for the elite choir of the Great Church) and the ***psaltikon*** (music for the soloist) manuscripts. In these collections we learn more about the structure of some of the chants, as, for instance, the *prokeimenon*, or gradual. Troelsgård draws our attention to how its cadence typically uses sets of double gammas (for instance, γγο- γγο-). His research includes an auralization of such a *prokeimenon* in an acoustic model of Hagia Sophia, produced by the team of Jens Holger Rindel from the Technical University of Denmark, showing how the double gammas elicit a specific tonal area effect; the acoustics of the Great Church frame the pitch by adding tones below and above, giving emphasis on the dominant sung tone.

The use of auralizations further strengthens Troelsgård's point that the Byzantine musically notated manuscripts are not equivalent to modern scores. Instead, a significant portion of Byzantine chant relied on the embodied memory of performance tradition and its "earcon" produced by prolonged exposure to listening in a particular acoustic environment. Thus, in order to "reconstruct" Byzantine chant, we need to "recompose" it according to the melodic formulas derived from the medieval sources in the diastematic Middle Byzantine musical notation and from a model system (largely preserved in the chant manuscripts known as *heirmologia*) and unique melodies gathered in the ***sticheraria***. The goal is to situate this recomposition in one of the surviving medieval churches. This process of recuperating Byzantine chant demonstrates the complexity of the modern process from transcription to performance.

The auralizations Troelsgård participated in were all done postproduction, using a virtual model created with the software Odeon, which was in turn designed on the basis of on-site measurements in Hagia Sophia gathered by the Danish–Turkish project Conservation of Acoustical Heritage by the Revival and Identification of Sinan's Mosque Acoustics (CAHRISMA, 2000–03). This meant that the performers could not interact live with the modeled acoustics of the building. "Icons of Sound" innovated by offering an aural venue for chanters to sing live in a virtual Hagia Sophia, a topic addressed by Jonathan Abel and Kurt Werner at the end of this volume.

Byzantine chant was one of the elements constituting the liturgical service, leading us to ask, What was the symbolism of the liturgy that the music helped to articulate? The interpretation of the semantics of the liturgy in Byzantium features in a special genre of texts, known as *mystagogiai*. Walter Ray's essay explores next this written tradition. Starting with the treatises of Theodore of Mopsuestia (d. 428) and Pseudo-Dionysius (fl. ca. 500), his essay traces echoes of these formative texts in the influential treatises of five Byzantine theologians, Maximus the Confessor, Patriarch Germanus, Nicholas and Theodore of Andida, Nicholas Cabasilas, and Symeon of Thessaloniki, spanning the mid-sixth to the fifteenth century. Attention is further placed on the meanings attached to the different spaces of the church and its altar and ambo.

Mystagogy attempts to explain how the liturgy confronts the divide between God and humans. Two theological positions emerged in Byzantium: one articulated by Pseudo-Dionysius and the other by Theodore of Mopsuestia. Pseudo-Dionysius defends a position known as "anagogical"; while it was produced slightly after Theodore's mystagogy, it is in fact older and closer to the perspective of the Early Christians. The anagogical model maintains that the liturgy can bridge the gap between the faithful and God in the way that it envisions how the terrestrial celebration mirrors the celestial liturgy, and thus draws the latter near. This enables humanity to be lifted toward a union with the divine in a kind of temporal *theōsis* (deification). The anagogical model is present most prominently in the mystagogy of Maximus the Confessor and Saint

Symeon of Thessaloniki, while that of Patriarch Germanus balances this vision of heavenly worship with the "life-of-Christ" model.

It is the anagogical exegesis that bears particular significance for the rituals in the Justinianic Hagia Sophia. The basilica's great dome articulates plastically this anagogical pull as a visual ascent, attracting the gaze to trace the vertical axis toward its luminous apex. At the same time, the acoustics of the cupola produce the effect of an aural waterfall, thus bringing "near" the bright disembodied voice of reflected sound waves. Similarly, the cathedral chant, characterized by its long **melismas** and the intercalation of nonsemantic syllables, expresses a tendency to push against the register of human speech toward an imagined celestial, cosmic sound.[22]

The second Byzantine mystagogical vision is identified as "historical"; it was articulated by Theodore of Mopsuestia and developed by several of the Byzantine mystagogy writers, including Patriarch Germanus, Nicholas and Theodore of Andida, and Nicholas Cabasilas, as well as Symeon of Thessaloniki. This "historical" model perceives that the gap between human and divine is unbridgeable in time. Such a union with the divine can only be a future expectation. Not until Christ had proven faithful even unto death did the union between Jesus and the Word become permanent. Humanity, which needs to imitate Jesus's historical actions, can hope to attain the closeness Christ now enjoys with the Lord only after the resurrection. This leads to the envisioning of the Eucharistic liturgy as an enactment of Christ's Passion.

If the "anagogical" model of liturgical exegesis corresponds to Hagia Sophia's architecture, its grand-scale, nonanthropomorphic decor, and a cathedral chant abounding in melismas and nonsemantic intercalations, the "historical" model, by contrast, responded to the scaling down of Middle Byzantine churches, which resulted in more intimate interiors now covered with figural programs illustrating Christ's terrestrial life. Viewed through the "historical" lens of Byzantine mystagogy, the liturgy becomes a rehearsal for the future expectation to meet Christ as judge at the end of time.

Ravinder Binning's essay addresses one such scenario unfolding in the small church of the Virgin Panagia ("All-Holy") at Trikomo (ca. 1130) on Cyprus. Christ materializes from the apex of the dome, his all-seeing eyes always following the viewer in space, no matter where the latter stands. Enveloped by the gaze from above, the visitors are forced into an uneasy bodily position of craning their necks. As the faithful stand, bare and exposed before the Judge, their conscience is troubled, their fear stirred. The inscription at Trikomo, surrounding the image of the Judge, reinforces this terrifying psychosomatic sensation of being watched and indicted. Fear leads to confession; confession to penance; and penance possibly to forgiveness, precipitating a vivid imagined dialogue unfolding between human and divine. A compelling record of one such intense interaction between a worshipper and the image of Christ (in the dome of the Holy Apostles in Constantinople) comes by way of a tenth-century edifying story, written by the bishop Paul of Monemvasia.

This dialogue is further rehearsed, Binning argues, by new elements emerging in the liturgical rite of the Stoudios Monastery in Constantinople.[23] Theodore of Stoudios (d. 826) steered this monastic reform, which adopted aspects of the Palestinian Sabaitic rite. At the heart of the Sabaitic rite is the desire to articulate in an affective way the presence of Christ to the faithful. But how could this be done in the framework of the existing liturgical texts? Up to this period, scripture and especially the Psalms formed the main body of performed poetry. Yet the Psalms inadvertently presented an Old Testament lens through which to experience this dialogue with Christ in the course of the liturgy. The Sabaitic rite in general, and the Stoudite reformed one in particular, introduced the composition of a new hymnography. The Triodion, a collection of hymns for the Lenten-Paschal period, is among this new poetry, and Binning turns specifically to its canon "On the Second Coming," which is credited to Theodore of Stoudios and performed on the eighth Sunday before Easter, the so-called Sunday of *Apokreas*, or Meatfare.[24] This canon vividly triggers trembling and fear, followed by penance and a plea for mercy. The poetry sung during this specific Sunday influenced the imagined exchange between mortal and divine at the end of time. Although it was performed only annually, the canon "On the Second Coming" acquired a daily reality produced by the spatial position of Christ's image in the dome and the psychosomatic effect it brought about in the faithful standing below.

The previous art historical literature on the image of Christ in the dome concerned itself predominantly with establishing a stable taxonomic designation for it, labeling it the Pantokrator. Yet most of these representations appear without such names. Instead of looking for taxonomies and fixed meanings, Binning finds these images to be works in progress, with semantics and affective power that shift according to the desires, practices, and emotions of the people who confront their visual and aural effects. That of Christ in the dome evokes a scenario of fear and penance resonant with the new liturgical poetry of the Triodion.

In the next essay Lora Webb leads us to consider another such liturgically shaped encounter with the monumental program of Byzantine churches: the Metamorphosis or Transfiguration mosaic in the mid-eleventh-century Monastery of the Theotokos at Nea Moni on the island of Chios. The ceremony it reflects, the Feast of the Transfiguration, celebrated on August 6, is firmly embedded in the cathedral liturgy of Constantinople. Webb turns to three typika—the tenth-century **typikon** of the Great Church, the eleventh-century example of the Theotokos Evergetis Monastery in Constantinople, and the twelfth-century one of Christ the Savior in Messina—in order to reconstruct the format of the liturgy for this feast. Her analysis uncovers how the passages that form the ***prokeimena*** (graduals), alleluias, and the *koinōnikon* (communion verse) for the day are those that are visualized in the scene of the Metamorphosis in the illustrated Byzantine Psalters. The semantics of these texts mold the perception of the image.

Yet, unlike the panoptic image of Christ in the dome, the Transfiguration does not control the apex of the nave in Nea Moni. Instead, it is placed in the southwest conch. In fact, for the duration of the feast, as Webb observes, this mosaic remains at the back of the faithful. The latter turn to face it at the end of the service, after receiving communion. As they ready themselves to leave the church and lock their gaze with the Metamorphosis, they may recognize how the circular shape of the Eucharistic bread they had encountered earlier in the Divine Liturgy coincides with the round mandorla of light from which the transfigured Christ emerges to his witnesses. The possibility of partaking in Christ comes to light from the visual field of the mosaic; the figure of Christ is pushed to the pictorial surface in a way that denies depth, so the representation projects into the physical space that surrounds and enfolds icon and beholder. Like a ladder of ascent, the physical space of the church extends into the pictorial Metamorphosis, lifting the beholder to heavenly heights of Tabor and Hermon (Ps. 88[89]:12–13), a phrase repeated in the *prokeimena* and alleluias sung for the liturgy of the Metamorphosis. The poetics of the visual confirm the aural promise unfolding in the chants, and this expectation becomes fulfilled in the olfactory, tactile, and gustatory act of consuming the Eucharist.

In the next essay, Laura Steenberge sustains our focus on the Eucharist. She studies how Late Byzantine compositions of the **Cheroubikon** hymn use melodic fragments from the **Trisagion** and the Sanctus in order to articulate mystagogy through music. She identifies three such motifs—a processional, a singing, and a descending formula—that shape through music a symbolic form, reminiscent of "text-painting" (the technique of writing music that reflects the literal meaning of a song). For instance, a singing formula, originating in the Trisagion and the Sanctus, is mirrored in the Cheroubika in order to enact the imagined celestial chant that the angels ceaselessly perform before the throne of God.

The quotations of melodic formulas through which the celestial sound of the angels, rooted in the Sanctus and the Trisagion, arises in the terrestrial Cheroubika express the anagogical model first articulated by Pseudo-Dionysius and that majestically unfolded in the Justinianic liturgy of Hagia Sophia.[25] Yet after the end of Iconoclasm in 843, this vision appears to have given way to the increasingly more influential "historical" model. In the Late Byzantine period, Symeon of Thessaloniki endeavored to correct this imbalance by producing a synthesis of the anagogical and historical mystagogies and reaffirming the role of *theōsis*. In order to show the mirroring of the celestial in the terrestrial performance, the Late Byzantine Cheroubika further employ prosody of nonsemantic syllables (***kratemata*** or ***teretismata***) and aspiratory sounds (*ho*, *he*, *ha*). They form musical events not linked to the register of human speech or recognizable lyrics; instead, in their nonsemantic character, they remind us of what in later Western music takes form under the term "leitmotif." Steenberge concludes with how the widespread use of the diastematic musical notation in the Late Byzantine period aids this flourishing of musical compositions, which leads to the development of a repertoire of chants attributed to named contemporary composers.

If the Cheroubikon imprints in its melody the angelic sound of the celestial liturgy, then how was this hymn originally perceived in the space, Hagia Sophia, for which it was created? Ekphraseis of the church building give us access to the medieval experience of aural architecture. The philologist Ruth Webb draws our attention to these texts and explores how they construct a multisensory experience of the space. Recent research on embodiment and agency informs this analysis. Ekphraseis are structured as a "tour around," a *periēgēsis*, that impels the listeners to imagine in their mind's eye being inside the building and encompassed by it. The periegetic literature uses the sensation of spatial envelopment to induce animation. Ekphrasis thus gives us another lens through which to view the same phenomenon of being enclosed that we encountered as the psychosomatic effect of the image of Christ confronting the faithful (as seen in the essays by Binning and Lora Webb). Similarly, this sensation of being encircled has an aural manifestation, as many of the domed churches produce an enveloping acoustics, Hagia Sophia being the most prominent example (as established by Woszczyk's essay in this volume). Envelopment emerges as a multimodal aesthetics structuring the Byzantine ecclesiastical space.

The space under the dome is perceived not as an inert void of air but, in the ekphraseis, as permeated by the energies of the Holy Spirit. The writers of these texts conceived of their creative process and inspiration as connecting to this greater force—*pneuma*—and channeling its energies through their bodies and mouths. In this model of poetic inspiration, I see a reverse mirror of the action of the chanters. By exhaling their energy in the ecclesiastical space, the singers activate its resonant acoustics. Reverberation in turn becomes an aural manifestation of the Holy Spirit as a bodiless voice. In both cases, be it the singers activating *pneuma* as reverberation or the poets channeling *pneuma* as inspiration, the creative act makes the metaphysical sensorially present to the faithful.

In some of the Byzantine ekphraseis, the building is understood as exercising agency over the writer and listener. In the same way, the acoustics modify the voice of the singer. Modern technology complicates this process. On the surface it may seem to reduce these buildings to instruments of the human voice; a space such as Hagia Sophia can be digitally reconstructed and offered to new performers to play in. At the same time, these virtual aural architectures constructed from interactive virtual environments become new sites synthesizing building, voice, and imagination. The next two essays turn to the acoustics of Hagia Sophia and the use of digital technology to model this building's enveloping soundfield. Wieslaw Woszczyk presents the results of his acoustic measurements on-site in 2013 and the potential to build a digital model for modern performances. The essay of Jonathan Abel and Kurt Werner traces the stages in the development of live auralizations produced as part of the Stanford's "Icons of Sound" project. They survey the process from the measuring of room acoustics on the basis of recorded balloons popping in Hagia Sophia to the concert of Cappella Romana in Stanford's Bing Hall in 2013, which was imprinted live with the acoustic signature of Hagia Sophia.

As readers explore the diversity of approaches, often with their own field-specific terminology, they may consult at their convenience a glossary that is placed at the end of this volume. Moreover, all the titles cited by individual articles can be easily found in one location in the bibliography.

It may be wondered why I have not written an essay for *Aural Architecture in Byzantium*. This is largely because my work on Hagia Sophia has been consolidated in a new book, *Hagia Sophia: Sound, Space, and Spirit in Byzantium*, which develops an integrated approach inspired by the Onassis seminar.[26]

Bringing together art and architectural history, liturgy, musicology, and acoustics, *Hagia Sophia: Sound, Space and Spirit* aims to uncover the spiritual and aesthetic principles that structure the sacred space of the Great Church. I argue that the *hieros* acquires presence in the interior of the Great Church in a series of temporal phenomena through two processes: mirroring (*esoptron*) and inspiriting (*empsychōsis*). Mirroring is plastically expressed in the synergy of ambo (the platform from which scripture was recited and from which the elite choir sang) and dome and in the reflective surfaces of the Eucharist chalices and patens. The concave form of the dome inverted in the silver bowls of the liturgical objects shows this intertwining, or *perigraphē*, of the macro- and microscales. The second process, inspiriting, is activated in the consecration ceremony and repeatedly enacted in the Eucharist ceremony. Its traces are present in the chiastic structure of some of the psalms and in the prosody of aspiratory syllables chanted in the service.

When Hagion Pneuma infuses matter, the inert becomes alive, and for the Byzantines this transformation constitutes a form of "performative" iconicity. A nonrepresentational image, it is produced through breath (*pneuma*) and is linked to the way the Lord inspirits the body of Adam (Gen. 2:7), transforming him into an image of the God, or *eikōn tou Theou*. I connect the performative iconicity to the nonfigural decorative program of the Justinianic church and to chant and reverberation: just as vital inbreathing makes the inanimate animate, so, too, the breath exhaled in song activates the acoustic return of the building, producing the reverberant and enveloping soundfield of the Great Church and thus activating a form of call and response between animate (singer) and inanimate (building). The Justinianic Great Church in the process of its liturgy created such a nonrepresentation image, which I call an "icon in the expanded field"; it unfolds as a temporal phenomenon. All the faithful are implicated in the *eikōn tou Theou* through their own participation in the chant and the Eucharist.

The process of inspiriting, furthermore, has a visual correlate in the glitter, or *marmarygma*, of the marble (*marmaron*) covering the walls and floor and in the coruscating live waters of the sea (*marmairousa thalassa*). The iterative *marmar*, vocalizing the murmur of quivering water, again draws attention to a mirroring that is a multimodal phenomenon. The same *esoptron* principle underlines the cathedral liturgy envisioned as a temporal mirror of the eternal celestial rite. Reflection and reverberation, which constitute the visual and sonic mirroring, and the process of inspiriting, invested in the chiastic structure of the psalmody and in the aspiratory prosody of the sung poetry, together orchestrate a multisensory experience that has the potential to destabilize the divide between real and oneiric. This process places the faithful in a space in between the terrestrial and celestial.

Hagia Sophia: Sound, Space and Spirit further introduces a short film on aesthetics and samples of Byzantine chant digitally imprinted with the acoustics of Hagia Sophia. Both are developed as integral segments of the project, allowing the reader to transcend the limits of textual analysis and experience the temporal dimension to the Byzantine process engendering the animation of the inert.[27] Thus, my monograph dovetails with *Aural Architecture in Byzantium*, mobilizing the different fields of music, architecture, liturgy, and art in order to offer access to the complex spiritual experience orchestrated by the sung office of the cathedral liturgy of the Great Church.

This sensual experience comes to life in performance. And in my closing lines I will return to the psychosomatic effect the live auralizations created at Stanford. The process of reattaching Hagia Sophia's "detachable echo" onto the live sound of Cappella Romana makes it possible to manipulate the experience of space. This can, in turn, produce the effect of several different spaces acoustically present in one location.[28] Media scholar Lev Manovich has defined such a flow of the digital in the real as an "augmented" reality.[29] The audience at the Bing lived ephemerally in a similar "augmented" reality in which the flow of the digitally resurrected medieval intersected with the real. It is my hope that such encounters of the in-between can elicit a much stronger empathy for the past.

The importance of *Aural Architecture in Byzantium* lies in the collective richness and variety of papers on a multitude of different topics orchestrated around the synergy among architecture, music, acoustics, and digital technology. The dialogue between medieval and modern created by these essays offers a paradoxical experience of both connecting to what is ancient and historical—the Byzantine liturgy, its music, art, and architecture—and the latest of technology, which has the potential to bring to the fore some of the sensorial aspects of the distant past. My hope is that *Aural Architecture in Byzantium* will stimulate an integrated approach to the study and experience of medieval art and expand the envelope of research in sound studies of the premodern period.

Notes

1 Jonathan Sterne, "Sonic Imaginations," in *The Sound Studies Reader*, ed. Sterne (Abingdon, UK: Routledge University Press, 2012), 1–17; Michael Bull and Les Back, Eds., *The Auditory Culture Reader* (Oxford: Berg, 2003); Mark Smith, Ed., *Hearing History: A Reader* (Athens, GA: University of Georgia Press, 2004); and Ros Brandt, Michael Duffy, and Dolly MacKinnon, Eds., *Hearing Places: Sound, Place, Time and Culture* (2007; reprint, Newcastle upon Tyne: Cambridge Scholars Publishing, 2009).
2 R. Murray Schafer, *The Soundscape: Our Sonic Environment and the Tuning of the World* (Rochester, VT: Destiny Books, 1994), 4, 10–11, 237–52.
3 Emily Ann Thompson, *The Soundscape of Modernity: Architectural Acoustics and the Culture of Listening in America*, 1900–33 (2002; reprint, Cambridge, MA: MIT Press, 2004), 1–2.
4 For the analysis of the wet acoustics of Hagia Sophia, see the essay by Wieslaw Woszczyk in this volume. Similar conclusions have been reached about the acoustics of the Late Antique churches in Ravenna; see David J. Knight, "The Archaeoacoustics of a Sixth-Century Christian Structure: San Vitale, Ravenna," in *Music & Ritual: Bridging Material and Living Cultures*, ed. Raquel Jiménez, Rupert Till, and Mark Howell (Berlin: ēchō Verlag, 2013), 133–46; and David Knight, "The Archaeoacoustics of San Vitale, Ravenna" (master's thesis, University of Southampton, 2010).
5 Chris Scarre, "Sound, Place and Space: Towards an Archaeology of Acoustics," in *Archaeoacoustics*, ed. Chris Scarre and Graeme Lawson (Cambridge: McDonald Institute for Archaeological Research, 2006), 1–10; and Igor Reznikoff, "The Evidence of the Use of Sound Resonance from Paleolithic to Medieval Times," in *Archaeoacoustics*, 77–84.
6 Scarre and Graeme Lawson, *Archaeoacoustics*, viii.
7 Barry Blesser and Linda-Ruth Salter, *Spaces Speak, Are You Listening? Experiencing Aural Architecture* (Cambridge, MA and London: MIT Press, 2007).
8 See the essay by Jonathan Abel and Kurt Werner in this volume. For the concert with live auralization, see www.youtube.com/watch?v=uKLkJJ3ftIw.
9 Blesser and Salter, *Spaces Speak, Are You Listening?*, 5.
10 For an extended discussion of "earcon" in Hagia Sophia, see Bissera V. Pentcheva, *Hagia Sophia: Sound, Space, and Spirit in Byzantium* (University Park, PA: Pennsylvania State University Press, 2017), chap. 4.
11 Deborah Howard and Laura Moretti, *Sound & Space in Renaissance Venice* (New Haven, CT: Yale University Press, 2010).
12 Bonna Wescoat and Robert Ousterhout, Eds., *Architecture of the Sacred: Space, Ritual, and Experience* (Cambridge and New York: Cambridge University Press, 2012).
13 Vasileios Marinis, *Architecture and Ritual in the Churches of Constantinople, Ninth to the Fifteenth Century* (Cambridge and New York: Cambridge University Press, 2014).
14 See http://iconsofsound.stanford.edu/. For an initial synthesis of this project, see Bissera V. Pentcheva, "Hagia Sophia and Multisensory Aesthetics," *Gesta* 50, no. 2 (2011): 93–111. Since its inception in 2008, my research has grown into a monograph, Pentcheva, *Hagia Sophia: Sound, Space and Spirit in Byzantium*.
15 Thompson, *The Soundscape of Modernity*, 167–235, 248–56. On the interconnection between materials and acoustics of churches, see also Ettore Cirillo and Francesco Martellotta, *Worship, Acoustics, and Architecture* (Brentwood, UK: Multi-Science Publishing, 2006).
16 See the essay by Jonathan Abel and Kurt Werner in this volume.
17 To view segments of Cappella Romana's concert at Stanford's Bing Hall, see www.youtube.com/watch?v=uKLkJJ3ftIw and www.youtube.com/watch?v=bHpOiX2sO-s. For a review of the event, see www.sfcv.org/reviews/stanford-live/cappella-romana-time-travel-to-constantinople
18 See http://auralarchitecture.stanford.edu/.
19 See http://auralarchitecture.stanford.edu/.
20 Anna Kartsonis, *Anastasis: The Making of the Image* (Princeton, NJ: Princeton University Press, 1986), 40–81; John Osborne, J. Rasmus Brandt, and Giuseppe Morganti, Eds., *Santa Maria Antiqua al Foro Romano cento anni dopo: Atti del colloquio internazionale, Roma, 5–6 maggio 2000* (Rome: Campisano, 2004); David Buckton, "Byzantine Enamel and the West," *Byzantinische Forschungen* 13 (1988): 235–44; Erik Thunø, *Image and Relic: Mediating the Sacred in Early Medieval Rome*, Analecta romana studi danci 32 (Rome: L'Erma di Bretschneider, 2002), chap.

6, 129–55; and Kathleen Corrigan, *Visual Polemics in the Ninth-Century Byzantine Psalters* (Cambridge: Cambridge University Press, 1992), 107–08.

21 Christian Troelsgård, *Byzantine Neumes: Byzantine Musical Notation* (Copenhagen: Museum Tusculanum Press, 2011).

22 Pentcheva, *Hagia Sophia: Sound, Space, and Spirit in Byzantium*. On the phenomenon of "acoustic waterfall" produced by the acoustics of the dome, see Wieslaw Woszczyk's essay in this volume.

23 Robert Taft, The Byzantine Rite: A Short History (Collegeville, MN: Liturgical Press, 1992), 53–66; and Thomas Pott, *La réforme liturgique byzantine: Étude du phénomène de l'évolution non-spontanée de la liturgie byzantine* (Rome: Orientalia Analecta Christiana, 2000), translated by Paul Meyendorff as *Byzantine Liturgical Reform: A Study of Liturgical Change in the Byzantine Tradition* (Crestwood, NY: St. Vladimir's Orthodox Seminary Press, 2010).

24 Derek Krueger, *Liturgical Subjects: Christian Ritual, Biblical Narrative and the Formation of the Self in Byzantium* (Philadelphia, PA: University of Pennsylvania Press, 2014), 164–97.

25 Bissera V. Pentcheva, "Mirror, Inspiration, and the Making of Art in Byzantium," *Convivium* 1, no. 2 (2014): 10–39.

26 Pentcheva, *Hagia Sophia: Sound, Space, and Spirit* and the connected Web site, http://hagiasophia.stanford.edu/.

27 See http://hagiasophia.stanford.edu/.

28 Jonathan Sterne, "Space within Space: Artificial Reverb and the Detachable Echo," *Grey Room* 60 (2015): 110–31.

29 Lev Manovich, "The Poetics of Augmented Space," *Visual Communication* 5/2 (2006): 219–40.

1 Aural architecture in Jerusalem, Rome, Constantinople, and Alexandria

Peter Jeffery

As the church emerged from persecution during the fourth century, two models of Christian community developed, each with its own shape of ritual: one type was formed in the great cities of the Roman Empire, the other in the monastic communities that rejected civil life. Subsequent "Christian history unfolds in an antithesis between the Empire and the Desert."[1] The model urban community was the one in Jerusalem, Christianity's holy city, whose ceremonies were widely imitated in much of the Christian world. But comparable urban rites emerged in the other great cities of the empire, different in both content and structure. Besides Jerusalem, we are best informed about the rites of Rome and Constantinople. In each of these three metropoleis, the known history of ritual life largely begins with architectural structures built by Emperor Constantine the Great (r. 324–37) or members of his family. Thus, the worship life of the three cities had something in common during the fourth century. By the fifth century, however, regional differences had already emerged. As we move to the eighth century and later, we can see the customs of each urban center developing in independent directions. Alexandria, which had no Constantinian buildings, offers a kind of parallel control, for here too we can observe gradual movement away from a shared late antique Greco-Roman culture. This essay explores this intensification of difference by tracing the interrelations among music, movable liturgical furnishings, and architectural form.

Jerusalem

In Jerusalem, the cult began with the recovery of the tomb of Jesus. Local tradition held that this had been covered over by the Temple of Aphrodite built under Emperor Hadrian after the Second Jewish War (135 CE), in a deliberate attempt to render it inaccessible. It was at just this time that all Jews were expelled from the city, and efforts were made to obliterate the memory of its history as a religious capital. But the Temple of Aphrodite may, instead, have had the opposite effect of marking the location of the site and thus preserving its significance. According to Eusebius, at any rate, the workers Constantine ordered to dig up the tomb had no trouble identifying what they were looking for: it emerged from the darkness into the light like an image of the savior returning to life.[2] Around it, Constantine built a complex of buildings: the rotunda of the Anastasis (Resurrection) surrounding the tomb itself, an open-air courtyard that incorporated the rock of Golgotha or Calvary, and a basilica known as the Martyrion or Martyrium.[3] Today, the Church of the Holy Sepulchre stands on the site of the Constantinian Anastasis.

As for the cultic practices that went on in these buildings, we first learn about them from a unique travelogue, written in the late fourth century.[4] This work never circulated widely; we have only two incomplete manuscripts of it, both discovered in modern times,[5] along with the indirect witness of a few medieval readers who had access to now-lost manuscripts. Thus, we lack the original title of the work, as well as its beginning and ending, and several sections in between. Only fairly recently have we come to agreement that the author's name was probably Egeria. Though she traveled with a group that is not described in the extant text, she seems to have been a celibate monastic, for she mentions no husband or family and clearly states that her record is intended for the "sisters" of her community back home. Indeed, her text is written in a peculiar dialect that reads like Latin turning into Spanish, and has therefore attracted much attention from Romance philologists.[6] It is just the sort of idiolect we might expect in a woman of that time, who would have had fewer educational opportunities than an upper-class man of her generation. This, in turn, tends to confirm that her travelogue is the most substantial Early Christian text known to have been authored by a woman.

Egeria gave an account of an entire liturgical year in Jerusalem, which must reflect the period from 381 to 384 CE.[7] As she tells it, on every day that commemorated an event in the life of Christ, the entire community went out to celebrate in the place where the event (according to local belief) had actually occurred. For example, on January 6, the birth of Jesus was celebrated at Bethlehem; on the last Saturday in Lent, the raising of Lazarus was celebrated in Bethany. Most events were commemorated at points within the city itself. At every location, the celebration included Bible readings about the original occurrence and the singing of psalms that were exegetically associated with it—a practice that Egeria greatly appreciated, though it seems to have been new to her:

> And what I admire and value most is that all the hymns and antiphons and readings they have, and all the prayers the bishop says, are always relevant to the day which is being observed and to the place in which they are used. They never fail to be appropriate.[8]

The readings and psalms that Egeria heard were collected into a liturgical book that survives only in Armenian translation, made at a time when the church in Armenia adopted the Jerusalem liturgy as its own. The oldest manuscript of this Armenian lectionary seems to translate a lost Greek original that must have been compiled between 417 and 438–39, that is, about two generations after Egeria's visit.[9] Its provisions are nevertheless very consistent with what Egeria related.[10]

The later development of the Jerusalem rite can be recovered from manuscripts in the Georgian language, translations of lost Greek books from about the eighth century.[11] With the passage of time and under the pressures of Islamic rule, the Greek community in Jerusalem eventually adopted the Byzantine rite. As a result, very few Greek manuscripts of the original hagiopolite rite have been preserved. Until recently only one Greek manuscript of the Jerusalem rite was known, containing just Holy Week and Easter Week and dating from the year 1122.[12] A few others have turned up among the new finds at the Monastery of St. Catherine's on Mount Sinai, which are just beginning to be investigated.[13]

Of all the rites described by Egeria, the most important has left a number of descendants; this is the Resurrection vigil that was celebrated every Sunday morning at the tomb of Jesus, "as if it was Easter."[14] According to Egeria, three psalms were sung, each followed by a prayer; then the bishop went into the tomb, where a light burned continually, and read a Gospel account of the Resurrection. The most dramatic survival of this is the rite of the Holy Fire, still celebrated every Easter.[15]

More important for music history was the evolution of the weekly celebration that took place every Sunday morning. The bishop's reading of the Resurrection account gradually solidified into a practice of reciting all four Gospels in regular succession, one per week. But since most Gospels preserve more than one narrative of the Resurrection, by the sixth century each Gospel account was split in half, producing an eight-week cycle of Resurrection readings for ***orthros***. It was during this period that we find the first evidence of the eight musical modes (***octoechos***), one for each week of the cycle.[16] Tenth-century Georgian manuscripts, thought to preserve the eighth-century state of the liturgy, give us lists of the three psalms arranged according to each musical mode.[17] The most developed form is preserved in the Byzantine rite, where there is now a sequence of eleven readings from the four Gospels, for which there is a set of eleven hymns or *troparia heothina* sung at *orthros*; this new hymnody has been attributed to Emperor Leo VI (866–912).[18]

Byzantine liturgical books still prescribe two of Egeria's three psalms at the Sunday morning vigil, though they are now reduced to one or two verses: for the first, there are eight possible texts, one for each mode.[19] For the second, often the only one performed today, the text is the last verse of the book of Psalms: "Let everything that breathes praise the Lord" (Psalm 150:6).[20]

Rome

Chronologically, the next city for which we have evidence is Rome. There, on about half the days of the year, the pope celebrated Mass at one of the city's churches. The selected church, where a stage of the liturgy would be celebrated, was known as the *statio*, a word that had a long association with fasting,[21] but also came to mean a stopping place on a procession. It is from this word that we derive the modern term "stational liturgy" for this type of peripatetic liturgical calendar.

Early evidence of stational churches survives in sermons of Popes Leo (r. 440–61) and Gregory (r. 590–604),[22] but the earliest complete list of stational churches is in the oldest register of epistle readings for the Roman Mass, the early eighth-century Epistolary or Comes of Würzburg.[23] Within the next few centuries, we also have station lists in Gospel lectionaries, sacramentaries containing the Mass prayers, and graduals containing the Mass chants. Variations between the lists are minor.[24]

While Jerusalem's ritual was concentrated on Golgotha and the Holy Tomb, Rome had two central churches that were used for the most important feasts: St. John Lateran, the pope's cathedral, and St. Peter's at the Vatican, where the first of the apostles was buried. On other days, the choice of stational church was determined by its relics, most of which were the bones of Early Christian martyrs. Thus, the Gospel of the raising of Lazarus (John 11:1–44), whom Jesus called out of his tomb, was read in the fourth week of Lent at the church of S. Eusebio, who was martyred by being locked in a small room. On the Sunday after Easter, the new converts who had just been reborn in baptism assembled to remove their white robes at the church of S. Pancrazio, a child martyr. The main Mass on Christmas was celebrated at S. Maria Maggiore, where the relic of the manger was held.[25]

The stational system was built atop an older classification of Roman churches. At the low end of the hierarchy were the deaconries, which had storehouses for distributingfood to the poor. Other churches served monastic communities, some of which also hosted pilgrims and cared for the sick. Pope Gregory the Great founded such a monastery in his palatial home on the Caelian Hill, overlooking the Circus Maximus. S. Saba (founded in Rome by monks coming from St. Sabas in Palestine) is the best known of the Greek-speaking monastic churches. At a more prestigious level were the ***tituli***, which were understood to have been built on the sites of house churches from the pre-Constantinian era. The *titulus*, or title, was a sign with the name of the presumed original owner of the home. The churches of S. Clemente and S. Cecilia were among this group. Largest and most important were the great basilicas of the Constantinian era: St. Peter's on Vatican Hill, S. Paolo fuori le Mura, St. John Lateran, S. Maria Maggiore, S. Lorenzo fuori le Mura, S. Croce in Gerusalemme.[26] In each of these basilicas, the daily liturgy was carried out by monks who lived in nearby monasteries. St. Peter's had four such monasteries, others had two or three, though we are not certain how many.[27] These monasteries could be described as pre-Benedictine, and it is from somewhere in or near this milieu that the Benedictine rule emerged. As Roman Christianity transitioned through the Carolingian era into the Middle Ages, however, only S. Paolo fuori le Mura became Benedictine. The clergy of St. Peter's and S. Maria Maggiore became canons; those at the Lateran became canons regular. However, it is the practices of the pre-Benedictine monks that were eventually enshrined in the medieval and Tridentine Roman rite. That is why the daily office of the Roman Breviary, though it was used by nonmonastic clergy, is of the monastic kind, structured around the weekly recitation of the 150 psalms.

A particularly interesting part of the Roman stational calendar is the sequence for Easter week, when the stations were held at seven ancient churches, starting with the Constantinian ones. We can see the beginnings of this arrangement in the sermons of Gregory the Great,[28] but the Roman **Missal** (the liturgical book that contains the texts and rubrics for the celebration of Mass) in its classic form is better organized.[29]

At the same time, a few manuscripts preserve a very ancient set of evening services for these days, the only survival in Rome of an urban or nonmonastic form of the daily office. Each night of this week, the clergy gathered at the great cross in St. John Lateran, singing "Kyrie eleison." Next, they sang some vespers psalms with alleluia refrains, then processed to the baptistry chapels singing melismatic alleluias with verses in Latin and Greek.[30] The boys' choir had a prominent role in this ceremony, alternating with the men.[31]

Gregory, Sermons 21–26 for April 15–21, 591	*Stations in the Missale romanum, 1570*
Sermon no. 21 Pascha: S. Maria Maggiore, Mark 16:1–7	Dominica Resurrectionis: S. Maria Maggiore
Sermon no. 22 Feria (Day) II: St. John Lateran, John 20:1–9	Feria II: St. Peter's
Sermon no. 23 Feria III: St. Peter's, Luke 24:13–35	Feria III: S. Paolo Feria IV: S. Lorenzo fuori le Mura
Sermon no. 24 Feria V: S. Lorenzo fuori le Mura, John 21:1–14	Feria V: SS. XII Apostoli (also known as SS. Philip and James)
Sermon no. 25 Feria VI: St. John Lateran, John 20:11–18	Feria VI: S. Maria ad Martyres *Rotunda* (=the Pantheon)
Sermon no. 26 Sabbato: St. John Lateran, John 20:19–31	Sabbato in Albis: St. John Lateran

What can we know about the organization of ritual space in the Roman church buildings during the period of stational liturgy? In a general way we know that early Roman churches were built in basilica form: rectangular, with two rows of columns, an apse at one end with a raised floor, containing the altar, the episcopal chair, and the synthronon, where the priests and deacons sat. In front of this area, on the level of the lower floor, was an enclosed area for the lower clergy and choir, and nearby a raised ambo from which the Gospel and other readings were proclaimed. But information more specific than that is difficult to obtain, for though many churches were built in very early times, unfortunately, no Roman basilica remains in anything like its eighth- to tenth-century state: all have been remodeled many times since the early medieval period. By the end of the Middle Ages, most were in poor condition, leading their cardinal patrons of the Renaissance and Baroque era to lavishly rebuild them. The nineteenth and early twentieth centuries saw a burst of antiquarian revision, an attempt to return many buildings to what was imagined to have been their Early Christian state. Another wave of restructuring after the Second Vatican Council (1962–65) sought to make these buildings more consistent with the council's liturgical reforms. Thus, many churches have experienced at least three phases of rebuilding since the end of the Middle Ages.

Some things we can know, however, either from archaeological and literary research on the history of the great basilicas or from the study of churches that were closed during the early Middle Ages and therefore never remodeled. The best preserved of these is S. Maria Antica in the Roman Forum, which seems to have housed a community where much Greek was spoken;[32] it was closed in the ninth century following an earthquake (Figures 1.1–3). Another is the old basilica of Ss. Nereo ed Acchileo at the entrance to the Catacomb of Domitilla. There are also the old churches beneath the present basilicas of S. Clemente and S. Marco in the Piazza Venezia. In every case, when these churches were closed, they were stripped of their altars, icons, marble, and anything else that could be reused.[33] These churches, therefore, give information only on their early medieval structural state. And in modern times these buildings underwent some degree of restoration before being reopened.

The restoration of S. Maria Antica was completed in 2013, though the building is open to the public only for limited periods due to the delicate state of the floors and frescoes. In the main building of the complex we can see the apse in the rear, with badly deteriorated frescoes (Figure 1.1). An altar installed in 1955, in the form of a flat marble slab resting on two short columns, stands in front of the apse. In front of the table are the remains of the balustrade that marked the altar area as a clergy space, the presbytery/presbyterium, off limits to the laity. The surviving wall is higher on the right, with partially preserved frescoes. In the center is the entrance to this area, still lined with marble, but now blocked with modern orange-and-white traffic barriers. One can see that the presbytery area was raised one step above the level of the floor in front of it. This area in front of the balustrade is itself one step up from the main floor level. There is a step down to the left of the Corinthian column on our right, and another one at the far end of the low brick wall on our left. This structure contains the remains of an enclosure for the choir and the lower clergy who were less directly involved in the service at the altar. The octagonal hole in the floor, now filled in with wood, seems to have held the base of a medieval fountain; presumably, it did not exist when the space was in use as a church. The remnants of an ambo can be seen to the left of the choir wall enclosure.

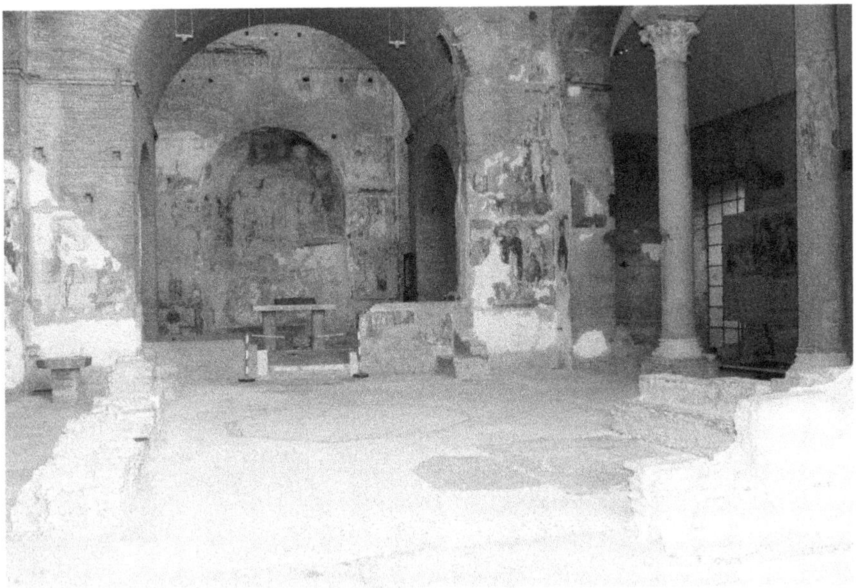

Figure 1.1 Santa Maria Antica in the Roman Forum, transformed into a church in the late sixth or early seventh century, nave. The low brick wall extending into the foreground marks the remains of an enclosure for the choir and the lower clergy.

Photograph: Peter Jeffery.

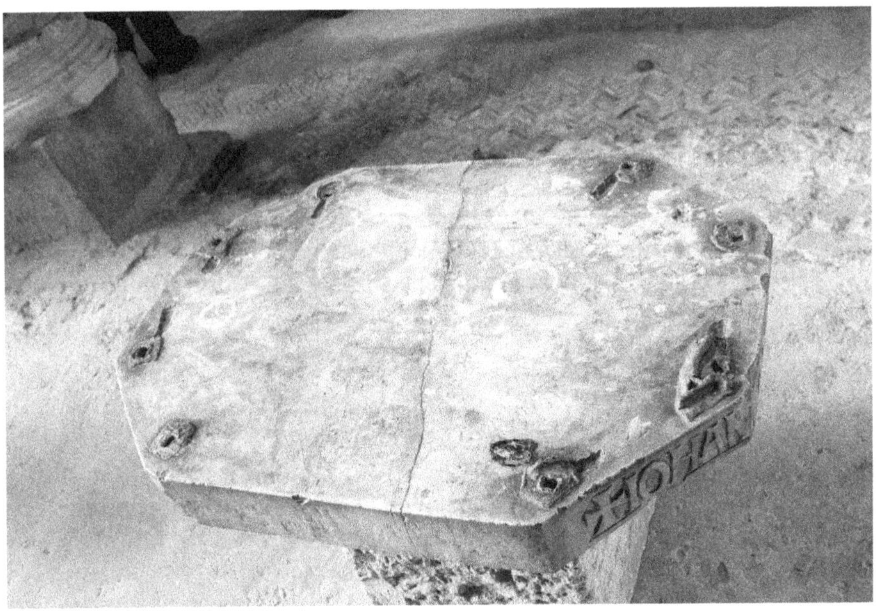

Figure 1.2 Santa Maria Antica in the Roman Forum, late sixth or early seventh century. The base of the ambo with Greek and Latin inscription and the holes on the surface suggest the previous existence of a metal balustrade.

Photograph: Peter Jeffery.

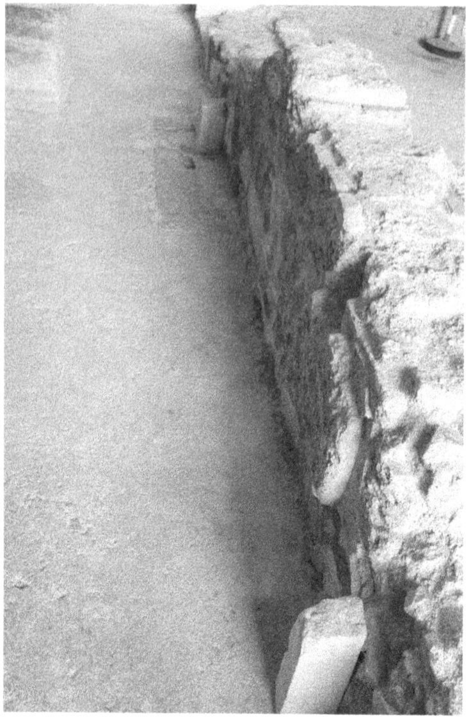

Figure 1.3 Santa Maria Antica in the Roman Forum, late sixth or early seventh century. Wall of the choir enclosure with the legs of the ambo embedded.
Photograph: Peter Jeffery.

It is the ambo that is most interesting for our purposes: this was a raised structure from which the scriptures were proclaimed and some chants were sung. What survives is the platform itself, which would have been raised and accessed by two staircases (Figure 1.2). The surviving structure today in the garden of the Hagia Sophia museum could help us envision the appearance of such an ambo (Figure 1.9).

The platform at Santa Maria Antica bears a bilingual inscription, identifying the donor as "John, servant of St. Mary, the Theotokos:"

+IOHANNES SERVV SCAE MARIAE +ΙΩΑΝΝΟΥ ΔΟΥΛΟΥ ΤΗϹ ΘΕΩΤΟΚΟΥ

This is, then, the base of the ambo donated by Pope John VII (r. 705–07), the son of a Byzantine civil official. The contemporary chronicle of his reign confirms that John "adorned with painting the basilica of the holy mother of God which is called *Antiqua*, and there he built a new ambo."[34] The ambo originally stood on four white marble legs directly over the low brick wall, at just the place where the wall becomes higher. The bottoms of two of these legs can still be seen on the outside of the wall as well as one on the inside (Figure 1.3). Above the base there was evidently an enclosure made of metal; one can still see the holes for attaching this framework to the base (Figure 1.2). However, there must have been a change in liturgical practice, for this ambo was dismantled even while the church building was still in use. When it was discovered, the ambo base was being used in the floor as a paving stone, directly in front of the step leading up to the altar area.

The layout and furnishing of the space was clearly somewhat different from the much better-known marble enclosure we find in the present church of S. Clemente, which has been called a schola cantorum since about the seventeenth century and is often presented in textbooks as the archetype of an early Roman or even Byzantine liturgical setting (Figures 1.4–6).[35] One reason for the confusion: the marble panels, evidently manufactured in the sixth century, were salvaged from the older church of S. Clemente, which was closed about the twelfth century but still lies beneath the present one and was excavated in the nineteenth century. We can date the panels because some of them bear a monogram of the name Johannes, now identified as Pope John II (r. 533–35), the first pontiff known to have chosen a papal name. Prior to being elected pope, he was identified as Mercurius, a presbyter of S. Clemente whose name also occurs on some of the marbles salvaged from the earlier basilica.[36]

S. Clemente's enclosure is noted for its large ambo to the left of the altar, with two staircases and a large candlestick with a decorative stone inlay, known as Cosmatesque (Figures 1.4, 5). On the right side are two smaller reading stands, one facing the altar, the other facing away (Figure 1.6). If one assumes, as many people do, that the Gospel was read at the large ambo on the left, but the epistle at one of the smaller stands on the right, this would seem to reproduce, or rather foreshadow, the Gospel side/epistle side arrangement of the Tridentine Roman Missal.[37] However, the altar is positioned "backwards," or *versus populum*, and has been for many centuries. That is to say, the

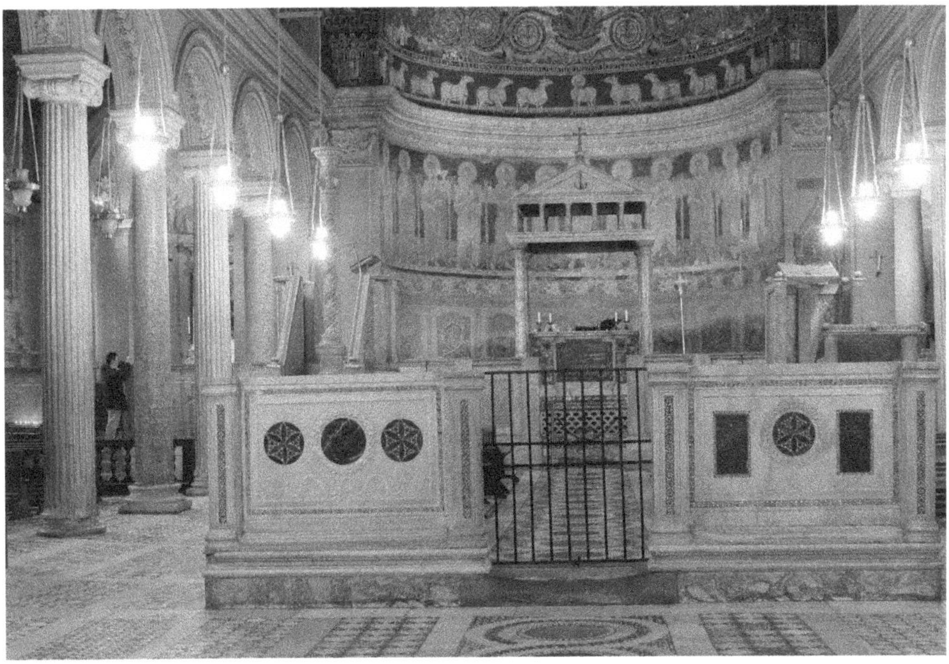

Figure 1.4 The church of San Clemente, Rome, ca. 1099–1120, the marble enclosure of the schola cantorum.

Photograph: Peter Jeffery.

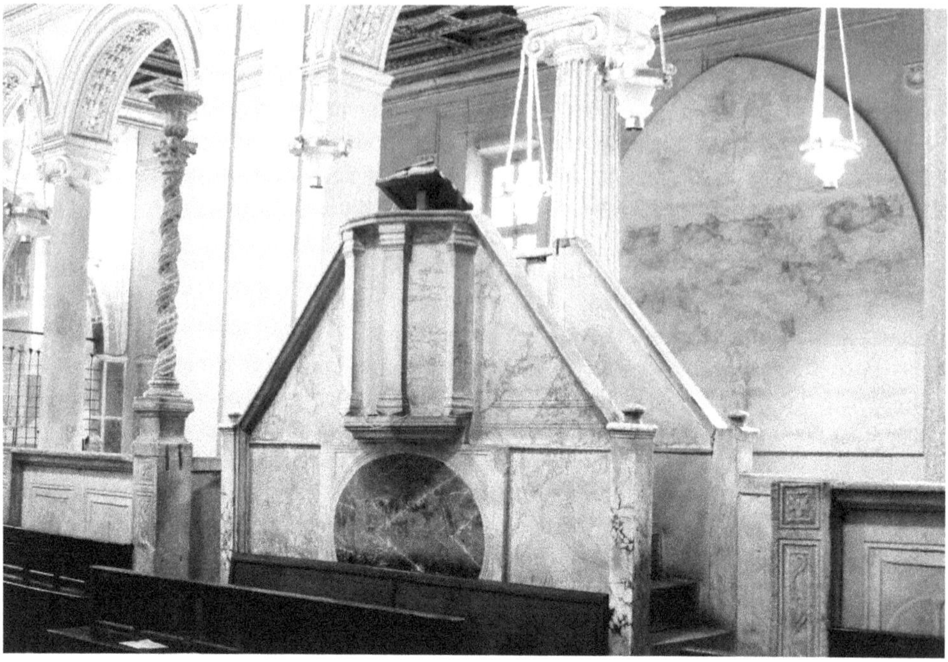

Figure 1.5 The church of San Clemente, Rome, ca. 1099–1120, ambo and candlestick.
Photograph: Peter Jeffery.

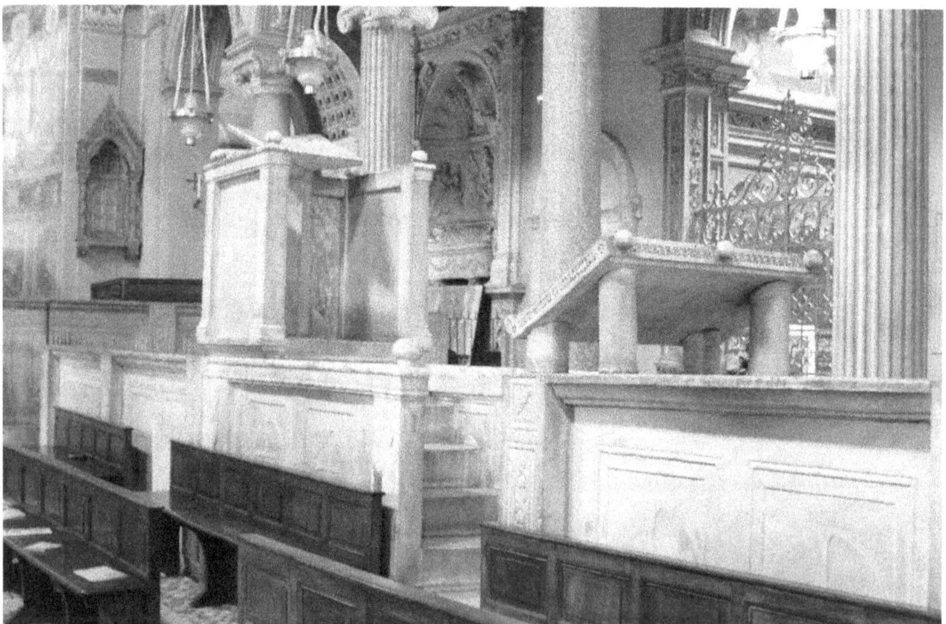

Figure 1.6 The church of San Clemente, Rome, ca. 1099–1120, second ambo and lectern.
Photograph: Peter Jeffery.

priest had no choice but to stand behind it when celebrating Mass, whether the intent was to face the people, in the modern way, or to avoid obstructing the worshippers' view of the tomb of Saint Clement directly below.

Constantinople

The stational cycle of Constantinople is preserved in the tenth-century typikon,[38] a book comparable to the ordinal of the medieval Western liturgy. The description of every feast begins by stating at what church in the city it will be celebrated, recording processions to many places in the city, especially to shrines of saints on their feast days. But on major feasts such as Christmas and Epiphany, we are told that "it is celebrated in the Great Church and in the churches in every place."[39] The Great Church here, as in Jerusalem, was the cathedral and ritual center, the Cathedral of Holy Wisdom, Hagia Sophia. The present building, of course, is the well-known edifice built by Emperor Justinian and consecrated in 537, but repaired several times following earthquake damage.[40] Its interior, too, is hardly in its original state, as it was converted into a mosque after the city fell to the Turks in 1453, and then into a museum during the rulership of Kemal Atatürk in the early twentieth century. But the unique design of the building, dominated by a massive dome, marks a step away from the shared antique culture that favored basilica plans in both East and West. Fragments of the earlier basilica, built by Theodosius II (r. 408–50) early in his reign, in fact remain on the site (Figure 1.7).[41] They reveal an iconography of the Christian disciples as sheep that is often seen in the basilicas of Rome—for example, at S. Clemente (Figure 1.4) and S. Cecilia (Figure 1.8). A two-staired ambo is also displayed in the museum's garden (Figure 1.9). Although it originally comes from a different church, it offers a good example of the typical shape; it is very similar to the one in the upper church of S. Clemente.

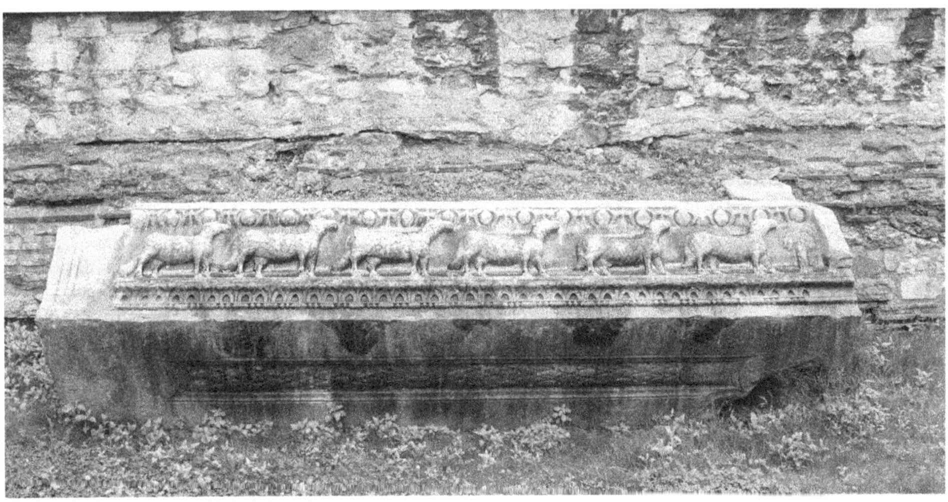

Figure 1.7 The basilica of Hagia Sophia, Constantinople, early fifth century, carved frieze showing sheep processing to a palm tree, from the architrave of the Theodosian basilica.

Photograph: Peter Jeffery.

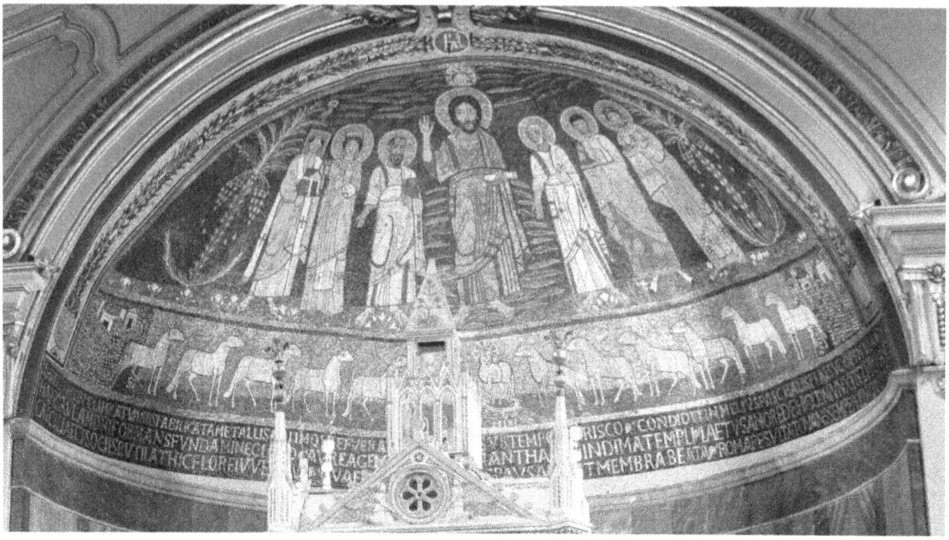

Figure 1.8 Apse mosaic of Santa Cecilia, Pope Paschal I (817–24), depicted at the extreme left, Rome.

Photograph: Peter Jeffery.

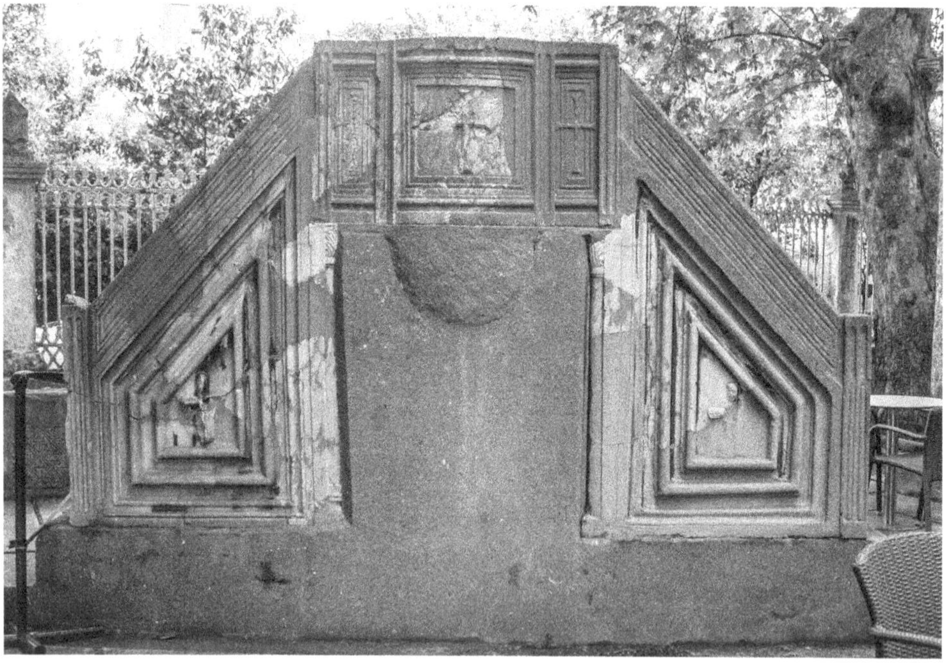

Figure 1.9 The basilica of Hagia Sophia, Constantinople garden. Portions of an ambo, early fifth century.

Photograph: Peter Jeffery.

Up until the Latin conquest, the liturgy carried on in Justinian's edifice was not the Byzantine rite used by all the Orthodox churches today but a nonmonastic cathedral rite known as the *ecclesiastes*, or "cathedral," or more commonly referred to as the "asmatic," or Sung Office.[42] A surviving element of this rite is the presence of green stripes or "rivers" in the marble floor, which helped guide the movement of clergy and processions around the building. One of them can be seen in Figure 1.10, the horizontal line running left to right in the floor, near the bottom of the photograph (Figures 1.10, 11).[43] When the Greeks retook the city of Constantine from the Latins in 1261, worship in Greek was restored, but the task was given to the monks, who followed a monastic rite of Palestinian origin that was related to the original Jerusalem practice. It was this that became the Byzantine rite used by Orthodox churches today. However, the Sung Office continued to be celebrated in other Greek cathedrals, notably, at Thessaloniki, where it lasted right up to the Turkish conquest of the fifteenth century.[44] For that reason, a small number of manuscripts still survive that preserve the repertory of the Sung Office.

The Sung Office did not include a Resurrection vigil like that of the monastic Byzantine rite. On Easter morning, the Eucharistic Divine Liturgy began with three antiphons, the third of which included one of the best known of all Byzantine chants, the Easter ***troparion*** Χριστὸς ἀνέστη ("Christ is risen").[45] A deacon read the first chapter of the Gospel of John in Latin, then the patriarch read it again in Greek. From high in the ambo, the deacon repeated each phrase that the patriarch read in a loud voice.[46]

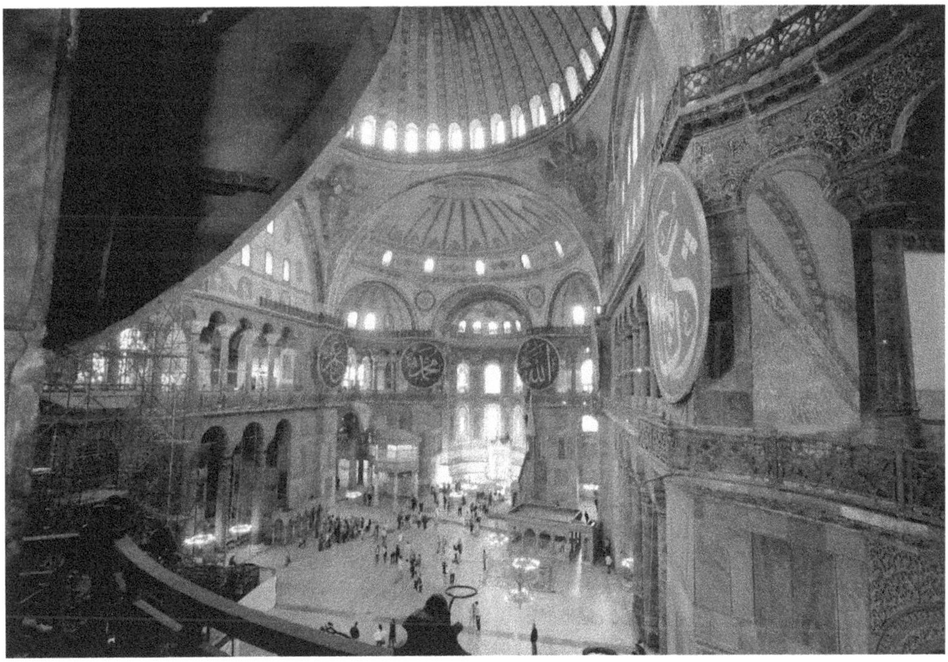

Figure 1.10 Hagia Sophia, Constantinople, 532–37 and 562, proconnesian marble showing one of the green stripes or "rivers."

Photograph: Bissera V. Pentcheva.

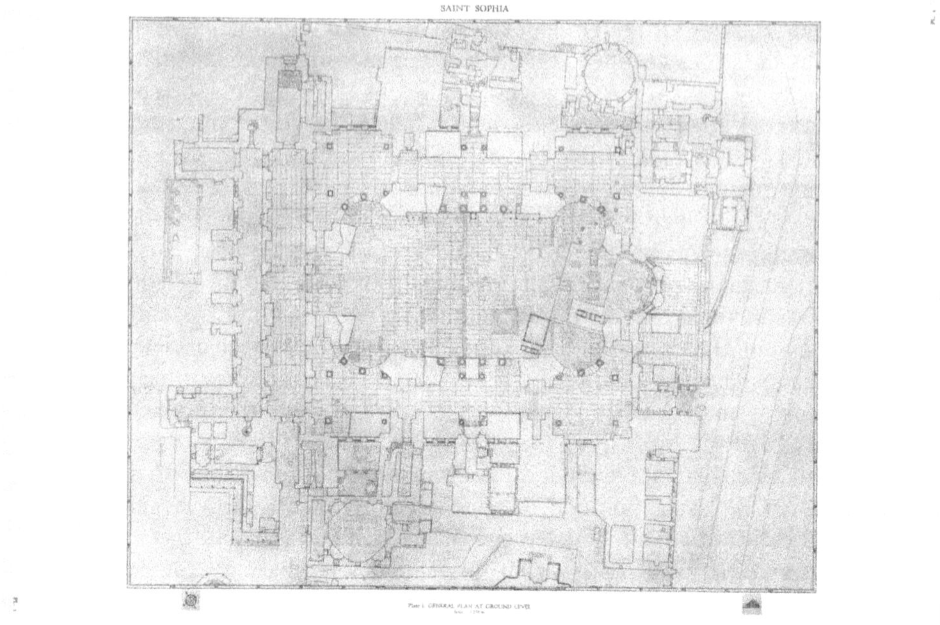

Figure 1.11 Hagia Sophia, Constantinople, 532–37 and 562, drawing of the pavement showing the four green-marble stripes, known as "rivers."

Photograph: © Robert van Nice, Image Collection and Fieldwork Archive, Dumbarton Oaks Field and Photographic Archive, Trustees for Harvard University, Washington, DC.

Alexandria

Rome, Constantinople, and Jerusalem were not the only cities in the Roman Empire, of course. Alexandria and Antioch were, in fact, the second- and third-largest cities after Rome. The bishops of these metropoleis still retain the title "patriarch" in both the Greek Orthodox and Roman Catholic hierarchies.

From the documents we have, it is not known if either Antioch or Alexandria ever had a fully developed stational liturgy, comparable to those of Jerusalem, Rome, and Constantinople. From late fourth-century Antioch we have many reports of street processions, both Christian and non-Christian.[47] But Antioch at the time was a raucous religious marketplace, with as many as three or four bishops of competing Christian sects.[48] By the time we have documentation of the stational calendars of Rome and Constantinople (from the eighth to tenth centuries), both Antioch and Alexandria had come under Muslim rule: the Greek-speaking Christians were slowly gravitating to the monastic Byzantine rite, while the former local liturgical tradition was preserved mainly by non-Chalcedonian or Monophysite Christians, translated into local languages. Thus, elements of the Alexandrian rite survive in Coptic and Ethiopic; elements of the Antiochene rite survive in Syriac and Armenian.

For the Alexandrian liturgical tradition, the most abundant early evidence is in the lectionary, which gives us the readings for the entire liturgical year. The oldest accessible manuscript (Pierpont Morgan Library, New York, MS M 573) dates to the ninth century.[49] There are many early fragments, and their relations to the main tradition continue to be underinvestigated.[50] The lectionary is notable for its comprehensive use of scripture: four New Testament readings at every Mass, with additional readings at the morning and evening office. Unlike in the Roman epistle and Gospel books, however, there are no indications of stational churches.

What do we know about the early church buildings of Alexandria? One problem we face here is that, following a series of earthquakes, much of ancient Alexandria is now under water. Archaeological investigations, not surprisingly, have focused on the more spectacular Pharaonic and pre-Christian sites, dating from an era when, as Philo Judaeus complained, the Egyptians worshipped animals.[51] The remains of a royal palace have been discovered,[52] as well as portions of the great lighthouse that was one of the seven wonders of the ancient world.[53]

Alexandria does not boast any structures built by Constantine or his family. Literary evidence seems to show that the first Christian edifices were built outside the city near the cemeteries. From the fourth century, synagogues and some pagan temples were converted and rebuilt into Christian buildings, beginning with the Caesareum, a large structure built by Cleopatra that had been dedicated to the imperial cult.[54] The only early church building that survives within the modern city is now the Greek Orthodox cathedral of St. Sabas (also identified with Saint Katherine), which was first mentioned in an eighth-century source. It is basically a rather short three-aisled basilica with a simple apse; ancient monasteries in the Egyptian desert preserve similar buildings, some with triconch apses, dating as early as the fifth century.[55] The non-Chalcedonian Coptic Orthodox cathedral of St. Mark is a twentieth-century building; for much of this community's history, the patriarchs lived in monasteries outside the city proper.

The Coptic Orthodox church still sings quite a few texts in Greek, among them the Χριστὸς ἀνέστη. The Coptic chant tradition has never adopted written notation, in part because many of the singers were blind. But the melodies of the oral tradition as we have it now do not especially resemble the medieval or modern Byzantine melodies, so it is difficult to say that, even when there is a shared text, there must be a common musical ancestor.[56]

Conclusion

It is not easy to reconstruct the aural architecture of the distant past. Even when ancient buildings survive, they have been much modified since their initial construction. The church of S. Maria Antica brings us about as close as we can get to the ninth century, when it was closed—but even at that date, the ambo of Pope John VII had already been dismantled and its base reused as a paving stone, and much else was removed at the time of closure. The church of S. Clemente preserves many of the sixth-century marbles, but they have been moved to the twelfth-century building and are no longer in their original configuration. In Constantinople, Hagia Sophia was converted into a mosque and then a museum, losing its Christian liturgical furniture in the process. Very little survives of the Constantinian complex that Egeria visited in Jerusalem, and much of ancient Alexandria is under water. Still, there are signs of a common ancestral culture, from which each tradition has grown in its own direction. The earlier Hagia Sophia of Theodosius II apparently looked much more like a Roman basilica, in both structure and iconography, than Justinian's building now does. The oldest churches in Egypt were also constructed on the basilica plan.

Architectural history shows a more or less common culture at the beginning, centered on the basilica structure, but gradually moving in different directions as we approach the Middle Ages. Musical evidence is not fully available until the eleventh and twelfth centuries in Latin and Greek, and until the twentieth century in Coptic, but there seems to be a similar trend of moving away, in independent directions, from an original, shared culture in Greek.

Notes

1 Georges Florovsky, *Christianity and Culture*, Collected Works 2 (Belmont, Mass.: Nordland, 1974) 126.
2 Eusebius of Caesarea, Eusèbe de Césarée. *Vie de Constantin*, ed. Friedhelm Winkelmann et al., Sources Chrétiennes 559 (Paris: Cerf, 2013), 3.25–40, pp. 384–401, with the excavation of the tomb at 3. 28, 388–89. The excavation must have happened between the years 327, when Constantine's mother Helen arrived in Jerusalem, and 335, when the first church on the site was dedicated. See Charles Odahl, *Constantine and the Christian Empire*, 2nd ed. (London: Routledge, 2010) 211–20.
3 Richard Krautheimer and Slobodan Ćurčić, *Early Christian and Byzantine Architecture*, Pelican History of Art, 4th ed. (New Haven, CT: Yale University Press, 1986), 60–67.
4 Egeria, *Itinerarium*. The newest edition of the Latin text is Egérie: *Journal de Voyage (Itinéraire)*, ed. Pierre Maraval, Sources Chrétiennes 296 (Paris: Cerf, 1982). The best translation and introduction is John Wilkinson, *Egeria's Travels*, 3rd ed. (Oxford: Aris & Phillips, 1999; reprinted with corrections 2006). An international effort to reconstruct her route and identify the monuments she saw can be found at www.egeriaproject.net.
5 The Arezzo Codex (or Codex Aretinus) VI.3, discovered and edited first by Gian Francesco Gamurrini in 1884, and some Madrid fragments published by Jesús Alturo, who also lists the indirect witnesses, in "Deux nouveaux fragments de l' 'Itinerarium Egeriae' du IXe–Xe siècle," *Revue Bénédictine* 115 (2005): 241–50.
6 For example, Veikko Väänänen, *Le journal-épître d'Egérie (Itinerarium Egeriae): Étude linguistique* (Helsinki: Suomalainen Tiedeakatemia, 1987).
7 Wilkinson, *Egeria's Travels*, 169–71, 35–45.
8 Wilkinson, *Egeria's Travels*, 163.
9 Athanase [Charles] Renoux, Ed., *Le Codex arménien Jérusalem 121*, 2 vols., Patrologia Orientalis, vol. 35, fasc. 1 and vol. 36, fasc. 2 (Turnhout: Brepols, 1969–71); C. Renoux, "De Jérusalem en Arménie: L'héritage liturgique de l'Église arménienne," in *Patterns of the Past, Prospects for the Future: The Christian Heritage in the Holy Land*, ed. Thomas Hummel, Kevork Hintlian, and Ulf Carmesund (London: Melisende, 1999), 114–23; and Christina Maranci's essay in this volume, "The Great Outdoors: Liturgical Encounters with the Early Medieval Armenian Church."
10 Wilkinson, *Egeria's Travels*, 175–76.
11 Stig Simeon R. Frøyshov, "The Georgian Witness to the Jerusalem Liturgy: New Sources and Studies," in *Inquiries into Eastern Christian Worship: Selected Papers of the Second International Congress of the Society of Oriental Liturgy, Rome, September 17–21, 2008*, ed. Bert Groen, Steven Hawkes-Teeples, and Stefanos Alexopoulos (Louvain: Peeters, 2012), 227–67.
12 Peter Jeffery, "The Earliest Christian Chant Repertory Recovered: The Georgian Witnesses to Jerusalem Chant," *Journal of the American Musicological Society* 47 (1994): 1–39; and Jeffery, "The Lost Chant Tradition of Early Christian Jerusalem: Some Possible Melodic Survivals in the Byzantine and Latin Chant Repertories," *Early Music History* 11 (1992): 151–90.
13 Tinatin Chronz and Alexandra Nikiforova, "Beobachtungen zum ältesten Tropologion-Codex Sinaiticus graecus MΓ 56+5 des 8.–9. Jhs. mit Erstedition ausgewählter Abschnitte," in *Σύναξις καθολική: Beiträge zu Gottesdienst und Geschichte der fünf altkirchlichen Patriarchate für Heinzgerd Brakmann zum 70. Geburtstag*, ed. Diliana Atanassova and Tinatin Chronz, Orientalia—Patristica—Oecumenica 6 (Berlin: LIT Verlag, 2014), 1:147–74.
14 Egeria, *Itinerarium* 24.8–11; see Wilkinson, *Egeria's Travels*, 144–45.
15 Credulous but well-documented surveys include Bishop Auxentios of Photiki, *The Paschal Fire in Jerusalem: A Study of the Rite of the Holy Fire in the Church of the Holy Sepulchre*, 3rd ed.

(Berkeley, CA: Saint John Chrysostom Press, 1999); and Haris Skarlakidix, *Holy Fire: The Miracle of the Light of the Resurrection at the Tomb of Jesus; Seventy Historical Accounts (4th–16th c.)* (Athens: Elea Publishing, 2011).

16 Stig Simeon R. Frøyshov, "The Early Development of the Liturgical Eight-Mode System in Jerusalem," *St Vladimir's Theological Quarterly* 51 (2007): 139–78; and Peter Jeffery, "The Earliest Oktōēchoi: The Role of Jerusalem and Palestine in the Beginnings of Modal Ordering," in *The Study of Medieval Chant, Paths and Bridges, East and West: In Honor of Kenneth Levy*, ed. Jeffery (Woodbridge, UK: Boydell Press, 2000), 144–206.

17 Peter Jeffery, "The Sunday Office of Seventh-Century Jerusalem in the Georgian Chantbook (Iadgari): A Preliminary Report," *Studia Liturgica* 21 (1991): 52–75.

18 Solomon (Solon) J. Hadjisolomos, *Byzantine Music: The Modal Structure of the 11 Eothina Anastassima Ascribed to the Emperor Leo (+912): A Musicological Study* (Nicosia, Cyprus: Holy Monastery of Kykko, 1986); and Christian Troelsgård, "The Exaposteilaria Anastasima with Round Notation in MS Athos, Iberon 953," in *Studi di musica bizantina in onore di Giovanni Marzi*, ed. Alberto Doda, Studi e testi musicali, n.s., 6 (Lucca: Libreria Musicale Italiana, 1995), 15–28. On the Byzantine *octoechos*, see Christian Troelsgård's essay in this volume, "Byzantine Chant Notation: Written Documents in an Aural Tradition"; and Peter Jeffery, "Oktōēchos," in *The New Grove Dictionary of Music and Musicians*, 2nd ed. (London: Macmillan, 2001), 18:370–73. In the Byzantine rite, one starts on Sunday after Easter with mode one in both the psalms and the readings. After the eighth week, one is back to mode one in the psalms, but on to reading nine in the gospel. Thereafter one is out of kilter for the rest of the year until the system starts again. The first eight *troparia heothina* are sung to modes one to eight, the ninth, tenth, and eleventh *troparia* are sung to modes five, six, and eight.

19 The full series of texts can be found in Isabel Florence Hapgood, *Service Book of the Holy Orthodox-Catholic Apostolic Church*, rev. ed. (New York: Association Press, 1922; reprint, New York, then Englewood, NJ: Antiochian Orthodox Christian Archdiocese of North America, 1956, 1965, 1975, 1983, 1996), 29–31. Part of a twelfth–thirteenth-century manuscript, preserving the beginning of the first two Gospels and the first two psalm verses, is the "Willoughby Heothina Fragment," now in the University of Chicago Library, Goodspeed Manuscript Collection, MS Gr. 1342. A description with images can be seen at http://goodspeed.lib.uchicago.edu/ms/index.php?doc=1342&view=description (accessed August 30, 2015).

20 One can see this ritual, celebrated by the archbishop of Thessaloniki, in the video posted at www.youtube.com/watch?v=Ctk6X3LbDJQ, time code 3:30–6:49 (accessed August 30, 2015).

21 It appears already in a second-century Roman text known as *The Shepherd of Hermas* 54:1–2 (=Similitude 5, chapter 1). See Michael W. Holmes, Ed., *The Apostolic Fathers* (Grand Rapids, MI: Baker Academic, 2007), 568–69.

22 John Baldovin, *The Urban Character of Christian Worship: The Origins, Development, and Meaning of Stational Liturgy* (Rome: Pont. Institutum Studiorum Orientalium, 1987), 110, 124–25, 147, 149, 157.

23 Germain Morin, "Le plus ancien Comes ou Lectionnaire de l'Église romaine," *Revue Bénédictine* 27 (1910): 41–74. For a facsimile of the manuscript, see Hans Thurn, Ed., *Comes Romanus Wirziburgensis: Faksimileausgabe des Codex M. p. th. f. 62 der Universitätsbibliothek Würzburg* (Graz: Akademische Druck, 1968). See Baldovin, *The Urban Character of Christian Worship*, 125–26.

24 Antoine Chavasse, *Les lectionnaires romains de la messe au VIIe et au VIIIe siècle: Sources et dérivés*, 2 vols., Spicilegii Friburgensis Subsidia 22/1–2 (Fribourg: Éditions Universitaires, 1993).

25 Hartmann Grisar, *Das Missale im Lichte römischer Stadtgeschichte: Stationen, Perikopen, Gebräuche* (Freiburg im Breisgau: Herder, 1925).

26 Authoritative archaeological studies of the surviving buildings are found in Richard Krautheimer et al., *Corpus Basilicarum Christianarum Romae*, 5 vols. (Vatican City: Pontif. Istituto di Archeologia Cristiana, 1937–77). A good guidebook is Matilda Webb, *The Churches and Catacombs of Early Christian Rome: A Comprehensive Guide* (Brighton, UK: Sussex Academic Press, 2001).

27 Peter Jeffery, "The Roman Liturgical Year and the Early Liturgy of St. Peter's," in *Old Saint Peter's, Rome*, ed. Rosamond McKitterick et al., British School at Rome Studies (Cambridge: Cambridge University Press, 2013), 157–76.

28 Grégoire le Grand, *Homélies sur l'évangile, Livre II: Homélies XII–XL*, ed. Raymond Étaix et al., Sources Chrétiennes 522 (Paris: Éditions du Cerf, 2008), table 17, texts, 25–159.
29 See *Missale Romanum: Editio Princeps (1570)*, ed. Manlio Sodi and Achille Maria Triacca, Monumenta Liturgica Concilii Tridentini 2 (Vatican City: Libreria Editrice Vaticana, 1998), 353–63.
30 See David Hiley, *Western Plainchant: A Handbook* (Oxford: Clarendon Press, 1995), 529, 538–39.
31 The rite is reconstructed from all known sources in John K. Brooks-Leonard, *Easter Vespers in Early Medieval Rome: A Critical Edition and Study* (PhD diss., University of Notre Dame, 1988). See also S. J. P. Van Dijk, "The Medieval Easter Vespers of the Roman Clergy," *Sacris Erudiri* 19 (1969): 261–363.
32 Stephen J. Lucey, "Art and Socio-Cultural Identity in Early Medieval Rome: The Patrons of S. Maria Antiqua," in *Roma Felix: Formation and Reflections of Medieval Rome*, ed. Éamonn Ó Carragáin and Carol Neuman de Vegvar (Aldershot, UK: Ashgate, 2007), 139–58.
33 For example, an ancient icon of the Virgin and Child that has been housed for centuries in the church of S. Maria Nova (now better known as S. Francesca Romana) is believed to have been transferred there from S. Maria Antiqua at the time it was closed. A complex but fascinating theory about the origins and history of this icon is spelled out in Margherita Guarducci, *La più antica icona di Maria: Un prodigioso vincolo fra Oriente e Occidente* (Rome: Istituto Poligrafico e Zecca dello Stato, 1989). See the review by Paul Corby Finney in *Journal of Early Christian Studies* 1 (1993): 104–05.
34 Raymond Davis, Ed., *The Book of the Popes (Liber Pontificalis): The Ancient Biographies of the First Ninety Roman Bishops to AD 715*, Translated Texts for Historians 6, rev. ed. (Liverpool: Liverpool University Press, 2000), 90.
35 Elaine De Benedictis, "The 'Schola Cantorum' in Rome during the High Middle Ages" (PhD diss., Bryn Mawr College, 1983).
36 Federico Guidobaldi, Claudia Barsanti, and Alessandra Guiglia Guidobaldi, *San Clemente: La scultura del VI secolo*, San Clemente Miscellany IV, 2 (Rome: S. Clemente, 1992), 13–16, figs. 3–5, 38–40, 78, 89, 100, 106, 120, 122, 205.
37 Adrian Fortescue, *The Ceremonies of the Roman Rite Described*, rev. J. B. O'Connell, 5th ed. (London: Burns, Oates, & Washbourne, 1934), 4–5.
38 Juan Mateos, Ed., *Le Typicon de la grande Église*, 2 vols., Orientalia Christiana Analecta 165–66 (Rome: Pont. Institutum Orientalium Studiorum, 1962–63).
39 Mateos, *Le Typicon, de la grande Église*, 1:155, 175, 185.
40 Robert Mark and Ahmet Ş. Çakmak, Eds., *Hagia Sophia: From the Age of Justinian to the Present* (New York: Cambridge University Press, 1992).
41 For the story of the excavation, see Sven Larsen, "A Forerunner of Hagia Sophia," *American Journal of Archeology* 41, no. 1 (January–March 1937): 1–5.
42 Stefano Parenti, "The Cathedral Rite of Constantinople: Evolution of a Local Tradition," *Orientalia Christiana Periodica* 77 (2011): 449–69; and Alexander Lingas, "Sunday Matins in the Byzantine Cathedral Rite: Music and Liturgy" (PhD diss., University of British Columbia, 1996).
43 George P. Majeska, "Notes on the Archeology of St. Sophia at Constantinople: The Green Marble Bands on the Floor," *Dumbarton Oaks Papers* 32 (1978): 299–308.
44 For a discussion of Saint Symeon of Thessaloniki, see Walter Ray's essay in this volume, "Understanding Liturgy in the Byzantine Liturgical Commentaries."
45 The melody can be heard at www.youtube.com/watch?v=8ZSyTUsdpu0.
46 Mateos, *Le Typicon de la grande Église*, 2:94–97.
47 J. H. W. G. Leibeschuetz, *Antioch: City and Imperial Administration in the Later Roman Empire* (Oxford: Clarendon Press, 1972) 208–19, 224–42, 278–80; Frans van de Paverd, *St John Chrysostom, the Homilies on the Statues: An Introduction, Orientalia Christiana Analecta 239* (Rome: Pontificium Institutum Studiorum Orientalium, 1991); and Peter Jeffery, "Philo's Impact on Christian Psalmody," in *Psalms in Community: Jewish and Christian Textual, Liturgical, and Artistic Traditions*, ed. Harold W. Attridge and Margot E. Fassler, Society of Biblical Literature Symposium Series 25 (Leiden: E. J. Brill; Atlanta: Society of Biblical Literature, 2003), 147–87, esp. 180–84.
48 Marcel Metzger, Ed., *Les constitutions apostoliques*, 3 vols., Sources Chrétiennes 320, 329, 336 (Paris: Éditions du Cerf, 1985–87) 1:54–62, 2:10–39, 105; and Gary Philippe Raczka, "The Lectionary at the Time of Saint John Chrysostom" (PhD diss., University of Notre Dame, 2015).

49 Leo Depuydt, *Catalogue of Coptic Manuscripts in the Pierpont Morgan Library* (Louvain: Uitgeverij Peeters, 1993), vol. 1, no. 51.

50 Peter Jeffery, *The Secret Gospel of Mark Unveiled: Imagined Rituals of Sex, Death, and Madness in a Biblical Forgery* (New Haven, CT: Yale University Press, 2006), 79–86.

51 Louis H. Feldman, "The Case of the Blasphemer (Lev. 24:10–16) according to Philo and Josephus," in *Heavenly Tablets: Interpretation, Identity and Tradition in Ancient Judaism*, ed. Lynn LiDonnici and Andrea Lieber (Leiden: Brill, 2007), 213–26, esp. 219–20.

52 Franck Goddio et al., *Alexandria: The Submerged Royal Quarters* (London: Periplus, 1998). Skin divers explore the underwater palace in a Discovery Channel video posted at www.youtube.com/watch?v=gBja7TUobVE.

53 William La Riche, photographs by Stéphane Compoint, *Alexandria: The Sunken City* (London: Weidenfeld & Nicolson, 1996); and Jean-Yves Empereur, *Alexandria Rediscovered*, trans. Margaret Maehler, photographs by Stéphane Compoint (New York: George Braziller, 1998).

54 Judith McKenzie, *The Architecture of Alexandria and Egypt c. 300 B.C.–A.D. 700* (New Haven, CT: Yale University Press, 2007), 177–78, 240–44.

55 McKenzie, *Architecture of Alexandria*, 236–38, 271–87; Elizabeth Bolman, "The White Monastery Federation and the Angelic Life," in *Byzantium and Islam: Age of Transition, 7th–9th Century*, ed. Helen C. Evans (New Haven, CT: Yale University Press, 2015), 75–77, with further bibliography; and Bolman, Ed., *Monastic Visions: Wall Paintings in the Monastery of St. Antony at the Red Sea* (New Haven, CT: Yale University Press, 2002).

56 For Coptic melodies of "Christ Is Risen," see http://tasbeha.org/hymn_library/view/372.

2 The great outdoors

Liturgical encounters with the early medieval Armenian church

Christina Maranci

Introduction

> Then Solomon said: the Lord said he would reside in a dark cloud. I have built for you a dwelling place, a seat of stability for you to reside in forever.
>
> 1 Kings 8:12[1]

> My Lord . . ., send to this church the grace of your holy spirit, and, in the manner of the temple of Solomon, adorn and ornament it with a spiritual cloud of your glory, as thick darkness.
>
> The Prayer of Prince Ĵuanšēr, bk. 2, chap. 25[2]

In his study of early medieval Armenian church inscriptions, Timothy Greenwood noted Solomonic themes in a seventh-century account of a contemporary church consecration by the prince Ĵuanšēr, a text contained in book 2 of Movsēs Dasxuranc'i's tenth-century *History of the Caucasian Albanians* (*Patmut'iwn Ałuanic'*). At the moment of the consecration, Ĵuanšēr prays to the Lord to fill his church with the Holy Spirit in the form of a dark cloud, in the way that God entered the Temple of Solomon. The use of the Prayer of Solomon for a consecration ritual is, of course, most appropriate, and indeed, as Greenwood observes, 1 Kings 8 is read as a lection in most Eastern Christian liturgical rites for the dedication of a church (including the Armenian).[3] The shared themes of the ecclesiastical rite and historical text, remarked on in passing by Greenwood, form the point of departure for the present study. Considering three mid-seventh-century Armenian churches, I examine relief sculpture, epigraphy, and architectural settings in relation to early Armenian ritual, with particular attention to the hagiopolite, or Jerusalemic, meanings produced through a liturgical encounter with the Armenian church facade.

In so doing, there are several challenges that I face. The inherent difference between an abstract representation of organized movement and a specific physical setting hamper any straightforward application of texts to monument. Nor can we be sure that the rites, as preserved in the texts, existed at the time that the churches were constructed. Yet in my view it is a graver error to cast these texts aside because they cannot be grafted perfectly onto the architectural evidence. As early, if not contemporary, documentation for the experience of the church building, they allow us precious insight into the symbolic meanings of the church, as evoked through prayer, hymns, and movement.[4]

That Jerusalem in particular should be evoked through the liturgy is not at all surprising. Scholars have filled many volumes dealing with the Armenian experience of the Holy City, whether as a real or an imagined place.[5] Jerusalem was home to a community of Armenians from at least the fifth century, and the sixth and seventh centuries saw increased Armenian pilgrimage to and settlement in the region, as attested by written sources, epigraphy, and archaeological evidence. The seventh-century *Geography* attributed to Anania Širakac‘i referred to Jerusalem at "the center of all," like many medieval geographies, and to Armenia as the "northern region."[6] Seventh-century Armenian sources chronicle events in the city and their reception at home; they also include extraordinarily detailed descriptions of the holy places and relics and an abiding concern for the monuments, their destruction at the hands of the Persians, and their subsequent renewal. The central role of Jerusalem in the Armenian liturgy is demonstrated by the Armenian Lectionary, a precious fifth-century text preserving in detail the rites celebrated in the Holy City. The subject of much scholarly attention, this text offered to congregations in early medieval Armenia an imaginative topography of Jerusalem in which they could commemorate and enact Christ's Passion.[7]

For these reasons, scholars have long understood the built culture of early medieval Armenia in terms of Jerusalem. Armen Kazaryan has drawn parallels between images of the tomb aedicula of Christ, with its peaked roof and twisted columns, and the design of the drum that crowned the Cathedral of Vałaršapat.[8] The same scholar drew a persuasive comparison between the aedicula and the liturgical furniture at the church of Zuart‘noc‘, as we will discuss below. The most thoroughgoing hagiopolite interpretation of Armenian architecture is *La Jérusalem nouvelle et les premiers sanctuaires chrétiens de l'Arménie*, in which Nazénie Garibian de Vartavan suggests that the layout of the churches in the holy cities of Vałaršapat and Mtskheta in Georgia is based on the topography of the principal holy sites in Jerusalem.[9]

The liturgical dimension of this discussion has received little attention. Yet in light of the nature of Armenian architecture and early medieval ritual directives, the opportunities for its examination are rich. The prominent exterior position of relief sculpture and epigraphy on Armenian monuments invites us to reflect on the possible role of the exterior facades, as well as the church interior, in shaping the experience of the early medieval churchgoer. As we will see, this encounter was inherently multi-sensory and kinetic, requiring seeing and reading, singing, climbing, carrying, and smelling (the anointed walls), and as such offers an important tool for interpreting the many engraved and sculpted church exteriors of the Armenian architectural tradition. In undertaking this task, I make use of the scholarship on the rite of the Armenian church consecration and the Hymns (or *šarakans*) of the Holy Cross, as well as the aforementioned fifth-century Armenian Lectionary.

Mastara

Mastara (also known as Mazdara) is one of more than seventy seventh-century churches preserved from the regions of historic Armenia, today divided among the Republic of Armenia, the Republic of Arc‘ax (Mountainous Łarabał), eastern Turkey, Azerbaijan, and northern Iran (Figure 2.1).[10] Located in the Aragacotn Province of the Armenian Republic, Mastara is dated by its epigraphy to between about 640 and 650. As is typical of Armenian and Georgian architecture, it is constructed of rubble masonry, consisting of a thick core of mortar and fieldstone faced by squared, well-joined slabs

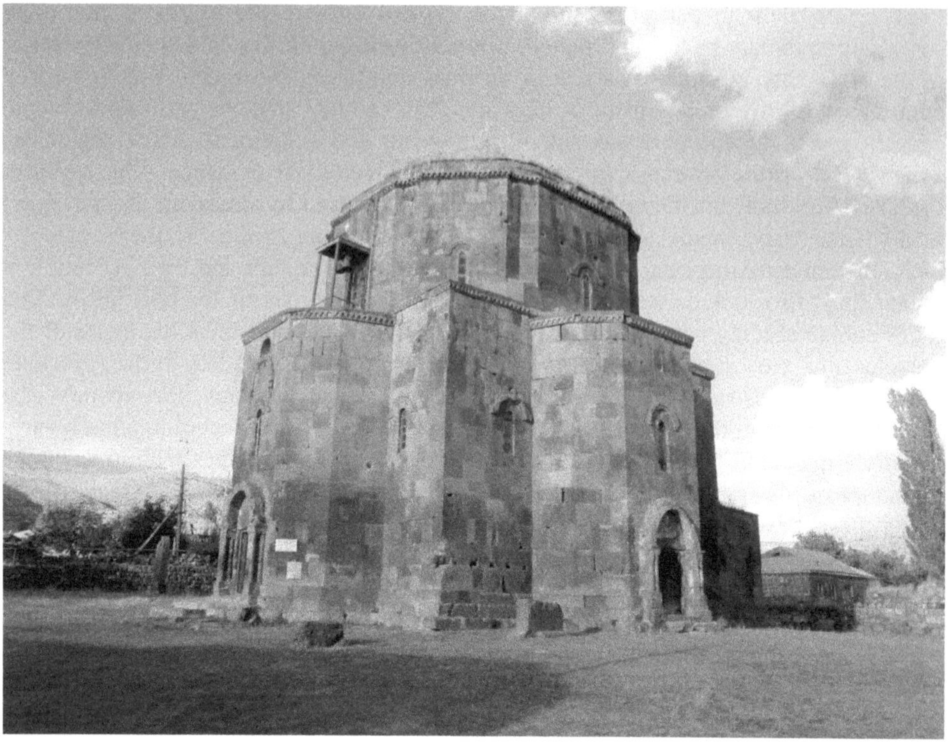

Figure 2.1 Church of Mastara, Republic of Armenia, ca. 630, view of the exterior from the west. Photograph: Wikimedia Commons.

of tuff. The exterior is strongly geometric, dominated by the tall central mass, and elevated on a stylobate. The plan is centralized, with a large dome set into the corners of a square bay, from which four conches project (Figure 2.2).[11] The dome is set on squinches, and the interior is defined by the rhythm of apsidal curvatures, squinches, and smaller squinches above.

Four inscriptions on the exterior of the church attest to the historical circumstances of its construction. On the western section of the southern elevation, a fragmentary text reads, "Of the month Arac‘ [day 14] at the consecration of this holy church and to the memory of bishop . . . [illegible words of uncertain number]."[12] Another is located on the southern facade, on the central (southern) facet of the projecting apse, above and on the arched frame of the window over the entrance: "In the years of Lord T‘edoros bishop of Gnunik‘ this holy house was built to expiate the unworthy Grigoras. Christ God, be compassionate to Grigoras sinner and to me Kep’[. . .] and [–]"[13] On the western elevation, on its southern part, a text reads:

> I thank God who permitted me Grēgoras Siwni and beloved nephew Grigor to build a house of glory and through this made me . . . [illegible words] bishop of Apahunik‘. This is a refuge [*apawēn*] for Mazdara, a place of prayer for the faithful, a place of expiation for sinners, and a memorial for me and for mine. And you who pray, remember us. . . .[14]

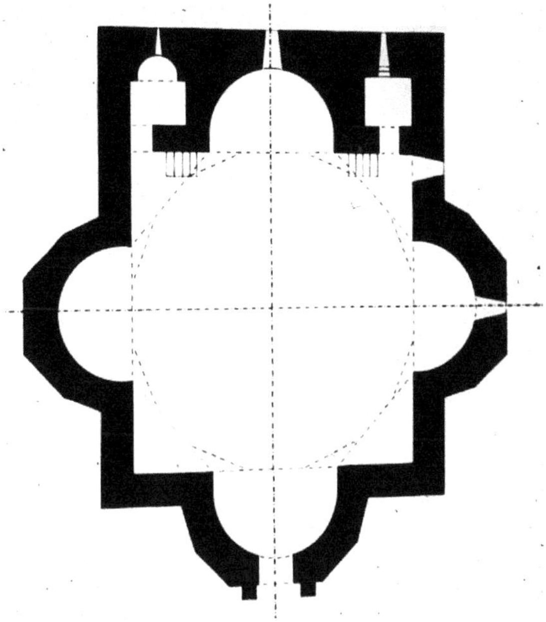

Figure 2.2 Church of Mastara, plan.
Photograph: Wikimedia Commons.

Putting together this information, historians have surmised that the church was constructed during the episcopate of T'eodoros Gnuni (c. 645) by the monk Grigoras Siwni and his nephew, for the expiation of their sins and as a refuge for Mazdara. We are also provided with a date of Arac' 14 (November 30) for the date of consecration.[15]

On the west facade, a fourth inscription has been given particular visual emphasis (Figure 2.3). It appears within a blind arcade over the west window. The text is arranged around and below a sculpted cross on a pedestal. Although the cross is badly weathered, we can see clearly its stepped podium, the flared ends of its arms, and what seem to be tendrils or wings extending from its base. The inscription reads as follows:

> Ա(ՍՏՈՒԾՈ)Յ ԱՃՈՂԵԼՈՎ ԳՐԻԳՈՐԱՍԱ(Յ) ՎԱՆԱԿԱՆԻ ՇԻՆԵՑԱՒ ԱՊԱՒԷՆ ՄԱԶԴԱՐԱՒ ԱՅՍ ԿԱԹՈՂԻԿԷ ՀԱՐՍ ՆԽԱՋ(ԱՆ)ՇԱՆ ԹԱԳԱՒ ՊՍԱԿԵԱԼ ՈՒՆԻ ՓԵՍԱՅ ԶՔՐԻՍՏՈՍ ՓԵՍԱՒԷՐ ԶԱՌԱՔԵԱԼՍ ՄԱՐԳԱՐԷՍ ԶՎԿԱՅՍ ՍԱ ՊԱՐԵՇԷՆ ՈՒՆԻ ԶՄԱՍՏԱՐԱ ԵՒ ՓՐԿԷ ԶԳՐ(ԻԳՈՐՈՍ)
>
> (Through God's augment of Grigoras the monk this cathedral was built as a refuge for Mazdara. The bride crowned with the cross-signed crown has as bridegroom Christ and as bridal companions the apostles, prophets and martyrs. Keep Mazdara prosperous and save Grigoras)

As Greenwood pointed out, the reference to the "bridegroom of Christ" is from John 3:29.[16] The same imagery, with greater emphasis, also occurs in two Armenian ritual contexts: the rite of consecration and, much more robustly, the sequence of hymns devoted to the Feast of the Exaltation of the Cross.

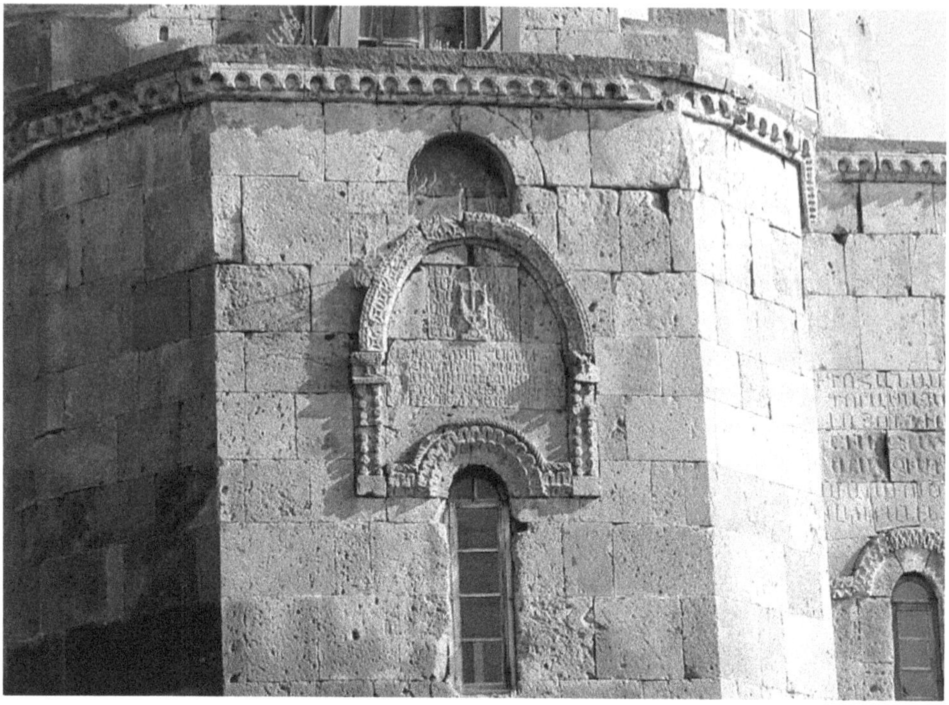

Figure 2.3 Church of Mastara, west facade inscription.
Photograph: Wikimedia Commons.

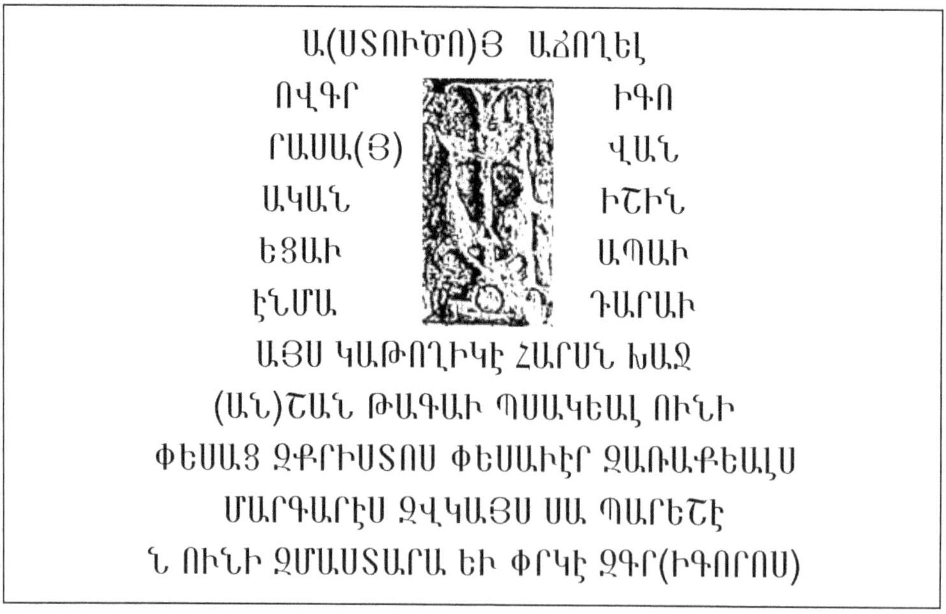

Figure 2.4 Church of Mastara, schematic drawing of west facade inscription with sculpted cross.
Drawing by author.

A 1998 study of the Armenian church dedication service by Michael Daniel Findikyan invites us to meditate on the liturgical imagery of the engraved portal and its potential ritual context. Findikyan collated three early textual accounts of this rite: a *maštoc‘*, or ritual, probably of the late ninth century, and two allegorical commentaries on the consecration rite, both dating to the first half of the eighth century, one by Yovhannēs Ōjnec‘i and the other attributed to Step‘anos Siwnec‘i.[17] All three of the texts prescribe the withdrawal from the church building and the performance of exterior services equipped with a cross. First, the altar table is carried out of the church, as the congregation gathers around it singing psalms, after which the altar table is reinstalled within the church and elevated to the bema. The next exterior unit is the "Naming of the Church" and the blessing of the exterior walls. At this point, the clergy and congregation depart the monument, and the bishop declares in whose name it has been erected, making a circuit around the church. The allegorical commentary attributed to Step‘anos Siwnec‘i further mentions "tracing the Lord with the cross" on the exterior and anointing the four sides of the building. This exterior moment is felicitous in light of the exterior epigraphy of Mastara.

We can find in these texts nuptial imagery. During the Introit, when the congregation and clergy approach the door of the church carrying the altar, the bishop traces the sign of the cross over the door and then opens it. Ōjnec‘i, observes the appropriateness of this action: "for by the cross Christ opened the entrance to paradise and to the heavenly bridal chamber [*erknayin aṙagasan*]." This interpretation of the Introit of the church consecration thus offers a fitting liturgical moment for the Mastara portal, which not only contains the nuptial and cross imagery but also is located over the western door of the church: on the threshold, therefore, of "the heavenly bridal chamber."

Another text warrants attention in this regard. Findikyan does not include it in his reconstruction of the rite, because presumably it did not occur in any of the three early medieval texts he used, but it is one of two hymns appended to the rite as recorded by Frederick Cornwallis Conybeare. Conybeare tells us that the text is drawn from three manuscripts of the *Ganjaran*, or *Book of Canticles*: the first, in BM Or. 2609, entitled "A Canticle of the Shołakat' [lit. effusion of light] of the Consecration of the Holy Church"; the second, in BM Or. 2608, where the text is called the "Canon of the Holy Ark and Ecumenical Church"; and finally in the third, Vienna Mekhitarists MS 133, where it is titled "A Canticle of the Holy Church." The index of this final manuscript, Conybeare comments, "ascribes this canticle to one Mkrtitch, who perhaps in the thirteenth century compiled it out of earlier material."[18] While we cannot be at all sure that this hymn in its preserved form dates from the early Middle Ages, nor that it was sung at the dedication rite, it is rife with bridal imagery:

> Daughter of Ancient Sion, receiver of the message, *to thee the Bridegroom Christ hath condescended, bringing thee an unfading wreath, by will of Father and of Spirit crowned*. Lo, the Bride gorgeously arrayed in her glory goes forth to meet the Lord the King who is come out to meet (her). Into the Holy pavilion invited, The Bridegroom Christ, the Sovereign, is arrived. The children of the Church encircle him and utter songs of praise . . .[19]

The crowned bride, we are then told in the following passage, is accompanied by the twelve apostles, the holy prophets, the holy pontiffs, and the blood of the holy martyrs—an ensemble of figures that brings to mind the Mastara inscription, with its crowned bride and her companions, the "apostles, prophets, and martyrs."[20]

We find the greatest liturgical parallels with the inscription, however, in the many hymns sung during the Feasts of the Cross as observed already by Patrick Donabédian in a footnote to his discussion of the church of Mastara in a study of 2008.[21] These hymns, studied in depth by Athanase Renoux, have much more recently formed the focus of publications by Findikyan, and I provide here a selection of pertinent passages from his English translations.

> Canon for the Dedication of the Holy Cross
>
> 4. At the newly-marvelous Dedication in Jerusalem [նորահրաշ նավակատիքն որ յերուսաղեմ] *your cross* was shown to us in radiant majesty, O Lord, God of our fathers. . . . *A queen stands on the right, the holy Church, crowned in gold braids, in the sign of your cross* [նշանաւ խաչի], O God of our fathers. . . .[22]
>
> 5. Bless the Lord and exalt him forever.
>
> *For the holy Church is betrothed to Christ. The heavenly bridegroom has crowned her with the cross; to the left and to the right it takes wing*, making heirs of nations.[23]
>
> 19. Faithful people, let us always sing a triumphant and new blessing in the highest to Christ the king. *Who came to illuminate his chosen, holy church. And he crowned her with his holy cross*. Let us sing his glory. Today we too celebrate the Dedication [of] the Holy Cross. And to the Saviour we offer glory and honor forever.[24]
>
> 39. *The Heavenly Bridegroom has come near you. Granting your salvation, he has crowned you with his wondrous glory*.[25]
>
> 41. *Rejoice O Holy Church, for Christ the king of heaven today has crowned you with his cross, and he has adorned your fortress with his wondrous glory. . . .* With the choirs of the heavenly hosts, we celebrate today and lift up unceasing glorification. *Be glad, immaculate Bride, in your inscrutable mystery*.[26]

As in John 3:29, the texts above refer to the bride (*hars*) and the bridegroom (*p'esa*); the hymns, unlike the biblical text, indicate that the Church is the bride. In no fewer than four of the verses, the bride is crowned with, or with the sign of, the cross. In the hymns this is rendered as "nšxanaw xač'i psakeal"; in the Mastara inscription, as "xač'anšxan t'agaw psakeal."

At Mastara, the bas-relief cross dominates the center of the composition both visually and thematically (Figure 2.4). The text is positioned in lines on either side of the form, requiring the viewer to pass over the cross in order to make sense of the names "Grigoras" and "Mazdara." In the latter case, the Z (Զ) is engraved into the base of the stepped podium, constituting a kind of decorative form.

By its position at the base of the cross and its circular shape, this letter might have also reminded the spectator of the "place of the skull," or, literally, Golgotha. As mentioned, the cross above it is of the Latin type, with flared arms, a form widely known from early Byzantine and Armenian metalwork and sculpture. Because of its base, it further resembles what is variously called a stepped, graded, or Calvary cross, the last term making clear reference to the mound on Golgotha where the shrine of the Crucixion is located. In seventh-century Byzantine art, this stepped-cross type is used on the solidi of the Byzantine emperor Heraclius in reference to his victorious restoration of the True Cross to Jerusalem in 630—an event to which we will return in a later section. In other ways, the Mastara cross resembles Armenian stone crosses of the same era, such as those from T'alin and Duin, from the base of which project a pair of tendrils or wings. A hymnographic reading of the Mastara cross, I would argue, invites the latter interpretation: recall the imagery of the bridegroom crowning the bride with a cross that "to the left and to the right . . . takes wing [*ew yaǰ ew yaheak t'řowc'eal*]." The coordination of the cross and text certainly support, in my view, Donabédian's passing observation on the source of the Mastara inscription. These correspondences allow us to imagine the inscription and, more broadly, the church's west and south walls as settings appropriate to the hymns sung on the Dedication of the Cross.

It is relevant, then, to consider Findikyan's further arguments regarding the date and origins of these hymns. While they are often dated by tradition to early eighth-century Armenia, Findikyan suggests that their content, arrangement, and vocabulary indicate a much earlier date and, in a related point, their associations with the Dedication of the Holy Places of Jerusalem (the Encaenia).[27] That is, in Findikyan's view, they were originally sung to commemorate the Encaenia and then later became associated with the Feast of the Exaltation of the Cross.[28] The evidence he brings forward is compelling; the hymns make mention, as we have seen, of the "newly-marvelous Dedication in Jerusalem" and are rife with ecclesiological and architectural imagery.[29] Further, the dates of the Feast of the Exaltation of the Cross coincide precisely with those of the Encaenia (seven days beginning September 13). Findikyan finally observes that the term "cross" that occurs in the hymns could have referred to the shrine of Golgotha —and not the cross itself as a relic—as it was used in the account of the fourth-century pilgrim Egeria, in the fifth-century Armenian Lectionary and in the tenth century Georgian Book of Hymns (***Iadgari***). [30]

We cannot know, of course, how much of this putative hagio-dedicatory meaning would have resonated in seventh-century Armenia nor, more particularly, at the church of Mastara. But it is tantalizing to think that these early festal hymns were sung on the day and subsequent anniversaries of the consecration of Mastara, and that they evoked the dedication of the holy places in Jerusalem. For the consecration of a church, such imagery would be both entirely appropriate and historically compelling, to judge from Ĵuanšēr's evocation of Solomon's Temple with which we began. What seems certain is that consideration of the early ritual sources opens up new ways to interpret Armenian church walls, and that the thematic parallels drawn above, interesting in their own right, also point toward a powerful experience of the church, not only in visual but also in aural terms.

Zuart'noc'

The church of Zuart'noc' was constructed as part of the residence of Nersēs III, patriarch of Armenia between about 641 and 661 (Figure 2.5). This structure collapsed in an earthquake in the early eleventh century but is well documented by contemporary sources and preserved archaeological remains. Built to commemorate a heavenly vision of Gregory, the patron saint of Armenia, the structure is closely tied to the country's sacred landscape. Yet, as many have pointed out, it also shows familiarity with the cultural traditions of Byzantine Constantinople, Syria and Mesopotamia, and the Holy Land. The program of epigraphy includes not only Armenian but also Greek, signaling,

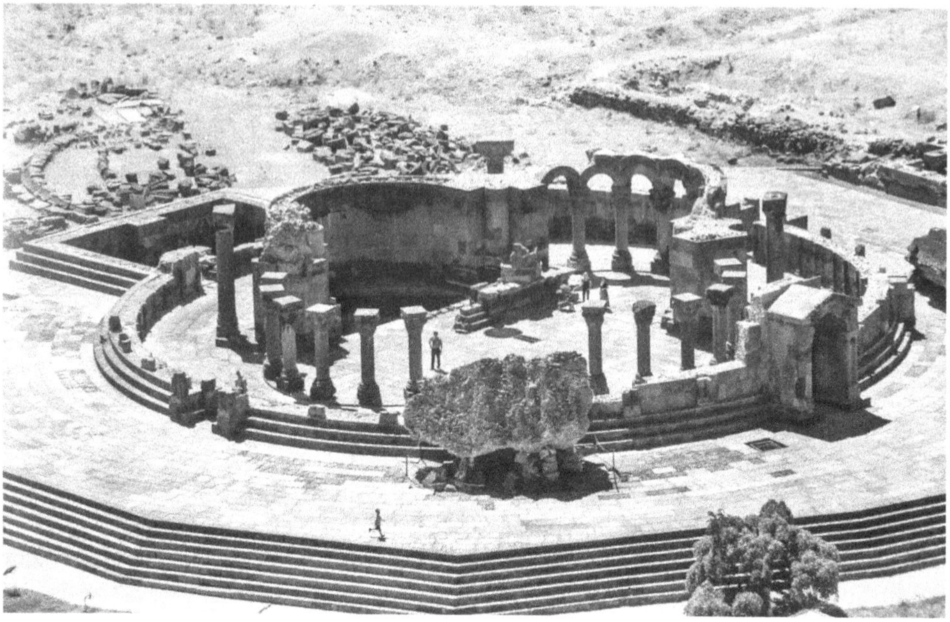

Figure 2.5 Church of Zuart'noc', Republic of Armenia, ca. 641–ca. 661, aerial view.
Photograph courtesy of Hrair Hawk Khatcherian.

many believe, the close relations Nersēs held with the Byzantine Empire. Most interesting in the present context is the unique plan of the church (Figure 2.6). The inner shell consists of a series of columnar exedrae (the earliest established example in the Southern Caucasus) joined by large, W-shaped piers, which once supported the dome. This tetraconchal shell was enveloped by a quasi-circular perimeter wall. The entire structure was elevated on a tall pedestal of seven steps, broad enough to accommodate a walkway around the exterior walls of the church.

While many scholars have drawn attention to the aisled tetraconchal shape of Zuart'noc' and its relations to the Syrian and Mesopotamian monuments of the same type, the circular plan of the monument finds its most obvious prototype in the martyria of the Holy Land, above all, the Anastasis Rotunda. Completed by 336 to shelter the traditional site of Christ's burial and resurrection, this structure formed the focal point of Christian Jerusalem and, indeed, of medieval Christendom more generally. The formal resemblances of the Anastasis Rotunda and Zuart'noc' have already elicited commentary from scholars either in passing or in more depth.[31] In the mid-twentieth century, Step'an Mnac'akanyan was the first to connect the two monuments; the relationship was recognized also by L. Durnovo, who in a 1952 essay speculated about the possibility of Jerusalemic imagery in the applied arcades of early medieval Armenian architecture (and also in the design of canon tables in manuscripts).[32] Recently, Dora Piguet-Panayotova, Zaruhi Hakobyan, Nazénie Garibian de Vartavan, and Armen Kazaryan have produced more comprehensive examinations of the problem. Piguet-Panayotova has suggested that the Anastasis Rotunda provided the "fundamental elements" of Zuart'noc'. Accepting the reconstruction of Zuart'noc' by T'oros T'oramanyan, she argued that both monuments shared the form of superimposed

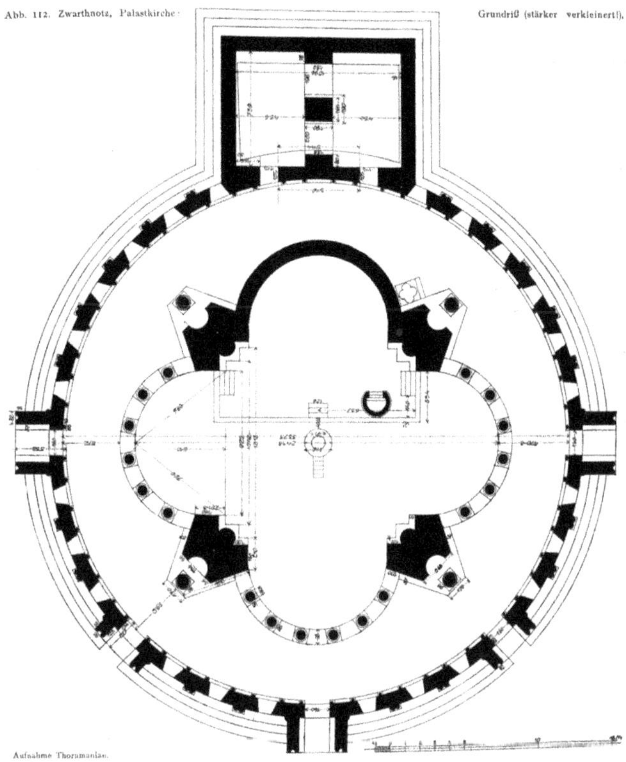

Figure 2.6 Church of Zuart'noc', plan.
T'oros T'oramanyan, in Josef Strzygowski, *Die Baukunst der Armenier und Europa* (Vienna: Anton Schroll, 1918), 113, fig. 112.

cylinders, the first enclosing the ambulatory, and the second enclosing a gallery level.[33] She also drew a correlation between the Ionic basket capitals at the rotunda and on the Temple Mount with those of Zuart'noc'. More recently, Kazaryan proposed a reconstruction of the liturgical space of Zuart'noc' based on existing archaeological materials at the church. He envisioned a partitioned enclosure under the domed space of the church and surrounding the cylindrical cavity at its center.[34] From the measurements of this crypt, Kazaryan surmised that a cylindrical stone object at the site, previously thought to be an ambo, was originally positioned over the cylindrical pit in the center of the church. This construction, in his view, was designed to mark the relics of Saint Gregory, and, judging from its form, was inspired by the Holy Sepulchre.

The exterior sculpture of Zuart'noc' has also been linked to the image of Jerusalem (Figure 2.7). The arcades of the first tier are composed of double colonnettes crowned with capitals from which spring molded arched frames adorned with a rinceau of grape bunches and leaves. Above the rinceau, a large sculpted field displays moregrape-vines, bunches of fruit, and trees with straight branches from which pomegranates hang. A horizontal stringcourse limits this zone, above which appear oculi framed by diverse ornamented moldings. The first tier of the building is crowned, finally, with a cornice that includes a running band of strapwork.

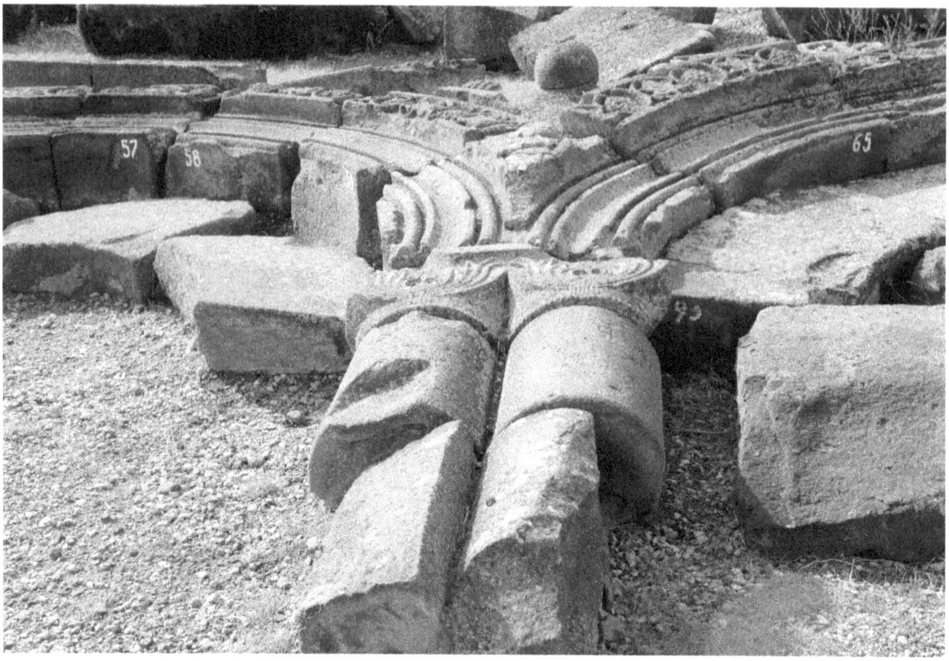

Figure 2.7 Church of Zuart'noc', fragments of the first tier laid out on the ground. Photograph by author.

Donabédian associates the decoration of the exterior arcade of Zuart'noc' with the Temple of Solomon.[35] He has put forward the descriptions of the temple in 1 Kings 7 and 2 Chronicles 4:12–13 as possible inspiration for the vegetal and interlace imagery of Armenian churches.[36] The Zuart'noc' arcade certainly provides an extraordinarily dense and copious array of associations with the temple, particularly as detailed in the book of Kings:[37]

> And he made two covering lattices for the capital and also two covering lattices for the second capital, and hanging work.
>
> And bronze pomegranates in a grating, a hanging work, row upon row. And in that way he made the second capital.
>
> And on the tops of the columns, there was lily work, four cubits long, near the arcade.[38]

Although the specific spatial relations of the building elements in this passage cannot be made to conform to the exterior of Zuart'noc', we should nevertheless recognize the coincidence of three motifs: the latticework, the pomegranate, and the lily form of the arcade capitals.[39]

If much ink has been spilled on the hagiopolite associations of the exterior facade of Zuart'noc', the degree to which certain ritual contexts would have activated these meanings has received little attention. As with the facade at Mastara, I propose a scenario in which this perimeter wall formed part of the physical setting for the performance of hymns sung during the rite of the church consecration. We have already mentioned the procession of entry into the church with the altar. This procession is accompanied by Psalms 119–21: Psalm 119, "In my distress I cry to the Lord"; Psalm 120, "I lifted up my eyes to the hills, from whence my help comes"; and Psalm 121, "I was glad when they said to me, 'Let us go to the house of the Lord!' Our feet have been standing within your gates, O Jerusalem!"[40] Findikyan points out that Yovhannēs Ōjnec'i refers to these three psalms as "gradual psalms [*sałmosk' astijanac'*]"and suggests that this may reflect the general belief that they were sung by pilgrims climbing Mount Zion to the Temple of Solomon.[41] The concept of ascent, he continues, is illustrated in the "crescendo from abject despair, through acknowledgement of God as protector to rejoicing for having arrived at Jerusalem."[42] One can imagine how effective such a psalmody would be while climbing the steep podium of Zuart'noc': the themes of the psalms, the built landscape, and the accompanying physical movement would have worked together to re-create the experience of pilgrimage to the Holy Land.

The bas-relief human figures within the spandrels of Zuart'noc' may also be understood within this context (Figure 2.8). Eleven are preserved, all holding tools of various sorts. Forming an unusual (although not unprecedented) iconographic subject for the era and region, they have been variously interpreted as holy persons or (mystifyingly) as patrons. Some have proposed that they represent Saint Gregory and the pagan king Trdat building

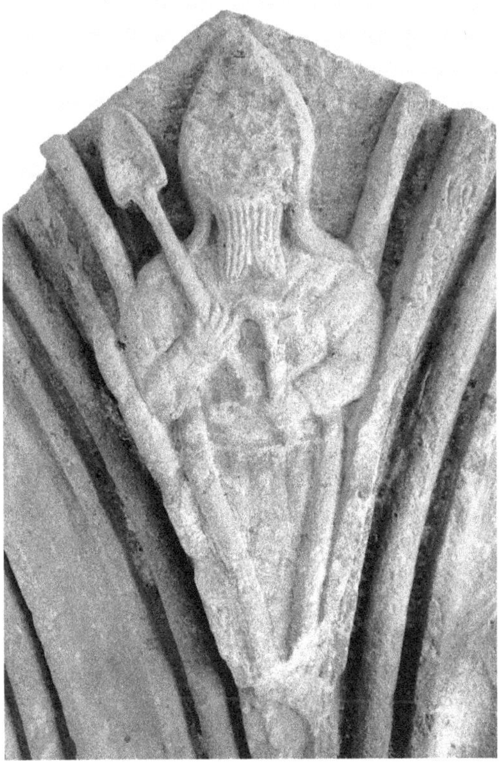

Figure 2.8 Church of Zuart'noc', fragment of figural bas-reliefs.
Photograph by author.

the first Christian shrines in Armenia—a passage that forms part of the fifth-century conversion story. Their costume, their lack of halos, the specific tools they hold, and the fact that they are shown actually in the process of working on the spandrels, encourages, in my view, a direct interpretation as a team of builders and workers.

But whether or not we wish to assign to them specific and stable identities, one ought also to consider how the performance of the consecration rite could have inflected their interpretation by the medieval churchgoer. Directly after the singing of Psalms 119–21, another Psalm, 117, is sung three times, after clergy and congregation have processed toward the church and arrived at its doors. The psalm is first recited outside the door, and then is repeated until verse 19: "Open to us the gate of mercy." This verse is repeated three times, the sign of the cross is made over the door, and the bishop intones verse 20: "This is the door of the Lord and only the righteous shall enter." Then, finally, Psalm 117 is repeated and the bishop and clergy enter. Verse 22 is of particular relevance to the Zuart'noc' imagery: "the stone which the builders rejected has become the cornerstone of the church."

Recognizing the general appropriateness of this psalm, with its imagery of entrance and its mention of builders, Findikyan has called it an "ideal accompaniment to the procession into the newly-built church." For Zuart'noc' in particular, the sung themes of lifting up the eyes, arriving at the gates of Jerusalem, and builders and stones would have accorded well with the visual program of the church. Unlike the larger, strongly projecting vegetal forms of the facade, the builders are carved in shallow relief, and indeed are most clearly apprehended from close to the wall surface. This suits the liturgical prescription to sing verse 22 just at the entrance to the church, following the gradual psalms describing the approach to Jerusalem. In this way, the eyes and ears of the early medieval participant were filled with the evocation of the Holy City just as their bodies, after a procession to any one of the five entrances of Zuart'noc'—whether undertaken by circumambulation around the building on its paved walkway or by climbing the steep steps of the podium—had undergone physical work akin to that of pilgrimage.[43]

Finally, it is noteworthy that Psalm 117 itself was sung by pilgrims in Jerusalem, as described by the fifth-century Armenian Lectionary. At the end of his discussion of Psalm 117 in the consecration rite, Findikyan mentions that it is prescribed with varying refrains in three contexts: during a procession from the house of Caiaphas to Golgotha during the vigil of Holy Friday, during the Paschal vigil preceding the lections, and, most relevant to the present argument, during a procession from the Mount of Olives to the Anastasis (*Surb Hariwt'iwn*—Holy Resurrection) on Palm Sunday. In this last processional, the Lectionary makes clear the action of the participants as "descending" and "psalm-singing [*sałmoselov*]."[44] How much of this was known to the churchgoer in early medieval Armenia cannot be ascertained. Yet it is possible to argue that when sung in the approach to Zuart'noc', Psalm 117 formed part of a coherent and multidimensional hagiopolite experience, generated not just by the round shape of the church, its sculpted walls, or the thematic content of the psalm but also by the association of the psalm with pilgrims processing to the Anastasis Rotunda.

The church of Mren and the liturgical image

The seventh-century church of Mren is located in what is now eastern Turkey in a military zone next to the closed Armenian border (Figure 2.9).[45] Mren is well known

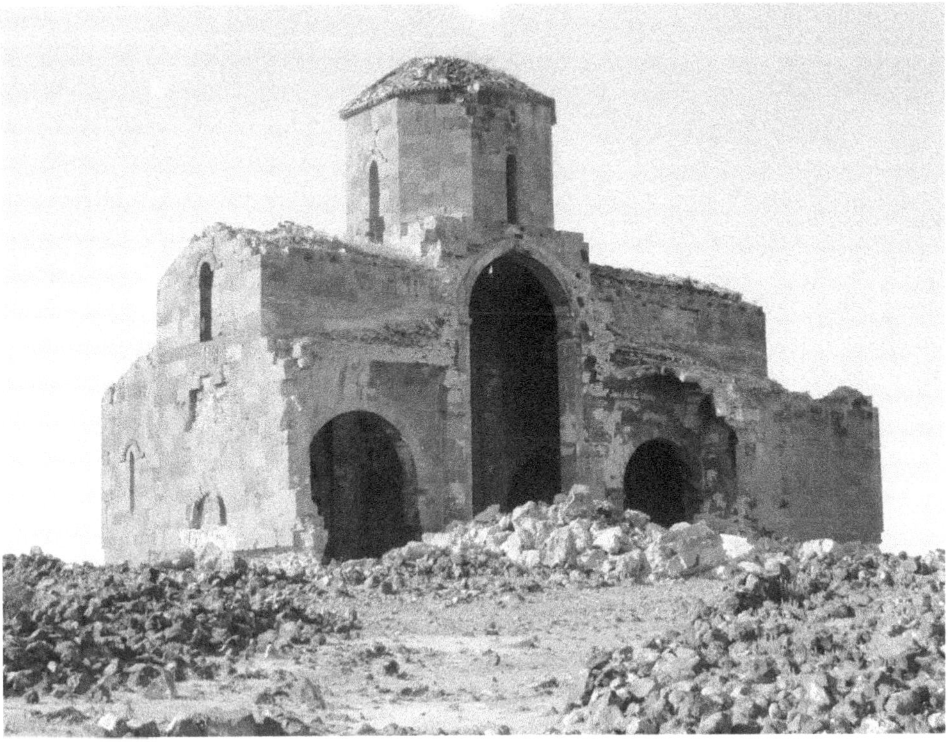

Figure 2.9 Church of Mren, Kars Province, Republic of Turkey, 638, view from the southwest. Photograph by author.

to historians of Byzantium and Armenia: dating to about 638, its epigraphy asserts interactions between the emperor Heraclius and the Armenian nobility and to the imperial goal of consolidating the eastern frontier against Persian attack.[46] Mren is additionally famous for its sculpted reliefs, one of which deserves special mention. On the north facade is a portal with a lintel (at present, unsupported and unsecured on its left side) bearing images of a horse, a tree, three human figures, and a central cross (Figure 2.10).

While each of the forms on the lintel is fairly easy to discern, deciphering the meaning of their combination has sustained decades of debate. Many of the earliest theories associated the lintel with a princely scene, making particular reference to the presence of the horse.[47] In 1966, Minas Sargsyan suggested that it depicts a church foundation, enacted by the cleric and nobles named and portrayed on the west portal.[48] In 1971 and more fully in 1997, Nicole Thierry detailed a series of problems with Sargsyan's argument and proposed instead that the scene represents the return of the True Cross to Jerusalem by Heraclius in 630.[49] Signaling the invocation of the "triumphant [*bareyałt'oł*] King Heraclius" on the west portal inscription, Thierry identifies the emperor as the left-hand figure on the lintel, honoring a cross intended to symbolize the relic. The larger censing figure at right represents, in her view, Modestos, bishop of Jerusalem, who received the relic from Heraclius. This interpretation has attracted the support of many Armenologists and Byzantinists, and I have recently adduced more evidence for this argument in the form of two early medieval Latin accounts of

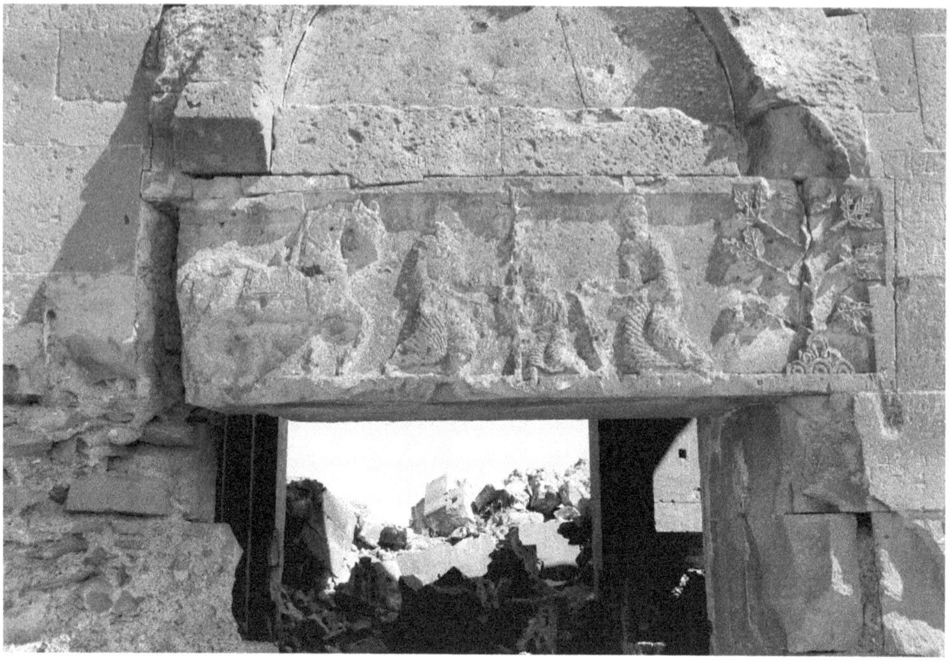

Figure 2.10 Church of Mren, north facade portal lintel.
Photograph by author.

the Return of the Cross, which offer a textual explanation for the unusual representation of Heraclius without crown or diadem and dismounted.[50]

Yet this identification does not account fully for the lintel at Mren or, more specifically, the strong ritual character of the scene.[51] The large incense burner, at the backswing of its movement, is a type well known from contemporary Byzantine examples in bronze. The composition focuses our gaze on the central cross, which is addressed by all the figures. With decorative branches at the corners of each arm, the cross, like that of Mastara, bears the morphology of late antique and early medieval examples from Byzantium and Armenia, known both from metalwork and from pictorial representation.[52] These features led Nicole Thierry to regard the scene at Mren as an "imaginative and reduced" image of the return of the Cross by Heraclius, one that fused the historical event with ritual meaning.[53]

Again, the dedication rite is of particular value in understanding the church facades, and especially the procession of reentry accompanied by Psalms 119–21.[54] For a worshipper approaching the north portal at Mren, the imagery of the psalmody would have been particularly germane. Singing the first-person lines, "I lifted up my eyes to the hills, from whence my help comes," the participants' gaze would have traveled from the altar stone, removed from the church interior, to the portal, where it would be met by the central cross, supplicants, and the magnificent sculpted tree on its mound. As the words were sung, the visitor's eyes would, truly, be "lifted" to the scene on the portal. While we cannot be sure that this rite was performed at Mren, it is nevertheless instructive that at an early date in the formation of the Armenian liturgy, a procession into the church was understood in terms of entry into Jerusalem.

The northern position of the portal also holds special significance in this context. Of the lateral sides of the church in early medieval Armenian architecture, the south, rather than the north, facade was typically preferred for access and epigraphy.[55] By contrast, at Mren, moving south through the north portal oriented the spectator not only toward the sacred space of the church but also, at least symbolically, toward Jerusalem.[56] This axis of approach could have evoked the arrivals of Christ and, later, Heraclius to the holy city. Such a procession would also have followed the southward progress of the Heraclian campaigns of 627 to 628, which descended via the Axurean river valley, quite near Mren, into Persian territory: an operation whose success led ultimately to the surrender of the holy relics to the Byzantines.[57] The north portal may thus have recalled, at once, memories of recent military campaigns, of the imperial *adventus*, and of the sacred narratives of the Holy Land. The city gate, although absent in the bas-relief, may be interpreted as the architectural threshold. While the medieval church portal has long been viewed as a topos for the gates to the holy city, the north portal at Mren, particularly when read together with the Armenian liturgy of dedication, presents an early and forceful expression of this concept.

Conclusion

I have sought to understand Armenian architectural facades through liturgical rite. At Mastara, the sculpted cross and surrounding inscription were presented in relation to the liturgy of church dedication, with its bridal themes, and also in relation to the hymns of the cross, themselves recently connected to the archetypal Christian dedication service: the Jerusalem Encaenia. At Zuart'noc', a monument already associated strongly by scholars with the Temple of Solomon and the Rotunda of Jerusalem, the Introit of the dedication rite would have compounded themes of the pilgrimage to and within the holy city. The participant approached the church while singing of the arrival at the gates of Jerusalem (Pss. 119–21), and at the doors of the church, he or she would sing of the "builders" in Psalm 117. In this context, the ritual directives would have attuned the participant to an experience of Jerusalem through the combined effects of processing, singing, and seeing. At Mren, a sculpted lintel on the north portal may also be interpreted in view of the triple psalmody of the Introit. In the case of this church, associated with the emperor Heraclius and the return of the True Cross, such a procession may be considered to contain a powerful timeliness, simultaneously recalling to the worshipper multiple periods in the history of Jerusalem, from the time of Christ to the contemporary moment.

Of course, these are not the only ways to understand the imagery of the churches. Indeed, the exercise undertaken here raises the question about what visual interpretation means and entails. A liturgical act takes place through time, which opens the possibility for a temporary or temporarily heightened meaning of an image. For example, the builders found on the arcade of Zuart'noc' might have been intended to represent Gregory and Trdat, or perhaps the actual building team who worked at the church. But when forming the backdrop for the rite of entry into the church, they would have echoed and, one may argue, been imbued with the architectural imagery of Psalm 117, inviting us to consider the process in the language of the psalms, and their performance in procession inflected, if for a few moments, the visual imagery of the church. When we remember that the verses were sung at the walls of the church and prescribed with modulations in volume, we can imagine a powerful moment of transformation, much like the liturgy of consecration as a whole, when the building yard became a holy place. Surely it is in just such a moment that archetypal holy places would be evoked in the mind, as they were for Prince Ĵuanšēr.

It is also worth underscoring the distinctive nature of the exterior ritual acts undertaken for the Armenian church consecration and, at the same time, the distinctiveness of the early medieval Armenian and Georgian exteriors. Regarding the former, Findikyan notes that the exterior canons are not derived from Byzantine liturgy, which contains no equivalent exterior movement. In his examination of the exterior wraparound inscriptions on Armenian churches, Greenwood was also unable to find parallels in early Byzantine or Umayyad architecture. I have also underlined the strong "exteriority" of Armenian monuments, produced not only through exterior relief sculpture and inscriptions but also through high podia, porticoes, exterior niches, paved walkways and plazas, and nearby stela monuments.[58] Could it be that these features reflect the desire to generate an exterior landscape akin to Jerusalem, or a place to perform processional rites such as those prescribed in the Armenian Lectionary of Jerusalem? One is struck by how many times pilgrims are described in this text as approaching, climbing, descending, and gathering and/or singing "before [*aṙajin*]" the church. These are tantalizing questions without ready answers. What is certain, however, is that the preserved liturgical texts are among the most powerful, if largely neglected tools for interpreting Armenian church exteriors.[59]

Notes

1 1 Kings 8:12 (Zohrab ed.), my translation.

2 *Patmut'iwn Ałuanic' Ašxarhi*, ed. Nikolai Emin (Moscow: Lazarean Institute, 1860), 149, trans. C. J. F. Dowsett as "The Prayer of Prince Ĵuanšēr," bk. 2, chap. 25 of, *History of the Caucasian Albanians*, by Movsēs Dasxuranc'i (London: Oxford University Press, 1961), 121.

3 Timothy Greenwood, "A Corpus of Early Medieval Armenian Inscriptions," *Dumbarton Oaks Papers* 58 (2004): 27–91 at 56n138.

4 Consideration of the multisensory experience of the Byzantine church interior has emerged in recent decades. See Bissera V. Pentcheva, "Hagia Sophia and the Multisensory Experience," *Gesta* 50, no. 2 (2011): 93–114.

5 See, generally, the scholarship of Michael E. Stone, Abraham Terian, and Robert W. Thomson. The last author has produced an essay with useful bibliography on the topic: "Jerusalem and Armenia," in *Papers of the Ninth International Conference on Patristic Studies, Oxford, 1983*, ed. E. A. Livingstone (Kalamazoo, MI: Cistercian Publications, 1985), 77–91.

6 *The Geography of Ananias of Širak: Ašxarhacoyc, the Long and the Short Recensions*, trans. and comm. Robert H. Hewsen (Wiesbaden: Dr. Ludwig Reichert, 1992), 46.

7 See, above all, Athanase Renoux, *Le Codex Arménien Jérusalem 121*, Patrologia Orientalis, vol. 35, fasc. 1 and vol. 36, fasc. 2 (Turnhout: Brepols, 1969–71).

8 Armen Kazaryan, "The Chancel and Liturgical Space in the Church of Zuart'noc'" [in Russian], in *Ikonostas: Proiskhozhdenie, Razvitie, Simvolika*, ed. Alexei M. Lidov (Moscow: Progress-Traditsiia, 2000), 85–117, at 99.

9 Nazénie Garibian de Vartavan, *La Jérusalem nouvelle et les premiers sanctuaires chrétiens de l'Arménie* (Erevan: Isis Pharia, 2009).

10 Any effort toward a precise count is, of course, fraught; various numbers can be derived, depending on the size of the territory considered, the ways in which the monuments are dated, the decision to include monuments known only from literary sources, and the states of renovation. My number is based in part on the catalog of Patrick Donabédian, focusing on the Armenian monuments, who suggests that between 630 and 690, "leur nombre pourrait s'élever à une soixantaine pour ces six décennies. . . ." (*L'âge d'or de l'architecture arménienne* [Marseilles: Parenthèsis, 2008], 275). When we add to this number monuments attributed to the sixth century, the number grows to more than seventy.

11 A type also known from the churches of Artik, Voskepar, and Hařič.

12 Greenwood, "A Corpus of Early Medieval Armenian Inscriptions," 86, inscription 10.4, fig. 10.4.

13 Greenwood, "A Corpus," 84, inscription 10.1.

14 Greenwood, "A Corpus," 84–85, inscription 10.2, fig. 10.2.

15 Greenwood, "A Corpus," 86.

16 Indeed, Greenwood ("A Corpus," 39) sees this as a reference to John, to whom the church is dedicated, "the apostle, prophet, and martyr, who is 'the companion of the bridegroom,' Christ himself." But this is contested by Patrick Donabédian (*L'âge d'or de l'architecture arménienne*, 156n222) who places apostle/prophet/martyr in the plural and thereby makes them the companions at the wedding. (In classical Armenian, the *-s* ending can indicate either a declined or a plural noun.)

17 Michael Daniel Findikyan, "The Armenian Liturgy of Dedicating a Church: A Textual and Comparative Analysis of Three Early Sources," *Orientalia Christiana Periodica* 64, no. 1 (1998): 75–121.

18 Frederick Cornwallis Conybeare, ed. and trans., and Rev. A. J. Maclean, trans., *Rituale Armenorum: Being the Administration of the Sacraments and the Breviary Rites of the Armenian Church Together with the Greek Rites of Baptism and Epiphany* (Oxford: Clarendon Press, 1905) 19.

19 Conybeare and Maclean, *Rituale Armenorum*, 20 (emphasis mine).

20 Conybeare and Maclean, *Rituale Armenorum*, 21.

21 "Dans les hymnes liturgiques liées aux célébrations de la croix, est plusieurs fois mentionée l'Eglise, épouse du Christ, couronnée par lui de la croix." Donabédian, *L'âge d'or de l'architecture arménienne*, 156n.222.

22 Michael Daniel Findikyan, "Armenian Hymns of the Church and the Cross," *St. Nersess Theological Review* 11 (2006): 63–105, at 74–75. "ի նորահրաշ նավակատիսն որ յերուսաղեմ պայծառազ զեաց վայելչոյ[ք] ցուցաւ խաչ քո տր ար—ծ հարցըն մերոց. Կայ դրօխոյ ընդ աջմէ սբ եկեղեցի. Յովկէհուրն պըսակեալ նըշանաւ խաչի քո. սստըւա: Եւ նորածնեալ քս մըկըրտու[ք] սբ աւազանին. Այսօր տօնեմք ուրախոյթք նըշանաւ խաչի քո. սստըւած հար" (emphasis mine, ellipses indicate my omission of words). Note that the English translation by Findikyan differs slightly from the Armenian above; he uses *Jaynk'ał Šarakan* [Hymns arranged by tone] (Jerusalem: St. James, 1914). I could not consult this at the time of writing; instead, I used a hymnal printed in 1833 in Ējmiacin, now in the Matendaran Institute of Ancient Manuscripts, Erevan (there the hymn appears on 387). See http://greenstone.flib.sci.am/gsdl/cgi-bin/library.cgi?e=d-01000-00---off-0armbook-armenian%2chajgirqn%2chaygirq%2carmbook%2cNo%5fDate%5fBooks-01-1----0-10-0---0---0direct-10---4-------0-1l--11-hu-50---20-help---00-3-1-00-0-0-11-1-0gbk-00-0-0-11-10-0utfZz-8-00&a=d&c=armbook&cl=CL5.35&d=HASHcee1ecf25e3766698aa98.

23 "Քանզի քի է հարսնացեալ Սբ եկեղեցի. Երգնաւոր փեսային պըսակեալ ըզսա խաչիւն. Եւ յաջ եւ յահեակ բոոյցեալ ըզհեթանոսս ժառանգելով բար" (emphasis mine).

24 Findikyan, "Armenian Hymns of the Church and the Cross," 79–80.

25 Findikyan, "Armenian Hymns of the Church and the Cross," 88.

26 Findikyan, "Armenian Hymns of the Church and the Cross," 89.

27 Michael Daniel Findikyan, "Armenian Hymns of the Holy Cross and the Jerusalem Encaenia," *Revue des études arméniennes* 32 (2010): 25–58.

28 For example, the patriarch Sophronius of Jerusalem (560–638) does not know why the Anastasis is celebrated first (as in the first day) and the cross second, since Christ first was crucified and then died. See Findikyan, "Armenian Hymns of the Holy Cross," 29.

29 Findikyan, "Armenian Hymns of the Holy Cross," 28. Findikyan notes that the hymns of days three, four, and five do not pertain to the cross so much as the church as a concept and as a structure, and some explicitly concern the consecration of buildings, such as the church of Holy Ējmiacin in Vałaršapat.

30 Findikyan, "Armenian Hymns of the Holy Cross," 34. It is relevant to our argument that Findikyan, and Athanase Renoux before him, noted the stational character of these hymns. See Findikyan, "Armenian Hymns of the Holy Cross," 48–50; and Renoux, *Le Codex Arménien Jerusalem 121*.

31 Dora Piguet-Panayotova, "Recherches sur les tétraconques à déambulatoire et leur décor en Transcaucasie au VIIe siècle," *Oriens Christianus*, 73 (1989): 166–212; Armen Kazaryan, "The Chancel and Liturgical Space in the Church of Zuart'noc'," 85–104; Garibian de Vartavan, *La Jérusalem nouvelle*; and Annegret Plontke-Lüning, *Früchristliche Architektur in Kaukasien* (Vienna: Verlag der Österreichischen Akademie der Wissenschaften, 2007).

32 Step‘an Mnac‘akanyan, *Zuart‘noc‘ə ev nuynatip hušarjannerə* (Erevan: Haykakan SSH Gitut’yunneri Akademiayi Hratarakč’ut’yun, 1971), 147; and Lydia Durnovo, “Stennaja Živopis’ v Aruc,” *Lraber hasarakakan gitut’yunneri*, no. 1 (1952): 49–66.

33 Piguet-Panayotova, “Recherches sur les tetraconques,” 176.

34 Kazaryan, “The Chancel and Liturgical Space in the Church of Zuart‘noc‘.”

35 Patrick Donabédian, “Les thèmes bibliques dans la sculpture arménienne préarabe,” *Revue des études arméniennes*, n.s., 22 (1990–91): 253–314, at 281.

36 Donabédian (“Les thèmes bibliques dans la sculpture arménienne préarabe,” 281–82) has also suggested that the basket capitals in the church interior may be related to the “bowl-shaped network capitals” mentioned in the construction of the Temple of Solomon.

37 Donabédian (“Les thèmes bibliques dans la sculpture arménienne préarabe”, 282) cites the tomb of Ałudi as presenting a juxtaposition of basketwork capital, pomegranate, and flower forms. I would venture that the example of Zuart‘noc‘ evokes the Temple of Solomon even more powerfully because of the multiplication of these themes around the perimeter of the church facade. But both sites certainly attest to the power of Solomonic imagery in the construction of early medieval Armenian monuments.

38 1 Kings 7:17–19, Titus Text Collection, Vetus Testamentum armeniace, from the so-called Zohrab Bible, http://titus.uni-frankfurt.de/texte/etcs/arm/zohrab/armat/armat.htm (my translation).: “Եւ արար երկուս վանդակս ծածկելէ զխոյակն, եւ արար երկոյս եւս վանդակս ծածկելէ զերկրորդ խոյակն. գործ կախաղանաւ: 18: Եւ երկոցունց սեանցն նռնաձեւս պղնձիս ‘ի վանդակապատսն, գործ կախաղանաւ, կարգ առ կարգ: Եւ այնպէս արար երկրորդի խոյակին: 19: ‘ի վ(ե)ր(այ) գլխոյ սեանցն, գործ շուշանագործ առ կամարաւ ‘ի չորից կանգնոց.”

39 Donabédian, “Les thèmes bibliques,” 280–83.

40 Findikyan, “The Armenian Liturgy of Dedicating a Church,” 91–93.

41 Findikyan, “The Armenian Liturgy of Dedicating a Church,” 90.

42 Findikyan, “The Armenian Liturgy of Dedicating a Church,” 90.

43 On circumambulation as one of the guiding principles in structuring ekphrasis of a church building, see Ruth Webb’s essay in this volume, “Spirituality, Embodiment, and Agency in Ekphraseis of Church Buildings.”

44 Renoux, *Le Codex Arménien Jérusalem 121*, 258–59.

45 See Donabédian, *L’âge d’or de l’architecture arménienne*, 108–10; Jean-Michel Thierry and Nicole Thierry, “La cathédrale de Mren et sa décoration,” *Cahiers archéologiques* 21 (1971): 43–77; and most recently, Samvel Karapetyan, “Mrenə ev Nra Hušarjannerə” [Mren and its monuments], *Vardsk’/Duty of Soul* 7 (2012): 31–63.

46 See *The Armenian History attributed to Sebeos*, trans. Robert Thomson, with historical commentary by James Howard-Johnston and assistance by Timothy Greenwood, 2 vols. (Liverpool: Liverpool University Press, 1999).

47 See for example Josef Strzygowski, *Die Baukunst der Armenier und Europa* (Vienna: Anton Schroll, 1918), 427–28.

48 Minas Sargsyan, “Mreni Tačarə Himnaderneri Patkerak‘andaknerə” [The images of founders on the church of Mren], *Patma-Banasirakan Handes* 35, no. 4 (1966): 241–50.

49 See Jean-Michel Thierry and Nicole Thierry, “La cathédrale de Mren,”; and by Nicole Thierry, “Héraclius et la vraie croix en Arménie,” in *From Byzantium to Iran: Armenian Studies in Honour of Nina G. Garsoïan*, ed. Jean-Pierre Mahé and Robert W. Thomson (Atlanta, GA: Scholars Press, 1997), 165–86.

50 Christina Maranci, “The Humble Heraclius: Revisiting the North Portal at Mren,” *Revue des études arméniennes* 31 (2009): 359–72. See also Stephan Borgehammar, “Heraclius Learns Humility: Two Early Latin Accounts Composed for the Celebration of Exaltatio Crucis,” *Millennium: Jahrbuch zu Kultur und Geschichte des ersten Jahrtausends n. Chr.* 6 (2009): 161–63.

51 Unlike, for example, the eleventh-century *Sacramentary of Mont-Saint-Michel*, in which the story is told in two registers of continuous narrative, the lintel presents a composition centered on the cross (Pierpont Morgan Library, New York, MS 641, fol. 155v).

52 See Susan A. Boyd and Marlia Mundell Mango, Eds., *Ecclesiastical Silver Plate in Sixth-Century Byzantium: Papers of the Symposium Held May 16–18, 1986, at the Walters Art Gallery, Baltimore, and Dumbarton Oaks, Washington, DC* (Washington, DC: Dumbarton Oaks, 1993). For Armenian crosses, see Jannic Durand, Ioanna Rapti, and Dorota Giovannoni, Eds., *Armenia Sacra: Mémoire chrétienne des Arméniens (IVe–XVIIIe siècles)* (Paris: Musée de Louvre, 2007).

53 Nicole Thierry, "Héraclius et la vraie croix en Arménie," 169. One might add that the proposed liturgical interpretation of the return of the Cross may be seen as in keeping with contemporary Heraclian ideology. See John Haldon, "The Reign of Heraclius: A Context for Change?" in *The Reign of Heraclius: Crisis and Confrontation*, ed. Gerrit J. Reinink and Bernard H. Stolte (Louvain: Peeters, 2002), 14–15.

54 Findikyan, "The Armenian Liturgy of Dedicating a Church," 91–93.

55 See Greenwood, "A Corpus of Early Medieval Armenian Inscriptions," 36.

56 Given the off-axis orientation of Mren, however, direct southward travel would actually bring one to the Persian Gulf.

57 See Robert H. Hewsen, *Armenia: A Historical Atlas* (Chicago, IL: University of Chicago Press, 2001), 90.

58 Christina Maranci, "Performance and Church Exterior in Early Medieval Armenia," in *Visualizing Medieval Performance: Perspectives, Histories, Contexts*, ed. Elina Gertsman (London: Ashgate, 2008), 17–32.

59 The exterior sculpture of Armenia and Georgia is physically fragile. At Zuart'noc', in the Armenian Republic, fragments of the facade now lie on the ground, exposed to weather, lichen, and damp. At the church of Djvari in Mtskheta, Georgia, the wall reliefs are weathering faster than they can be protected. The case of Mren is the most urgent, because the isolation of this church, and its unstabilized condition, mean that the north facade, and its portal, may soon go the way of the collapsed south facade. Tied to the fate of the monuments they adorn and exposed to the elements, the carved bas-reliefs discussed above require not only the attention of the scholar but also the care of the preservationist, and one hopes they will receive both.

3 Byzantine chant notation

Written documents in an aural tradition

Christian Troelsgård

In memory of Kenneth Levy (*New York 1927–†Princeton 2013)

Introductory remarks

Byzantine chant may be defined as the music used for the celebration of the Byzantine rite, which also has historical links to musical traditions of the Byzantine Empire. Such a definition is not exclusive in regard to language, geographic area, or ecclesiastical affiliation, and it focuses on the functional nature of ritual music. In addition, a close relation between music and text is stressed.

I shall in these pages try to concentrate on themes regarding the creation and transmission of chanted melodies that pertain to some of the various periods, regions, and places such as monasteries and cathedrals in which Byzantine chant has been cultivated. This approach matches the transmission of the source material, both geographically and chronologically, and reveals great lacunae. Thus, we are forced to realize how much—and, in many instances, how little—can be known about the medieval traditions of Byzantine chant.

Memory, writing, and the relationship between literacy and aurality/orality in the transmission of Byzantine chant

During his 1996 visit to Copenhagen, the late Princeton music professor Kenneth Levy read a lecture on the transmission of Byzantine chant. On this occasion, he insisted that in studying oral tradition simply as the opposite of written tradition, we run the risk of overlooking the important role memory played in the transmission of chant. At that time, the term "aural" had not yet gained traction in scholarly discourse as it has today; "oral" was still prevailing as the sole central concept when dealing with traditional musical cultures.

Levy, who had studied Byzantine chant for decades, had been introduced to the subject by Oliver Strunk, later the director of the Monumenta Musicae Byzantinae, or MMB (a project for publishing Byzantine chant documents and transcriptions established by the Western musicologists Carsten Høeg, H. J. W. Tillyard, and Egon Wellesz in 1931). Strunk was perhaps the first scholar to present a coherent theory on the nature of the "hidden interplay between oral and written tradition" in the transmission of Byzantine music.[1] Though he nourished interests in any Byzantine chant document from any period and location, he soon realized that the study of the Paleo-Byzantine sources was of key importance, as they preserved the earliest evidence from which rules of the "game" might be deduced. Such a deduction is indeed necessary, since very little explicit music theory from that period can be associated with the actual chant tradition. Beginning with these sources, more and more parts of the liturgical chant repertory were successively transferred into notated format.

At first, only parts of the hymnography of the daily offices were associated with **neumes** (early systems of musical signs). Later, a selection of elaborate settings for the cathedral rite was added, and in the Late Byzantine period, we find for the first time written documentation of the music associated with many of the daily, common chants. Many Byzantine chants known to be ancient through the recording of their texts—such as the age-old evening hymn Φῶς ἱλαρόν ("O gladsome light"), which is documented in its text-only version from about 300 CE and quickly became a set element in Byzantine vespers, or *hesperinos*—are not found with musical notation in sources earlier than Post-Byzantine ones.[2]

Models of chant transmission

How are musical traditions created, developed, and maintained? Each type of music, genre, period, and geographic region most likely had its own specific characteristic blend of elements in its transmission cycles. It means that the mix of these elements may look different even within such a roomy concept as "the Byzantine chant tradition." To account for this phenomenon, I will use the terms "aural" and "oral." The concept "aural tradition" (or "aural transmission") emphasizes the special dimension of sound: the acoustic environment in which chant is performed and the recipient's act of listening. The concept "oral tradition" (or "oral transmission") emphasizes the performance element in music and the active role of the singer(s) and/or musician(s). But either side of the musical activity presupposes in effect the other one (Figure 3.1).

On the one hand, there is passive reception. At an early age, Byzantine children were exposed to musical experiences: their mothers most probably mixed lullabies with elements derived from popular, "para-liturgical," or liturgical chanting. I consider

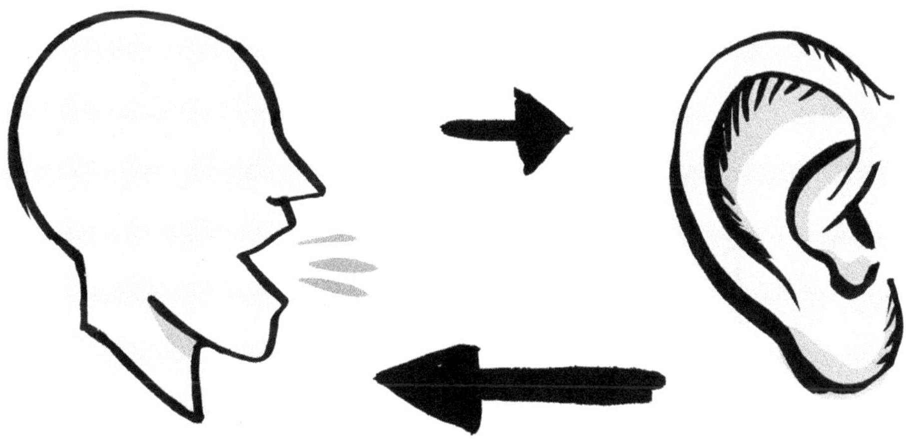

Figure 3.1 Oral–aural transmission: Model 1.

this aural exposure an important contribution to establishing a basic musical identity. As they grew older, the young children may have heard the chanting of an ecclesiastical procession passing by or attended different types of services in a local church or even in a major cathedral. Such "aural" musical activities define an acceptable musical performance to members of the group, for whom such music is said to be "traditional."

On the other hand, there is the active transmission ensured by the people performing the music. Their group includes lay people, members of the various orders of the clergy or monks, and professional chanters. They have all experienced the local blend of Byzantine chant from their childhood and, through memory and imitation, they produce acceptable (in the linguistic sense of "grammatical," as hinted in the paragraph above) performances of music. Their knowledge of the chant enables them to participate passively or actively by listening for liturgical moments, participating in congregational singing of whole hymns, or just responding to soloists/choirs.

Clergy and singers have in addition received specialized training in chant performance, since it was an obligatory part of their professional activities. Training likely contributed to stabilizing the memory of musical texts; at the same time, the knowledge and mastery of these texts—especially of the rules governing them—may well have inspired innovation and creativity. This last assumption has its analogy, also, in the slow yet steady development of language systems. The memory works not as exact reproduction, as a human "hard drive," but relies on the interpretation of rules, systems of scales, recurrent melodic formulas, and so on, which again is used for the "re-creation" or "recollection" of the music (Figure 3.2).

The decoding-encoding rules that are active in human memory may be partially supported by:

A) nonwritten music theory, for example, the grouping of melodies in specific genres, with specific contents, functions, tempos, or complexity (such as the placement

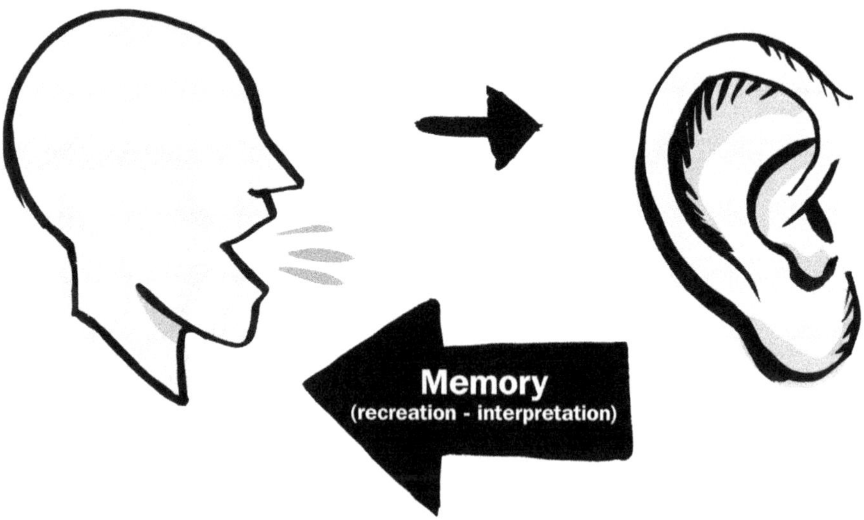

Figure 3.2 Oral–aural transmission: Model 2.

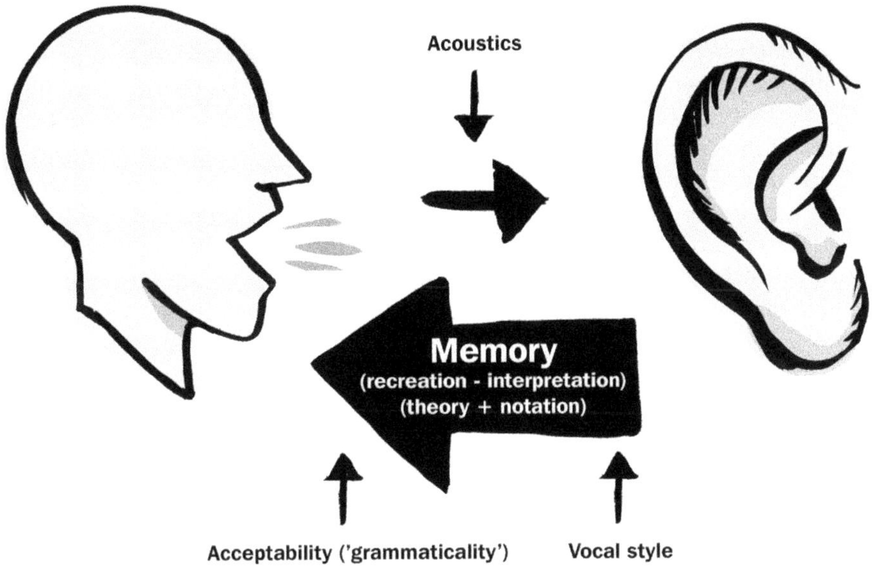

Figure 3.3 Oral–aural transmission: Model 3.
© Kim Broström.

in the continuum from simple syllabic to florid melismatic chant), considered to belong to a specific melody type or modal category; or by

B) written music theory (verbalized documentation of the above) plus musical notation. Further, the actual aural perception is dressed in the specific vocal qualities of individuals and choirs, affected by the number of performers and the architectural/topographic staging of the performance (Figure 3.3).

Oral–aural versus written chant administration

Orality and literacy have often been taken as opposites, belonging to two different worlds: orality has been associated with the collective and the focus on the group, literacy with the person and the focus on the individual. The two different "worlds" have been characterized in the following dichotomies:[3]

Oral	*Literate*
Intuitive	Objective
Integrated	Fragmented
Multicompetent	Specialized
Human being	Competent citizen
Wisdom	Knowledge
Kairos	*Chronos*
Repetition of archetypes	Innovation
Traditional local Spiritual	Literate global
Secular	

However, in a medieval context, such oppositions have to be understood in a specific context, namely, that of literacy in the era of manuscripts. In 1982 Walter Ong introduced a clear distinction between cheirographic (manuscript-based, which tends to be open, thus variable) and typographic (print-based, which tends to be closed, thus fixed) literacy.[4] The study of manuscript cultures since about 1985 has yielded more subtle descriptions of the specific ways in which oral and written procedures interacted in various areas of and activities in medieval culture. The role of memory and medieval concepts of memory have been explored, too, though mostly with regard to the medieval West. Mary Carruthers's *The Book of Memory* from 1990 inspired a whole series of studies with such an approach.[5] In this perspective, manuscripts are seen as just one type of aid in organizing a complex memory-based system of conceiving and retrieving knowledge.

Musical literacy might be studied from this angle, too; the written documents cannot replace or represent an oral tradition but should rather be looked at as memory aids or memory "technology." Hence, the study of the particular context of documents associated with chant reproduction becomes very important. In which institutions did they serve? for what purposes? and who used them? We hope that producing plausible answers to such questions can shed some light on how mouth, ear, memory, eyes, and pen worked together. We must also be ready to accept that the Byzantine chant documents in many respects deviate radically from the contemporary experience of "musical literacy," usually thought of as composing, reproducing, and performing music using original texts and printed scores.

I shall in the following sections try to present and interpret a number of different patterns of interaction between chant documents and the actual chant culture around them.

Musical manuscripts without notation?

Some would say that written musical notations may go back to what has paradoxically been termed "oral notation," defined as formalized systems of signaling melodic or other musical qualities between musicians. It may also refer to systems of memorizing and teaching music by means of spoken key words or phrases indicating melody, tempo, dynamics, repetitions, and so on. Finally, it can refer to cheironomic or hand gestures used by the singers and/or "conductors." But even without claiming that cheironomy would predate manuscripts with musical notation, we might consider the relation between the music performed and written elements in the oldest preserved text-only hymnals.

Let us begin by looking at a very ancient example of a written rendition of pieces of Byzantine chant: Papyrus Vienna G. 19.934, published by Kurt Treu and Johannes Diethart in 1993 (Figure 3.4).[6] It consists of seven fragmentary folios, both sides displaying an inclining majuscule script, which, according to paleographers, is typical of the sixth century.[7] The page presents the text of a hymn known as a *troparion* for a confessor bishop, beginning "Ὡς φωστὴρ ἐν κόσμῳ" (Like a star in this world), placed at the top of this fragment of a papyrus sheet. The contents and textual form of the fragment appear similar to what later on became known as the ***sticheron***, or a hymn genre performed at the morning (*orthros*) and evening (*hesperinos*) services as inserts in the recitation of Psalms.[8] If the sixth-century dating is correct, that is, in the century before John of Damascus, then this *troparion* definitely belongs to the formative

Figure 3.4 Papyrus Vienna G. 19.934, fol. 2r.

period of Byzantine chant. The provenance is unknown, but it probably surfaced in Egypt; it is likely that it originated in the province of Alexandria.

Leaving aside the textual analysis, I will focus instead on the musical evidence that can be deduced from the source. The *sticheron* was intended to be accompanied by music. As noted above, the earliest examples of Byzantine melodic chant notation that could be transcribed in modern staff notation did not appear until about the middle of the twelfth century.[9] Until then, therefore, Byzantine chant was said to be primarily transmitted orally, a concept that calls for qualification. It is not likely that the singer or singers, if a chant was intended for a choir, were left to improvise freely in their chanting. However close this document was to the origins of an Alexandrian branch of early Byzantine chant, we must presume that a tradition creating a framework for the singing had already been established for generations, and that certain "rules" for singing were supposed to be respected in the performance. Even if the document is older than the earliest preserved examples of Byzantine chant notation (it is reasonable here to disregard the third- or fourth-century Papyrus Oxyrhynchus 1786, which contains a hymn, furnished with an ancient Greek-letter notation, totally alien to the later Byzantine ones), it still gives us some evidence about the melody.[10] I shall point to two features of the "written" indications about the "musical" features of the piece. The first concerns the ascription to a specific mode (there are eight modes, forming the Byzantine *octoechos*; each mode is determined not on the base of a specific scale structure but on the application of a specific set of melodic formulas), in this case, an abbreviation of second plagal [mode] in the left margin, that is, "Πλβ"—the same modal key that is more clearly visible at the lower left corner (in the margin). This modal indication would tell the singer how to approach the performance. It is as if the following explanatory information were written down: "In our tradition we have divided the repertory of chants into eight modes [what later became known as the *octoechos*], and this piece belongs to the group of chants we call second plagal, as it shares many features with the other melodies of this same class." In fact, it seems from the sequence of modal indications in the seven fragmentary folios of Papyrus Vienna G. 19.934 that this is one of the earliest documents recording the *octoechos* system.

However, we do not know exactly what "second plagal" meant to this particular *psaltes* (cantor) of that distant past and that particular area. The point is that merely the use of lists of incipits and classes of songs that resemble each other shows that a concept of "pure oral tradition" might be misrepresenting early Byzantine chant. If we were to recompose the music of this *troparion sticheron*, we should hypothetically use the musical formulas from the sticheraric genre and an interpretation of the modal category as found in later, but more explicit written sources, those using the so-called Middle Byzantine diastematic or intervalic notation from about 1170 on. With our knowledge of the slow yet steady shifts of many details in the Byzantine chant documents over the centuries, we can be sure that we cannot re-create "the original piece," but we can "recompose" it according to the melodic formulas derived from the Middle Byzantine sources and a modal system whose age remains unknown.

The status of the document is also important: Was this an official hymn or an expression of private devotion? Was it a widespread hymn/tradition or something of restricted distribution? From church and monastery inventory lists (from the fifth through the seventh centuries) we have indications of institutional possession of manuscripts:

> Βιβλία διάφορα μεμβράϊνα καὶ χάρτινα ε' (various parchment and papyrus books: five), in *Papyri Graecae Wessely Pragenses (PPrag. II) 178.5–6*[11]
>
> Βιβλία δερμάτινα κα' (parchment books: 21; and likewise of papyrus: 3), in Bernard P. Grenfell and Arthur S. Hunt, eds., *New Classical Fragments and Other Greek and Latin Papyri, vol. 2, 111.27–28*[12]

Yet many of the early scriptural papyrus fragments and a number of hymnographic documents from the same area belonged to individuals, such as *anagnostai* (lectors), *psaltai* (cantors), or teachers.[13] The manuscripts are, in my view, often best interpreted as professional tools. The Vienna papyrus above seems neither to form part of a previously known hymnal nor to be organized according to any known liturgical system; the only applicable unifying feature is the eight-mode cycle. The constellation fits, however, a situation where a *psaltes*/teacher would like to teach his apprentice a variety of chants in all modes. We may suspect that certain variations in both text and music would arise when the chant repertories were administered in part by private entrepreneurs and not controlled by any specific authority. This can be inferred from a decree issued by the Synod of Laodicea (fourth century) that forbids unauthorized cantors, suggesting that such unofficial and perhaps innovative activities were in fact undertaken:

ιε' Περὶ τοῦ μὴ δεῖν πλὴν τῶν κανονικῶν ψαλτῶν, τῶν ἐπὶ τὸν ἄμβωνα ἀναβαινόντων καὶ ἀπὸ διφθέρας ψαλλόντων, ἑτέρους τινὰς ψάλλειν ἐν ἐκκλησίᾳ. (15. Concerning that, with the exception of the regular cantors, who ascend the ambo and chant from parchments, no others should chant in the church.)

νθ΄ Ὅτι οὐ δεῖ ἰδιωτικοὺς ψαλμοὺς λέγεσθαι ἐν τῇ ἐκκλησίᾳ, οὐδὲ ἀκανόνιστα βιβλία, ἀλλὰ μόνα τὰ κανονικὰ τῆς καινῆς καὶ παλαιᾶς διαθήκης. (59. Let no private psalms nor any uncanonical books be read in church, but only the canonical ones of the New and Old Testament.][14]

Well over sixty thousand Greek *troparia* are counted in the preserved standard repertories—according to Enrica Follieri's *Initia hymnorum Ecclesiae Graecae*—and with the addition of thousands of *apokrypha* (hymns not found in the later Byzantine standard repertory, such as the text I show here), such a mass of texts calls for some kind of written organization in a religious tradition, which definitely can be described as "script-based" in many other respects.[15]

The development of such a piece of music theory as the modal category (or *octoechos*) that is transmitted to some degree independently of specific chants seems to be the second step toward stabilization of the tradition. In practice, the singer would probably have recalled typical features of one or two well-known pieces of a specific group from memory. This recalled model would indicate how to intone the new piece, how to keep a phrase going, and how to make a medial or final stop. Nevertheless, the ascription to classes is administered in writing.

Another written "musical" feature of the chant collections without musical notation is punctuation. The singer understood the structure of the chanted text, and the punctuation signs in most cases strengthen natural syntactic divisions. Such is the case here, where *ano teleia* (·) and *comma* (,) are used.

However, in other cases, musical punctuation that goes against the natural syntax of the language can be seen in early chant manuscripts, especially in passages where musical elaboration such as a melisma is recommended. We might encounter such cases if we compare the actual passages marked with the Greek letter *theta* in the *stichera* and *heirmoi* (*troparia*, or hymns, with a fixed rhythm and melody that are used as the standard rhythmic and melodic pattern for other *troparia* in the canon of the morning office in the Eastern Church). We find, for example, a melodic elaboration on the word διό (therefore) before the traditional prayerlike conclusion of a *troparion*.[16] The musical elaboration, therefore, has a meaning related to the structural level of the chant.

Text accentuation and formulas

In addition to these features, knowledge of the natural accentuation of the language should also be considered. Stressed syllables may require a special treatment, and this practice seems to be inherited in many chant types from the simple psalmody, where text and music come together in a simple, yet efficient type of performance. Thus, we know from Byzantine psalmody as documented from the beginning of the ninth century through the amazing chain of evidence presented by Strunk that the initial phrase and recitation was governed by accent (pitch accent or sustained accent), while the cadence was mechanically adapted to the last four syllables of the verse.[17]

The application of such simple rules results in musical realizations of the same psalm text that, while similar, can take a variety of different forms. One must imagine the same variations in performance within a certain set of rules, which were supposed to be well known by the experienced singer. According to this model, it is the memory of the singer and his or her internalization of the rules—in combination with the memory of specific musical features of earlier performances of the same chant/text—that are evoked in the performance situation. From this point of view, performance acquires the status of composition or, rather, "re-composition." In historical studies of oral–aural traditions it is, however (in contrast to ethnomusicological field studies), impossible to confirm our suppositions by means of recordings of actual performances and interviews with active and passive members of the chant community.

The study of the formulaic structures in medieval and postmedieval Byzantine chant has now engaged scholars for almost a century. All known repertories have been recognized as being predominantly formulaic, as each mode in each genre is furnished with its own specific selection of formulas: initial, transitional, medial, and cadential. Paleo-Byzantine as well as Middle Byzantine musical notations operate with formulas as constituent and recurrent elements of the notation, since the so-called group signs, or the *megalai hypostaseis* (big entities), traditionally translated as "phrasing signs," are identified as formulas. Their names derive probably from chant teaching in monasteries and at cathedral schools, and they again testify to the close collaboration between writing and the human memory.[18] However, not all of the formulas were given idiomatic names for mnemonic reasons. For example, in the didactic chants ascribed to John Glykys (ca. 1260–1319) and John Koukouzeles (ca. 1280–1360), illustrating the notational signs and formulas associated with them, one of the stereotypical final cadences is given only with the description "τέλος τοῦ δευτέρου ἤχου" (ending in the second mode), since no particular sign is applied with it. I shall return to this material briefly below, but I can already disclose here that I see these written documents containing a mixture of didactic material and musical settings (dating from the later centuries of the Byzantine era) as primarily linked to chant teaching rather than to actual performance situations.

The "ordinary" and the "proper"

Specifically regarding the Byzantine tradition, both the early collections of chant texts without notation and the later books with notation seem to have been concerned primarily with chants very seldom sung, for example, once a year in connection with the celebration of a saint's day (the "proper" of the saint). Chants that were sung more often or daily entered only very late into the sphere of the written tradition, as these were known by heart by the clerics and the congregations and belonged to the "ordinary" or the "common." Although these terms have been borrowed from Western chant and liturgy studies, they are useful in detailing the distinction between the often sung and less often sung items in Eastern chant as well.

The uses of chant manuscripts

Musical manuscripts seem to have had a status in the Middle Ages different from a score or transcription of today; not every member of a choir would have a copy. The manuscripts are generally too small to be read by more than one person in a dark room. At the same time, they contain material that is supposed to be divided for performance by two separate choirs, alternating in the singing of psalm verse with its intercalated refrain or *troparion*. Yet no identical or even near identical pairs of handwritten chant books have been found.

Neither was it customary that every member of a choir had a copy of the text; much of the daily office and Divine Liturgy was, of course, known by heart. But to support the memory of the singers during performance of less-known chants, the institution of the *kanonarches* was established. Described in documents as early as the eleventh century, the *kanonarches* (a kind of prompter) of the Monastery of the Theotokos Evergetis in Constantinople, for instance, recited each text phrase immediately before the singing (guided by a chant manuscript with "musical punctuation"). He thus offered help for the singers in the performance of the *stichera idiomela* (hymns that have their own melody, as opposed to *stichera proshomoia*, which used a melody from a limited repertoire of well-known models) and of many other chants for the daily services.

A second type of help was offered by the notated manuscripts for choral music. They were probably regarded more as reference tools and teaching materials in the hands of the *protopsaltai* (soloists), presenting acceptable, even beautiful re-compositions of the melodic line of a given chant in a given mode of a given genre. The hymnals were used by experienced singers to show specific details to novices.

A third type of assistance was offered by the work of the ecclesiarch, or liturgy master, whose task it was (according to the Evergetis typikon, a document that prescribed the liturgical practice of the Theotokos Evergetis Monastery in Contantinople), to plan the specific schedule of each service. He made many decisions regarding the music, such as the number of stanzas, or *oikoi*, of the ***kontakion***, or sung sermon, that could be completed in the time allowed, adding yet another element of flexibility to the performance of chant.

Variation, change, and development

In practice, there might have been more than one accepted performance version, and it was likely that chants would differ slightly from day to day and from performance to performance. The multiplicity of acceptable versions is reflected by the presence of variant notation between the lines in the manuscripts. In this way, two, sometimes three, melodic versions are written in the same codex. These "scholia" (in practice, often variants written in red ink) indicate that multiple variants were taken into account. It is probable that more than one performance could be accepted by a congregation and *psaltai*—although I am sure that many singers would proudly contend that their own version was the only right one!

In a study of the formulaic structure in Ethiopian chant, Kay Kaufman Shelemay, Ingrid Monson, and Peter Jeffery introduced the word "substitutability" in order to express the phenomenon of substituting one formula with another under certain circumstances.[19] Thus, the analysis of chant traditions with strong elements of "oral–aural" qualities can lead to the recognition of formulas with similar musical functions. The flexibility and variability inherent in the very nature of transmission and the constant acts of re-composition of the traditional Byzantine repertories have led to the slow and organic development of many chant forms over years, decades, and centuries.

Such phenomena can be observed in many sections of the Byzantine repertories, for example, in the *heirmologia* (collections of *heirmoi* stanzas, which served as melodic-metrical models for other chants), a type of manuscript that is far less widespread than the *sticheraria* (a manuscript type collecting stichera *idiomela*, that is, those that have their own melody).[20] Despite the faithful copying of the notated manuscripts, the sung tradition of the *heirmoi* seems to have gotten out of step with the actual chant culture as time went by.

In "real" life, the *heirmoi* were used regularly because of their status as model melodies, and they thus were subject to the previously described judgment of common approval or rejection at each new performance. This situation seems to have led, as noted, to slow and organic development. At a certain time (about the thirteenth to fourteenth centuries), new *heirmologia* were notated in revised and homogeneous versions, predominantly in a style slightly simpler than the written versions of the earlier period. This apparent shift in the written record does not necessarily mean that the tradition of heirmological singing was abruptly broken. Instead, the written evidence suggests that the growing tension between the actually accepted versions and the written ones was finally resolved.

Yet the redaction of whole repertories by individuals could also take place now and then. The revision of the *Sticherarion* ascribed to John Koukouzeles (fourteenth century) might present one such instance.[21] With sufficient institutional backing, it is possible that sudden changes of a tradition could be effected. Even if we do not know the exact circumstances, a manifest reduction and standardization of the *Sticherarion* repertory was carried out about 1050 CE and quickly disseminated as the "Standard Abridged Version."[22] Such reforms of chant may have been connected with the introduction of new notational standards.

Automela and proshomoia, or model melodies versus contrafacta

As with the new *heirmoi*, also some sets of ***automela***, or "model melodies," were notated at quite a late period, not until the fourteenth to fifteenth centuries. These melodies belong to what Strunk named a "marginal repertory."[23] The term "marginal" here applies as seen from the perspective of the modern student of the written tradition, whose point of departure was the massive and stable tradition of the *sticheraria* and, to a lesser degree, that of the early ***heirmologion*** found in the musical manuscripts. In reality, the *automela* had a much wider usage than the proper hymns, being applied as model melodies to the proper texts of the daily service. Indeed, they form a very central repertory, based on quite a limited set of model melodies in a limited number of instances.

Contrafacta, the concept of writing new chants to preexisting and well-known melodic patterns, in stanzas with a specific number of lines and syllables and a specific distribution of the main accents (the principles of *isosyllabia* and *isotonia*), offers another principle of organizing a vast chant repertory, which in some ways competes with the principle of modal ascription. This principle of organization is very old, too. It is, for example, present in the *kontakion* repertory and documented in papyrus fragments, such as Universiteitsbibliotheek, Louvain, Pap. Khirbet Mird P.A.M. 1 (eighth–ninth century).[24] The persistent formula indicating the model melody is <ψάλλεται> πρὸς τὸ . . . (<sung> to . . . [followed by the incipit of the model melody]).

It presents again a combination of oral and written elements, since the indication of the model melody in the text manuscript is the main issue, while the indication of mode is of secondary importance. The *automelon-proshomoion* principle, that is the creation of a large number of chant texts (***proshomoia***) molded on a limited set of well-known model melodies (*automela*), has permeated almost all the offshoots of the Byzantine chant tradition.

Chant manuscripts with partial notation

Some notations are designed to give a broad range of information about pitch, rhythm, tempo, dynamics, and so on; others give only a small part of what would be needed by the uninitiated musician or singer. In the case of the latter, the "missing" pieces of information are withheld either because they belong to the common knowledge of a given musical culture and are therefore considered unnecessary or because of a desire to keep it secret, perhaps for professional or religious reasons.

No musical notation, however, can be identical or synonymous with the music itself. A "complete" notational system might be theoretically possible, but it would eventually become so complex that it would have no value as a tool to signal musical characteristics within the framework of human perception, whether to insiders or outsiders of a given musical culture. An early class of Byzantine melodic notation, the so-called *Oxeia*, *Diple*, or *Theta* notations, have been termed "partial musical notations." The manuscripts furnished with notational elements of this type embrace all the written parameters mentioned above; in addition, certain words or syllables are distinguished by notation in the form of a single acute accent (*Oxeia*), a double accent (*Diple*),[25] or the letter *Theta*.[26]

The older stages of the Paleo-Byzantine melodic notations are called archaic according to Strunk's terminology,[27] or stages 1–3 according to Constantin Floros.[28] These varieties of partial notations leave a number of blank syllables to be filled out by the singer according to his knowledge of the tradition. Only the syllables considered the more important, more difficult, or more ambiguous received musical signs to guide the singer in his re-composition.

Recently, partial notation has also been pointed out in a Strasbourg papyrus fragment from about 800 CE.[29] Though slightly different in appearance, it seems to be derived from an archaic stage of the Paleo-Byzantine notations. It can perhaps be described as a "crossover" between the *Theta* notation and the oldest variety of melodic notation. It gives signs for a complex, multisyllable melisma, as can be confirmed by later notated versions of the same piece in more developed Paleo-Byzantine and Middle-Byzantine notations on parchment. Still, it is a "partial notation," pertaining only to the melismatic passage.

From Paleo-Byzantine to Middle Byzantine notations

In the later and most developed stages of Paleo-Byzantine neumatic notations, all syllables came to be furnished with at least one sign, and in other respects, too, the notation became more and more explicit, leaving fewer elements to be supplied from memory. For example, these notations add signs to indicate "high" or "low" positions to melodic elements and to the modal signatures or the intonations (model signatures that carry information about the mode, also known as *ēchēmata*). Further, they begin to apply the ison and *oligon* signs for repetition and counting one step upward, respectively. The Chartres and the Coislin musical notations fall into this branch of Paleo-Byzantine systems, both named after French library collections with representative examples of these types. The Chartres notation was the first to combine consistently one neume with an indication of relative pitch, namely, in the form of the "straight" ison.

The transition from Paleo-Byzantine notation to Middle Byzantine notation, which probably took place about the middle of the twelfth century, consists in a precise indication of the number of steps, so-called *phonai*, to perform within a given mode. The actual interval relations within this mode are not pointed out, however, and the scale system was still to be supplied through the mechanisms of the "oral–aural transmission." Nevertheless, the application of a quite elaborate system of modal signatures in the musical manuscripts might give some important hints and often almost conclusive evidence, as studies by Christian Thodberg and Jørgen Raasted have shown.[30] The precise intonation of single notes and other performance niceties are, however, not revealed even in the most developed type of Byzantine notation and must be supplied from elsewhere to obtain a plausible reconstruction of the melody. This gap must be filled by the performers with guidance from traditional performance practices, interpretations of the music treatises, and, finally, personal taste. The same notation may thus result in quite different interpretations regarding such characteristics as rhythm, intonation, and vocal aesthetics, even if they all respect the melodic skeleton offered by the written document.

These twelfth- and thirteenth-century transnotations from Paleo-Byzantine to the Middle Byzantine supply the "melodic skeleton" that can be traced back to the earliest and archaic phases of the so-called Chartres variety of the Paleo-Byzantine neumes. Some contours of the melodies, such as the division of the text into phrases and melodic positions considered apt for melodic elaboration, even go back to the *Theta* notations and often display a remarkable stability. If we want to be confident in the succession of melodic steps, the transcription of pieces documented in the oldest Byzantine chant sources can be done only as long as a parallel exists in Middle Byzantine notation that can serve as a control for this process of "backward comparison."

In a list of holdings of books in the library of the Monastery of St. John the Evangelist in Patmos from about the year 1200, one of the *sticheraria* is described as "palaiotonon," or "furnished with the old [that is, Paleo-Byzantine] notation." After the invention and early dissemination of Middle Byzantine notation, Paleo-Byzantine musical manuscripts often continued to be used in many places, and even new manuscripts in Paleo-Byzantine notation were produced. This smooth transition from one notational syntax to another attests to the autonomy of the chant tradition in its entirety, drawing to a great extent on memory.

Working out the Cathedral Chant from the Psaltikon and Asmatikon manuscripts

In this section I will continue the discussion of oral–aural procedures in Byzantine chant transmission and present a case study that focuses on formulas, the relationship between music and text, and the roles of performance practices in the chant of the cathedral rite of Constantinople, the rite cultivated in Hagia Sophia and imitated in perhaps less rich forms in the major churches of the Byzantine area.[31]

The chants of the Byzantine Cathedral are primarily known from two complementary chant collections, the *asmatikon*, or the book with chants for the elite choir, and *the psaltikon*, also known as the ***kontakarion***, the book for the precentor or soloist. These collections represent most likely the usage in the church of Hagia Sophia in Constantinople between the eleventh and thirteenth centuries.[32]

Written documents of the cathedral repertories with musical notation certainly existed in the tenth and eleventh centuries, and these are given in Paleo-Byzantine notation. This can be ascertained from the Slavic reception of the *kontakarian* repertories. In Slavonic Church chant tradition, parts of the *psaltikon* and *asmatikon* are combined into one collection and furnished with a specific variety of the Old Church Slavonic notations appropriated from Byzantine models, the so-called ***kondakar*** notation.[33] Byzantine chant manuscripts offer few and fragmentary remnants of the Paleo-Byzantine tradition of the cathedral rite chant notation, that is, pertaining to the specific musical tradition of the patriarchal liturgy in Constantinople. My next example shows such a fragment of a partially notated version of the Good Friday *kontakion* "Τὸν δι' ἡμᾶς σταυρωθέντα" / (The one crucified for us) (Figure 3.5). It is unclear why the scribe began to fill in neumes for the first part of the *kontakion* but suddenly gave it up for the rest of the *prooimion* (that is, the introductory stanza) and the first *oikos* (the first stanza of the "body" of the *kontakion*, serving as musical model for all subsequent ones); these are the two stanzas found with notation in the *psaltika* manuscripts from a slightly later period (see an example in Figure 3.6). This *kontakion* is preceded by a *sticheron* in a very different and less complex notation, a moderately developed version of the Paleo-Byzantine Coislin notation (one of the two main branches of Paleo-Byzantine melodic notations, the other being the Chartres notation).

Relatively few complete or almost complete copies of the *psaltikon* and even fewer *asmatika* furnished with Middle Byzantine notation have been preserved. From the eastern part of the empire, perhaps only six remain. The four *psaltika* are in the Monastery of St. John the Evangelist, Patmos, MS Gr. 221 (dated to ca. 1162–79); Monastery of St. Catherine, Sinai, MS Gr. 1280 (13th–14th century) and MS Gr. 1314 (14th century); and National Library, Ochrid, MS Gr. 59 (13th–14th century). The two *asmatika* are in the Cathedral Library, Kastoria, MS Gr. 8 (14th century) and Library of the Great Laura, Athos, MS Gr. γ 3 (15th century). The remaining manuscripts of these classes, about four times as many, are of Italo-Greek origin, and they probably testify to the adoption of parts of the Constantinopolitan Cathedral repertoire in the newly founded and reformed Byzantine monasteries of southern Italy under the protection of the Norman kings in the twelfth century.

The *psaltikon* is the soloist's book containing: chants for the *prokeimena*, or graduals (psalm verses sung as an introduction to the readings from the Scripture) performed at the Divine Liturgy and at other offices; the verses of the so-called great *troparia*, or *hymns;* the *allelouia* verses for the Divine Liturgy; the ***hypakoai***, or great responsories, sung among other occasions at laudes on Great Feasts and Sundays; the *kontakia* for the entire year; and, as a special case, the full *Akathistos* hymn, the most famous poem dedicated to the Virgin Mary and performed at the feast of Annunciation (March 25) and at special Theotokos services on Fridays during the Great Lent. The functional division between the two collections—*asmatika* and *psaltika*—is so strictly observed

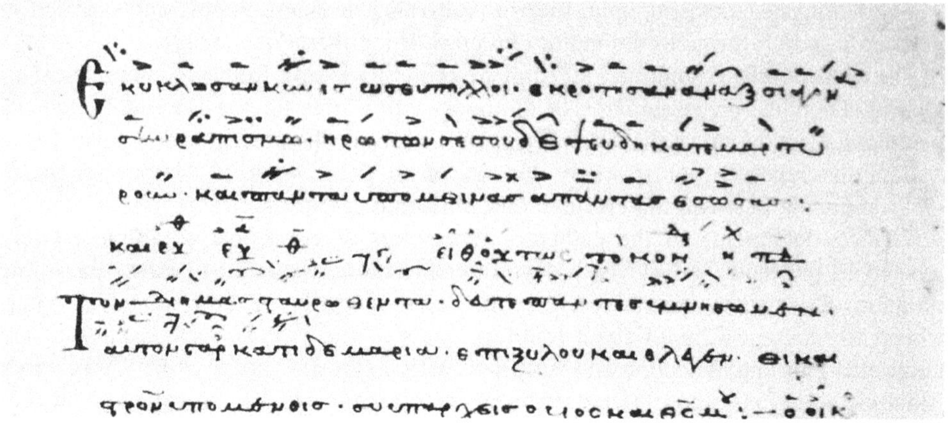

Figure 3.5 The *Kontakion "Τὸν δι' ἡμᾶς σταυρωθέντα"* for Good Friday in fourth plagal mode. From monastery of St. Catherine, Sinai, MS Sinai gr. 1214, early twelfth century, fol. 186v.

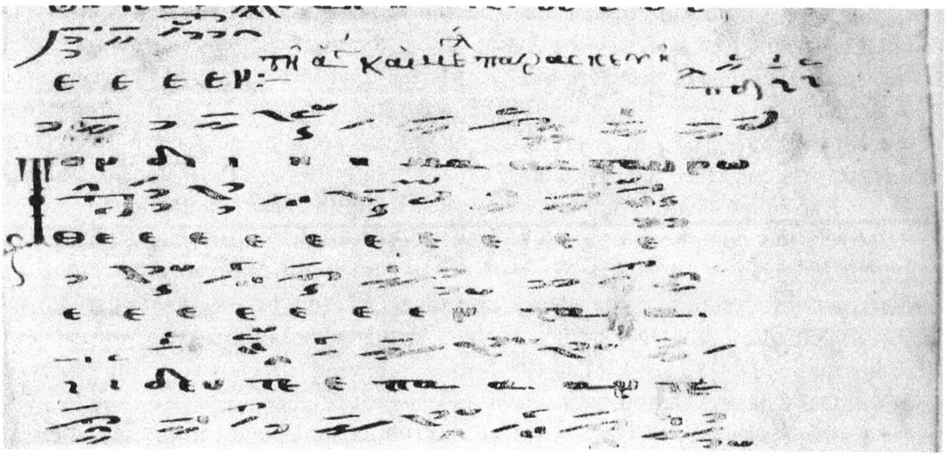

Figure 3.6 The *psaltikon* (Florence, Ashburnhamensis 64 (written at Grottaferrata, Italy, (AD 1289), fol., 124r) shows the Middle Byzantine counterpart for the Good Friday *kontakion* rendered in Middle Byzantine notation.

that in the case of responsorial chants such as the *prokeimena*, which are performed in part by the soloist, in part by the elite choir, the solo sections appear only in the *psaltikon*, the choral sections only in the *asmatikon*. To reconstruct the cathedral chants, both books are required. That division extends to style, too. The *psaltikon* has its own characteristic, melismatic formulas that differ in some respects from those seen in the melodies of the *asmatikon*. For example, chants such as the *hypakoai* and *kontakia* may occur in both books, and in these instances, the music follows the main style of the collection in which they are found.

The *prokeimenon*, sung before Apostle readings, demonstrates that even in the *psaltikon* cathedral rite repertories there are formulas governed by text accent.[34] The *prokeimena* are possibly derived from archetypes in the Early Christian psalmody, where the congregation responded with refrains to the verses of psalms sung by a soloist.[35] In the development of the early rites, the psalmody seems to have been modified in various ways. A successive reduction in the number of verses to be sung might correspond to the institutionalization of the liturgies in the ancient Christian metropoles. We see the earliest sources—in prose, that is, sources without notation—prescribing a kind of modified responsorial psalmody. The selection of particular psalm verses for particular liturgical occasions soon became a set tradition in Jerusalem, Rome, Alexandria, and Constantinople.[36]

In the *prokeimena*, the verses of the Psalter were not quoted slavishly and in a biblical sequence. Instead, they were selected to furnish a prophetic justification of the actual liturgical feast. For this reason, the verses were edited—cut and recombined to underline the intended theological interpretations. But these chants add up to more than a cento, or patchwork of scriptural texts; the words and their music contributed to the totality of liturgical action as performed by priests, congregation, and chanters together.

The prokeimena settings

At the same time, the melismatic character of the *prokeimena* chants clearly signals a musical approach in which formulaic melodies might work with the texts so as to not only beautify them but also add to the message of the words. At first sight, the most striking feature of the cathedral rite music of the *prokeimena* is the use of the so-called double-gamma endings (γγ-). The two gammas represent a sung syllable, as an extension of the final syllable. The double gammas are known also from the *asmatikon*. They seem to have a connection with a performance practice different from that of contemporary classical music. In today's performance practice, the giving of intonation tones to the choirs and the tuning of instruments—although these features might have clear musical qualities in themselves—are not considered part of the performance. By contrast, the intonation melodies and double gammas could be seen as a practical means to administer intonations. The "regular" and modally defined final tone is the one sung on the vowel immediately before the first group of gammas, the gammas probably representing a nasal melodic extension of the word, reaffirming the modality by its characteristic leaps and steps, leading to what is to be sung next. The presence of the double gammas in the musical manuscripts is thus linked to the practices of professional soloists/choirs at Hagia Sophia. Eventually this feature became a standard idiom of cathedral chanting, wherever the chant manuscripts were copied (Italy, for example).

In the *prokeimena*, the melody generally matches text syntax and accentuation and is structured by the placement of the melismas. The following table shows the melismas in bold (Figure 3.7). The text accents are expressed musically by sustained and prolonged tones as *diple* or *kratema* (marked in Figure 3.7 by "s"), but in addition, they are often supplemented by a tonic accent in the opening or the middle of the phrase, most often neumed with the dynamic interval sign called the *oxeia* or *petasthe* (marked in Figure 3.7 with a "t"). In opposition to simple psalmody, this principle of musical accentuation is extended also to the final cadences, embodied in an extended and (for

Prokeimenon *This is the day*, for Easter, Divine Liturgy, fourth plagal mode

	Text from Patmos 221	LXX	KJV
PS 117.24	Αὔ(st)-**τη** ἡ ἡ-μέ(ts)-**ϱα**, ἣν ἐ-ποί(st)-η-σεν ὁ **κύ-ϱι**-ο-γγ-ο-γγ-ος.	1 Αὕτη ἡ ἡμέϱα, ἣν ἐποίησεν ὁ κύϱιος· 2 ἀγαλλιασώμεθα καὶ εὐφϱανθῶμεν ἐν αὐτῇ.	This is the day which the Lord hath made; we will rejoice and be glad in it.
PS 117.1	Ἐ-ξο-μο-λο-**γεῖ**-σθε τῷ κυ-ϱί(s)-ῳ, **ὅ**-τι ἀ-γα-**θό**-γγ-ο-γγ-ος.	1 Ἀλληλουϊα. 2 Ἐξομολογεῖσθε τῷ κυϱίῳ, ὅτι ἀγαθός, 3 ὅτι εἰς τὸν αἰῶνα τὸ ἔλεος αὐτοῦ·	O give thanks unto the Lord; for he is good: because his mercy endureth for ever.
PS 117.2	**Εἰ-πά**-τω δὴ οἶ(t)-κος Ἰσ-ϱα-ὴλ(st) ὅτι ἀ-γα-**θό**-γγ-ο-γγ-ος:-	1 εἰπάτω δὴ οἶκος Ἀαϱὼν ὅτι ἀγαθός, 2 ὅτι εἰς τὸν αἰῶνα τὸ ἔλεος αὐτοῦ·	Let the house of Aaron now say, that his mercy endureth for ever.

Figure 3.7 Distribution of tonic and sustained accents in the Easter *prokeimenon* from the *psaltikon*.

the *prokeimenon* repertory) standard melisma. This melismatic setting has been accomplished at the expense of the biblical chant text, which appears only in a truncated form. Some have interpreted this abbreviation as a degeneration of the original, proto-Christian tradition of responsorial psalmody. Yet we see the skeleton of the age-old psalmody in the refrain selected for the celebration of the Resurrection (Ps. 117[118], verse 24) and in the fact that the subsequent selected verses at least appear in numeric order.

Another theory to explain the absent last parts of the verses—that either the choir or the congregation should complete them—finds little support in the manuscripts and liturgical orders. For some of the *prokeimena*, the refrain is taken up and brought to its "scripturally correct completion" by the so-called *Dochai* (receptions or responses) found in the *asmatikon* collection. For other *prokeimena*, this is not the case, and the chant text is left as a torso or a half-verse. Because of the original parallelisms inherent in the poetic composition of the Psalter, in many instances it is possible to leave the second half out without losing much of the meaning (the "missing" words from Septuaginta [Ps. 117(118): 22, 1–2] are supplied in the example shown here in the right column with gray shaded letters). Even if performed only in their truncated textual form, the *stichoi* in their musical form might nevertheless convey a plurality of meanings to those participating in the liturgical act.

The ownership of the psaltika-asamatika and other chant manuscripts

When it comes to repositories of *psaltika* and *asmatika*, southern Italy preserves a rich record. The monasteries in this area, while no longer part of the Byzantine Empire by the late eleventh century, emerged as important centers of Byzantine culture and cults.[37] Because they celebrated major feasts according to the Constantinopolitan cathedral rite, they most likely ordered the production of *psaltika* and *asmatika* with musical notation, as seen in Figure 3.6.

Their offices mixed the cathedral with the monastic Stoudite-Sabaite rite.[38] The eleventh-century Theotokos Evergetis Monastery in Constantinople exhibits a model for such a composite rite; its offices are recorded in its typikon, or liturgical order document, which presents the order of services. A southern Italian equivalent of the Evergetis typikon is seen in the monastery of the Santissimo Salvatore at Messina, Sicily (founded 1131 CE).[39] The liturgical rubrics in the Messina typikon present more aspects of the cathedral rite. It uses the term ὁ λαός ("the people," not "the monks") and mentions the choirs and church officials as if this was a veritable copy of the Constantinopolitan cathedral rite. *Psaltika* and *asmatika* associated with the Messina monastery are preserved in the Messina University Library.[40]

The previously mentioned list of manuscripts from Patmos also includes a *psaltikon* manuscript that may very well be the one we know as Patmos 221 (see above). These two examples, from southern Italy and Patmos, indicate that the music of the Constantinopolitan cathedral rite had found a place in the standard repertory and was eventually performed in many other contexts aside from Hagia Sophia, though still supported by written documents derived from copies of the cathedral's musical manuscripts.

The papadikai and akolouthiai manuscripts (from ca. 1300)

While many chant manuscripts of the eleventh to the fourteenth centuries were copied for and held by institutions for the maintenance of their local version of the Byzantine rite, it seems that the nature of the manuscripts generally changed during the last centuries of the Byzantine era. The ***akolouthiai*** manuscripts record the order and performance of the chants in the liturgy. This type, often associated with one of its early composers/editors, John Koukouzeles, exhibits a great variety; the contents differ from one manuscript to another to a greater extent than is seen in the traditional chant manuscript types discussed above.[41] Around some traditional blocks of simple ordinary chants and a few remnants of the *psaltikon* and *asmatikon*, each scribe of the *akoulouthiai* manuscripts gathered a more individually selected body of kalophonic settings (ornamented and expanded versions of traditional Byzantine chants). The designation "by my teacher" occasionally appears in the rubrics, and pieces composed or edited by the very scribe (and declared as such in the rubric) are not rare.

Moreover, the *akolouthiai* manuscripts are regularly prefaced by a brief collection of materials for teaching the basics of Middle Byzantine notation and exercises to improve reading skills. The adjective *papadikos* in the title of the earliest versions of these collections of didactic material is derived from the word *papas* (from the root *papad-), presumably not in the sense of "priest," but rather as an equivalent to *psaltes* (soloist or precentor). Finally, the manuscripts are often so small that the designation "pocket book" is well deserved.

It is thus reasonable to conclude that these types of chant manuscripts were the personal tools used by singers and chant teachers, who earned money from accepting chant pupils from outside the monastery. Ioannis Markouris has found and published a number of contracts of chant apprenticeships in Crete from the fourteenth and fifteenth centuries.[42] That chant teaching was increasingly privatized in the later period may also be deduced from the following ironic chant text, found in a number of *papadikai*, set to an extremely challenging melody modulating through all eight modes:

> Ὁ θέλων μουσικὴν μαθεῖν καὶ θέλων πενέσθαι
> θέλει πολλὰς ὑπομονάς, θέλει πολλὰς ἡμέρας
> θέλει καλὸν σωφρονισμὸν καὶ φόβον τοῦ Κυρίου
> τιμὴν πρὸς τὸν διδάσκαλον, δουκάτα εἰς τὰς χεῖρας
> τότε νὰ μάθει ὁ μαθητὴς καὶ τέλειος νὰ γίνῃ.

> The one desiring to learn music must be desiring to be poor,
> to desire [to suffer] many hardships, to desire [to work] day after day,
> to desire to exercise self-control, modesty, and fear from the Lord,
> to pay honour to the teacher, [and] with silver coins in his hands, [only] then the student will learn and reach perfection!

What consequences might such a development have for the stability of chant traditions and the whole concept of chant transmission? It is difficult to answer these questions now, but it can be assumed that such wide-ranging changes in the milieu surrounding the chant culture probably also changed the premises of the oral–aural transmission mechanisms as seen in earlier Byzantium. This degree of individualization and privatization may have led to faster development and innovation, in contrast to earlier conservatism in the reproduction of the traditional repertories and chant book types, controlled by institutions.[43]

Chant, location, and architectural space

Seldom do the older Byzantine chant manuscripts contain instructions as to the exact location where a chant is to be sung or the placement of the singers.[44] The chant collections were copied into books that could, in principle, be used anywhere in the empire. Relatively few individuals had competence in musical writing, and the ones with these skills often traveled from one place to another, bringing with them manuscripts for copying or instruction. Because of their local character, however, the typika give many useful insights into the locations and staging of the liturgical acts described, including chanting. In addition, literary sources, travelers' and pilgrims' accounts, deliver further information about the specific places in the city, including streets, squares, and positions inside churches, in which chant performance occurred.

It is highly likely that the physical surroundings influenced the maintenance and the development of the chant traditions and genres used in specific locations, although the same chants (that is, with the same text and sequence of melody) occasionally were performed in different places, depending on the liturgical demands of the specific services. Thus, the feedback effect between the performers and the soundscape, resulting in specific behaviors and practices, are hard to discern directly in the written sources, given the way these documents work (according to the line of thought presented above). But since the performance of specific services and chants recurrently took place in the same spatial contexts, the physics most likely left an imprint, over time, on specific chant genres and traditions.

Singing in streets, squares, and courtyards

The stational liturgy, a term referring to the processional character of Late Antique liturgy in which stages of the service would be performed en route at different churches and public spaces, has gained a new scholarly interest since John Baldovin's publication *The Urban Character of Christian Worship* (1987).[45] Many chants, such as the *Trisagion*, a famous hymn with a specific text ("Thrice Holy"), litanies, and groups of *troparia*, have been identified as processional hymns.[46]

Vocal performances set outdoors face a number of technical problems: reverberation times constantly vary and are often very brief, which has the effect of muffling the sound production and decreasing audibility, while wind, rain, and other weather phenomena interfere with the performance. Nevertheless, the very act of walking together in procession while singing produces an intense experience that perhaps reduces somewhat the demand for uniform performances or high artistic standards. It is the participation, solo or group, that matters.

Processional singing is documented in monastic contexts as well, as, for instance, in the Evergetis typikon. On a number of major saints' feasts, the celebrant's entrance procession at *hesperinos* (vespers) is adorned by the singing of "Φῶς ἱλαρόν" (Oh, Gladsome Light), and this was also the custom throughout the Triodion (Lent) period. At the Feast of the Transfiguration, August 6, a procession circled around the monastery, visiting all the cells, while singing the *troparion* Σῶσον, Κύριε, τὸν λαόν σου, a processional piece also sung by the brethren carrying the cross at the Adoration on the fourth Friday in Lent, and by the priest, the deacon, and the ecclesiarch (liturgist) while they prepare the cross for the ceremonies on September 14, the Exaltation of the Holy Cross. The melody of this processional hymn was first documented in the *akolouthiai* manuscripts from the fourteenth century.

Finally, on Palm Sunday a more extensive singing of different *troparia* took place during the procession with palm branches, leaving the main church for the Chapel of Our Savior after *orthros*. At the regular λιταί (processions) at the vespers office, prayers or litanies seem to have been performed without processional *stichera* or other *troparia*, consisting of a very simple musical form.

It may be presumed that in order to enlarge participation and make disruptions en route manageable, melodies and responses for outdoor singing were simplified to some degree. The traditions of processional chanting may have been created mostly by experience and maintained without the use of written notation until the late or post-Byzantine period. During the earlier period, though processional chant formed an essential part of the totality of the Byzantine chant *corpus*, many processional chants may be said to have belonged to the category of "marginal repertory." Another term should probably be used today, since scholars have begun to perceive musical notation not as the primary vehicle of the tradition but as one of the technologies integrated in its oral–aural transmission. However, it is difficult to find another equally brief term to describe the occasionally and heterogeneously notated "ordinary" chants.

Singing in churches

In 2009 and 2010, in collaboration with Professor Jens Holger Rindel from the Technical University of Denmark, I worked on an auralization experiment on three Constantinopolitan churches.[47] We studied the following churches:

- The Church of Sts. Sergius and Bacchus: volume = 15,000 m3 (cubic meters), RT (reverberation time) (1 kHz) (that is, the time passed during 60 dB of room decay of a well-defined tone at 1 kHz (meaning a pitch close to b″), though only measured in the interval –5 dB to –35 dB and doubled to match measurements over a decay of 60 dB. It could thus be described as RT T_{30}) = 3 seconds (empty church);
- Hagia Irene: volume = 39,000 m3, RT(1 kHz) = 4.3 seconds (empty church);
- Hagia Sophia: volume = 255,000 m3, RT(1 kHz) = 10 (empty church).

Acoustic 3-D models had already been produced in connection with the European–Turkish project CAHRISMA (Conservation of Acoustical Heritage by the Revival and Identification of Sinan's Mosque Acoustics). Using **anechoic** recordings of a chant reconstructed from the *psaltikon* repertory, we produced auralizations in each church with different numbers of singers (some of the recorded tracks were doubled artificially in order to make experiments with a greater number of singers) in different positions, a simulated listener in a variety of placements, and a simulated audience of various numbers of people as a virtual congregation. The impact of the surroundings is definitely much more predictable in confined spaces (or in semiconfined spaces, such as the ancient theaters) than outdoors, and the contrast between the soundscapes and the experience of nuance in the three different-sized churches was very enlightening.

In order to compare various types of input, we compared the reconstructed melody with auralizations in the same surroundings made from anechoic recordings with Leonidas Asteris, the then *protopsaltes* of the Great Church, representing the contemporary chant tradition of the Ecumenical Patriarchate. In this chant style, many fine ornaments were clearly audible in the Church of Sts. Sergius and Bacchus, while they were blurred or almost disappeared in the larger churches with longer reverberation time. The reconstruction of the *psaltikon* in a style without such ornamentation, however, seemed to create a specific and more clearly discernible sound in the larger churches. Tempo differences would, of course, have an impact, and further experiments with real-time auralization as opposed to auralization as a postproduction process, as employed by our project, will in the future bring further insight.[48]

In 1996, we learned something of the effect of architecture on music from a program of Byzantine music that we had prepared for the congress of Byzantine chant at the 1996 International Congress of Byzantine Studies in Copenhagen. The conductor Lykourgos Angelopoulos reported that the concert lasted considerably longer than planned, due to the feedback from the resonant acoustics in the Church of Our Lady, Copenhagen Cathedral.[49] In our auralization experiment, the use of ison (a drone note) in singing made interesting use of the reverberation. The constant repetition of the same frequency in the churches with longer reverberation times created the impression for both singers and listeners that it was almost "running by itself." It is clear from the historical information that the rather small-size choirs in Hagia Sophia would not prevent them from creating impressive performances in this extremely reverberant cathedral. Even if this drone practice is not described in sources from before about 1500, the experiment might nevertheless hint at the plausibility and characteristics of using such a performance practice in a cathedral context.

Another musical feature—one in fact present in the musical manuscripts, the so-called double-gamma endings discussed above—gave a specific tonal effect in the larger church, creating the sensation that several tones were sounding together at cadences, producing a tonal area. A tonal area—more than one tone sounding together—produces a tonal feeling of space. In the case of the double-gamma endings, the "final" tone is framed by other tones (below and above), emphasizing rather than blurring it. I cannot claim that the mystery of the function and meaning of this musical feature is now solved, but it has at least been illuminated from a new angle. The double-gamma endings pertained exclusively to the cathedral genres. They are occasionally seen also in later documents, in the *akolouthiai* manuscripts, where these include a few selections from the *asmatika* and *psaltika*, but in the post-Byzantine chant traditions this feature disappears. If I was asked to point to a specific element in which the cathedral soundscape became traceable in the neumated manuscripts, it would be these endings—together with the related phenomenon of the "**asmatic syllables**," non-semantic vocalisms inserted to extend and beautify the chanted texts of the cathedral rite, that is, in the *asmatika* and *psaltika*.

Conclusion

This is a plea for relinquishing the idea of categorizing chant traditions in a one-dimensional continuum from "oral" to "written." Instead, I point to the importance of studying the mechanics of oral–aural chant transmission and the variety of roles played by the written sources in the totality of "chant administration" in its sociological, ritual, and physical contexts. This approach looks at traditional musical cultures according to what is received, how it is perceived, in which surroundings it is heard, by which means it is recognized, how it is remembered and its meaning deduced, how it is re-created, and, finally, how it is passed on to the next generations through practice and performance.

Notes

1 Oliver Strunk, "P. Lorenzo Tardo and His 'Ottoeco nei mss. Melurgici'" (1967), reprinted in Strunk, *Essays on Music in the Byzantine World*, ed. Kenneth Levy (New York: W. W. Norton 1977), 267.

2 This pertains to many of the simple and daily chant items, such as the priest's parts, choral and congregational responses, and the so-called marginal repertories; see part 4.2 below ("*Automela* and *Proshomoia*, or Model Melodies versus *Contrafacta*"). For "ordinary" and "marginal" repertories, see, for example, Neil K. Moran, *The Ordinary Chants of the Byzantine Mass* Hamburger Beiträge sur Musikwissenschaft, 2 (Hamburg: Wagner 1975); and Christian Troelsgård, "Melodic Variation in the 'Marginal' Repertories of Byzantine Musical MSS, Exemplified by Apolytikia/Kontakia and Exaposteilaria Anatasima," in *Cantus planus: Papers Read at the 7th Meeting, Sopron, Hungary, 1995/International Musicological Society Study Group*, ed. Laszlo Dobszay (Budapest: Hungarian Academy of Sciences Institute for Musicology, 1998), 601–09.

3 Anne Pym, "Writing Oral Traditions: Clash between Worlds," *Review of Communication* 7 (2007): 303–10, esp. 304. Pym refers to a presentation by Michael Oleksa, *Communicating across Cultures*, 4-pt. video series, pts. 1, 2, KTOO and Capital Community Broadcasting, Juneau, AK, 1994. Such dichotomies were introduced in the discussion of oral traditions by Walter Ong. Here, they are given in a more popularized form, and from the sociology of literature rather than from chant research, see Walter Ong, *Orality and Literacy: The Technologizing of the Word* (New York: Routledge, 1982).

4 Walter Ong, *Orality and Literacy*, 5–77, 139–80. Ong studied a range of medieval and popular literary genres and found certain common traits that could be linked to their oral transmission. He opposed these to the concepts of literature and the "work of art" in the world of printing.

5 Mary Carruthers, *The Book of Memory: A Study of Memory in Medieval Culture* (Cambridge: Cambridge University Press, 1990).

6 Kurt Treu and Johannes Diethart, *Griechische Literarische Papyri christlichen Inhaltes*, vol. 2, Mitteilungen aus der Papyrussammlung der Österreichischen Nationalbibliothek (Papyrus Erzherzog Rainer), n.s.,17th ed. (Vienna: Brüder Hollinek, 1993).

7 I have dealt earlier with the contents of the codex and its importance for the study of the early *octoechos* system. See Christian Troelsgård, "A New Source for the Early Octoechos? Papyrus Vindobonensis G. 19.934 and Its Musical Implications," in *Byzantine Musical Culture: Proceedings of the 1st International Conference of the ASBMH*, 668–79, www.asbmh.pitt.edu/page12/Troelsgard.pdf (accessed May 20, 2015).

8 For an introduction to the sticheron, see Egon Wellesz, *A History of Byzantine Music and Hymnography* (Oxford: Clarendon Press 1949), 302–17; and Nina-Maria Wanek, *Sticheraria in spät- und postbyzantinischer Zeit: Untersuchungen anhand der Stichera für August* (Vienna: Praesens, 2013), 7–22.

9 Christian Troelsgård, chap. 4 in *Byzantine Neumes: A New Introduction to the Middle Byzantine Musical Notation*, Monumenta Musicae Byzantinae, Subsidia, vol. 9 (Copenhagen: Museum Tusculanum Press, 2011), 35–40.

10 Charles H. Cosgrove, *An Ancient Christian Hymn with Musical Notation: Papyrus Oxyrhynchus 1786; Text and Commentary* (Tübingen: Mohr Siebeck, 2011), 6–10.

11 Rosario Pintaudi et al., Eds., *Papyri Graecae Wessely Pragenses (PPrag. II)*, Papyrologica Florentina, vol. 26 (Florence: Gonnelli, 1995).

12 Bernard P. Grenfell and Arthur S. Hunt, Eds., *New Classical Fragments and Other Greek and Latin Papyri*, Greek Papyri, 2nd ed. (Oxford: Clarendon Press, 1897).

13 Harry Y. Gamble, *Books and Readers in the Early Church* (New Haven, CT: Yale University Press, 1995), 203–42.

14 Excerpt from a decree of the Synod of Laodicea, Greek text from Karl Joseph von Hefele and Charles Leclercq, *Histoire des conciles d'après les documents originaux*, vol. 1, pt. 2 (Paris: Letouzey et Ané, 1907), 737 ff.

15 Enrica Follieri, *Initia hymnorum ecclesiae graecae*, vols. 1–5, Studi e testi (Città del Vaticano: Biblioteca Apostolica Vaticana, 1960–66).

16 See below, section dealing with partial notation and *Theta* notation.

17 Oliver Strunk, "The Antiphons of the Oktoechos" (1960), in Strunk, *Essays on Music in the Byzantine World*, 165–90.

18 See Maria Alexandru, "Studie über die 'grossen Zeichen' der byzantinischen musikalischen Notation, unter besonderer Berücksichtigung der Periode vom Ende des 12. bis Anfang des 19. Jahrhunderts," 3 vols. (PhD diss., University of Copenhagen, 2000).

19 Kay Kaufman Shelemay, Peter Jeffery, and Ingrid Monson, "Oral and Written Transmission in Ethiopian Christian Chant," in *Early Music History* 12 (1993): 55–117.

20 See M. Velimirović, "The Byzantine Heirmos and Heirmologion," in *Gattungen der Musik in Einzeldarstellungen: Gedenkschrift für Leo Schrade*, ed. Wulf Arlt et al. (Bern: Francke, 1973), 192–244. For an introduction to the "Sticherarion," see Wellesz, *A History of Byzantine Music and Hymnography*, 302–19; and Wanek, *Sticheraria in spät- und postbyzantinischer Zeit*, 7–22.

21 See Jørgen Raasted, "Koukouzeles' Revision of the Sticherarion and Sinai gr. 1230," in "Laborare fratres in unum: Festschrift Laszlo Dobszay zum 60. Geburtstag," ed. J. Szandrei and David Hiley, special issue, *Spolia Berlinensia* 7 (Hildesheim, 1995), 261–77; and Raasted, "Koukouzeles' Sticherarion," in *Byzantine Chant: Tradition and Reform; Acts of a Meeting Held at the Danish Institute in Athens, 1993*, ed. C. Troelsgård (Athens: Danish Institute at Athens; Aarhus: Aarhus University Press, 1997), 9–21.

22 Wanek, *Sticheraria in spät- und postbyzantinischer Zeit*, 18–22; and Christian Troelsgård, "A List of Sticheron Call-Numbers of the Standard Abridged Version of the Sticherarion," pt. 1, "The Cycle of the Twelve Months," *Cahiers de l'Institut du Moyen-Âge Grec et Latin* 74 (2003): 3–20.

23 Strunk, "Melody Construction in Byzantine Chant" (1963), in Strunk, *Essays*, 191–201.

24 J. van Haelst, "Cinq textes provenant de Khirbet Mird," *Ancient Society* 22 (1991): 297–317.

25 Raasted, "Theta Notation," 57–62.

26 Troelsgård, *Byzantine Neumes*, 27; and Jørgen Raasted, "Theta Notation and Some Related Notational Types," in *Palaeobyzantine Notations: A Reconsideration of the Source Material*, ed. Raasted and Christian Troelsgård (Hernen: Brediusstichting, 1992), 57–62.

27 Oliver Strunk, *Specimina notationum antiquiorum*, Monumenta Musicae Byzantinae, Subsidia, vol. 7 (Copenhagen: Ejnar Munksgaard, 1966), Pars suppletoria, 1–7.

28 Constantin Floros, *Universale Neumenkunde* (Kassel-Wilhelmshohe: Bärenreiter, 1970), vol. 1. On Paleo-Byzantine notation, see further Gregorios Stathes, *Hē exēgēsis tēs palaias vyzantinēs sēmeiographias: kai ekdosis anōnymou syngraphēs tou kōdikos Xēropotamou 357 hōs kai epilogēs tēs Mousikēs technēs tou Apostolou Kōnsta Chiou ek tou kōdikos Docheiariou 389* (Athens: Institute for Byzantine Musicology, 1978).

29 Bibliothèque Nationale et Universitaire de Strasbourg, P. inv. 1185; see Geneviève Husson, "P. Strasb. inv. 1185: Hymne pour la fête de l'Hypapantè (2 février)", in *Atti del XXII Congresso internazionale di Papirologia, Firenze 1998*, ed. Isabella Andorlini (Florence: Istituto Papirologico G. Vitelli, 2001), 681–87. For a more recent discussion and color photographs, see Alan Gampel, "The Origins of Musical Notation," in *Age of Transition: Byzantine Culture in the Islamic World*, ed. Helen C. Evans (New Haven, CT: Yale University Press, 2015), 144–56, esp. figs. 7a–b.

30 Christian Thodberg, *Der byzantinische Alleluiarionzyklus: Studien im kurzen Psaltikonstil*, Monumenta Musicae Byzantinae, Subsidia, vol. 8 (Copenhagen: Ejnar Munksgaard, 1966), 48–63; and Jørgen Raasted, *Intonation Formulas and Modal Signatures in Byzantine Musical Mansucripts*, Monumenta Musicae Byzantinae, Subsidia, vol. 7 (Copenhagen: Ejnar Munksgaard, 1966), 48–63.

31 For the musical features of the rite, see Oliver Strunk, "S. Salvatore di Messina and the Musical Tradition of Magna Graecia" and "The Byzantine Office at Hagia Sophia," in *Essays on Music in the Byzantine World*, ed. Kenneth Levy (New York: W. W. Norton, 1975), 45–54 and 112–50. See also Alexander Lingas: "Festal Cathedral Vespers in Late Byzantium," *Orientalia Christiana Periodica* 63 (1997): 421–59. For the historical development of the Byzantine rite, see Robert Taft, *The Byzantine Rite, a Short History* (Collegeville, MN: Liturgical Press, 1992).

32 A brief introduction to these two book types is found in Strunk, "S. Salvatore di Messina and the Musical Tradition of Magna Graecia," 45–54.

33 Constantin Floros, *The Origins of Russian Music: Introduction to the Kondakarian Notation; Revised, Translated, and with a Chapter on Relationships between Latin, Byzantine and Slavonic Church Music by Neil K. Moran* (Frankfurt: Peter Lang, 2009).

34 G. Hintze, *Das byzantinische Prokeimena-Repertoire: Untersuchungen und kritische Edition*, Hamburger Beiträge zur Musikwissenschaft, 9 (Hamburg: Verlag der Musikalienhandlung Wagner, 1973).

35 Robert Taft, "Christian Liturgical Psalmody: Origins, Development and Decomposition," in *Psalms in Community: Jewish and Christian Textual, Liturgical, and Artistic Traditions*, ed. Harold Attridge and Margot Fassler (Atlanta, GA: Society of Biblical Literature, 2003), 7–32.

36 Christian Troelsgård, "The Prokeimena in Byzantine Rite: Performance and Tradition," in *Cantus Planus, Papers Read at the 6th Meeting, Eger, Hungary 1993*, ed. Laszlo Dobszay (Budapest: Hungarian Academy of Sciences,1995), 1:65–77.

37 Annalisa Doneda, "I manoscritti liturgico-musicali bizantini: Tipologie e organizzazione," in *El palimpsesto grecolatino como fenómeno librario y textual*, ed. À. Escobar, Collectión Actas, Filologia (Zaragoza: Institución "Fernando el Catolico," 2006), 103–10.

38 The Sabaitic liturgical tradition incorporates elements from the Mar Saba Monastery in the Judean Desert, mixed with local elements in the Stoudios Monastery in Constantinople. After the turbulent Iconoclast period, the reformed liturgy of the Stoudios Monastery was imitated by other cenobitic communities.

39 Miguel Arranz, *Le Typicon du monastère du Saint-Sauveur à Messine (Codex Messinensis gr. 115, A.D. 1131)*, Orientalia Christiana Analecta 185 (Rome: Istituto Pontificio Orientale, 1969).

40 Donatella Bucca, *Catalogo dei manoscritti musicali greci del SS. Salvatore di Messina (Biblioteca Regionale Universitaria di Messina)* (Rome: Comitato Nazionale per le Celebrazioni del Millenario della Fondazione dell'Abbazia di S. Nilo a Grottaferrata, 2011), 161ff.

41 See Edward V. Williams, "John Koukouzeles' Reform of Byzantine Chanting for Great Vespers in the Fourteenth Century" (PhD diss., Yale University, 1968); and Anne Elizabeth Pennington, "Seven *Akolouthiai* from Putna," *Studies in Eastern Chant*, ed. Egon Wellesz and Miloš Velimirović (Oxford: Oxford University Press, 1979), 4: 112–33.

42 Ioannis Markouris, "Apprenticeships in Greek Orthodox Chanting and Greek Language Learning in Venetian Crete (14th–15th Century)," in *I Greci durante la venetocrazia: Uomini, spazio, idee (XIII–XVIII); Atti del Convegno Internazionale di Studi, Venezia, 3–7 dicembre 2007*, ed. Chryssa Maltezou, Angeliki Tzavara, and Despina Vlassi (Venice: Istituto Ellenico di Studi Bizantini 2009), 233–49.

43 Laura Steenberge pursues this line of research in her essay for this volume, "*We Who Mystically Represent*: Text-Setting in the Medieval Cheroubika."

44 In the later chant manuscripts, however, a richness of descriptions begins to appear in the rubrics; see the documentation in Evangelia Spyrakou, *Οι χοροί ψαλτών κατά την Βυζαντινή παράδοση* [The Choirs of Singers according to the Byzantine Tradition] (Athens: Institute for Byzantine Musicology, 2008).

45 John F. Baldovin, *The Urban Character of Christian Worship: The Origins, Development, and Meaning of Stational Liturgy, Orientalia Christiana Analecta, 228 (Rome: Pontificio Istituto Orientale, 1987).* This study has been frequently referred to in later studies, for example, in the studies of Byzantine liturgical history by Robert Taft, The Byzantine Rite; and Vasileios Marinis, Architecture and Ritual in the Churches of Constantinople, Ninth to Fifteenth Centuries (Cambridge: Cambridge University Press, 2013).

46 Processional chants and rituals have been identified in connection with the study of imperial books of ceremonies, as suggested by Baldovin, *The Urban Character of Christian Worship*; and Taft, *The Byzantine Rite*. Other chants have been interpreted as closely related to the processional context; see, for example, Rainer Stichel, "Die musizierenden Hirten von Bethlehem," in *Beiträge zum Symposion zur byzantinischen Lexikographie*, ed. Wolfgang Hörandner and Eric Trapp, Byzantina Vindobonensia, 20 (Vienna: Verlag der Österreichische Akademie der Wissenschaften, 1989), 249–82.

47 Christian Troelsgård and Jens Holger Rindel, "Byzantine Hymns in Churches of Constantinople: An Archaeological Soundscape," www.odeon.dk/byzantine-hymns-churches-constantinople (accessed May 20, 2015).

48 For real-time auralization, see the essay by Jonathan Abel in this volume.

49 For a description of the concert, see Alex Lingas, "Performance Practice and the Politics of Transcribing Byzantine Chant," *Acta Musicae Byzantinae* 6 (2003): 56–76.

4 Understanding liturgy

The Byzantine liturgical commentaries

Walter D. Ray

Liturgical commentaries emerged at the beginning of the Middle Ages in both East and West, though arguably they played a greater role in the theology and spirituality of the Eastern churches. Developing out of the late fourth-century mystagogies that provided instruction for neophytes on the rites of initiation,[1] liturgical commentaries promised to reveal the significance of the liturgical rites for those who participated in them. In this brief survey I will look at the five principal commentaries on the Byzantine Divine (Eucharistic) Liturgy from the time of its first flowering in the sixth century in Justinian's church of Hagia Sophia until the fall of Constantinople in 1453, by which time the liturgy had virtually reached its full development. First, however, we must look at some modern attempts to evaluate these commentaries, in order to find the proper background against which to view them. I will then introduce each of the commentaries, placing it in its historical and theological context and presenting its main ideas. In keeping with the theme of this volume, I will pay specific attention to the commentaries' treatment of the church building.

Background: earlier commentaries and modern perspectives

A few significant studies deal with the five Byzantine liturgical commentaries as a group. René Bornert, *Les commentaires byzantins de la Divine Liturgie du VIIe au XVe siècle*, treats these commentaries as literary monuments, analyzing questions of authorship, textual integrity, genre, main ideas, and literary relationships.[2] Hans-Joachim Schulz's *Die byzantinische Liturgie*, first published in 1964,[3] revised in 1980 with a new section taking into account the work of Bornert and others,[4] and fully reworked in a third edition,[5] approaches each commentary as characteristic of a period in the history of the Byzantine liturgy, placing it within its historical, liturgical, and theological contexts. Hugh Wybrew's *The Orthodox Liturgy: The Development of the Eucharistic Liturgy in the Byzantine Rite* is similar to Schulz's work but offers a somewhat more systematic history of the liturgy.[6] All three remain foundational to any study of the Byzantine commentaries.

Since Bornert's volume came out, it has been customary to describe the immediate background to the Byzantine commentaries by contrasting the liturgical writings of Theodore of Mopsuestia and the late fifth-century author who wrote under the name Dionysius the Areopagite. Theodore (ca. 350–428), the last of the great Antiochene exegetes and author of the last of the great fourth-century mystagogies, is the first to apply historical, or "life-of-Christ," symbolism to the liturgy, in his *Catechetical Homilies* (Cat. Hom.).[7] In the Eucharistic liturgy, the Passion of Christ is represented by the transfer of the bread and wine to the altar before the Eucharistic prayer. The deacons bringing the gifts stand in for the angels that accompanied Christ both to his Passion and to the tomb. The altar denotes the tomb, and the altar cloth, the winding sheet in which Christ was buried. The deacons fanning the gifts to keep flies away are the angels who were at the tomb (Cat. Hom. 15.25–27). The Resurrection is represented by the *epiclesis* (invocation) of the Holy Spirit in the Eucharistic prayer (Cat. Hom. 16.11). The distribution of the elements is Christ's postresurrection appearance to his disciples (Cat. Hom. 16.17).

In *The Celestial Hierarchy* (CH) and *The Ecclesiastical Hierarchy* (EH),[8] the last of which gives a commentary on church rituals from baptism to the funeral, including the *synaxis* (the gathering, that is, the Eucharist), Dionysius (fl. ca. 500) finds in the earthly liturgy a representation of the heavenly liturgy. The church's hierarchy, from bishop to catechumen, mirrors the angelic (CH 1.3). Through these hierarchies, light descends from God, and God, "the gathering Father [*ho synagōgos Patros*]," draws people up into unity within themselves, with one another, and with God (CH 1.1–2). This movement also describes the Incarnation, since Jesus is the true light (CH 1.2). The outward movement and return to the sanctuary of the bishop as he incenses the church before the liturgy presents an image of God's movement through the hierarchies, outward into multiplicity, and returning into unity (EH 3.3). This movement is also seen in the distribution of Holy Communion, first to the bishop, then from bishop to the people, through which all are made one in communion with God (EH 3.13).

For Bornert, Theodore and Dionysius embody, respectively, "Antiochene realism" and "Alexandrian symbolism," which are rooted in the exegetical methods associated with these two Christian centers.[9] Robert Taft understands the contrast this way:

> The Antiochenes, more attentive in exegesis to the literal sense of scripture, favored a mystagogy that saw the liturgical mysteries chiefly as a portrayal of the historical mysteries of salvation. The Alexandrines, following the Origenist exegetical penchant for the allegorical, interpreted liturgy by a process of anagogy, whereby one rises from the letter to spirit, from the visible rites of the liturgical mysteries to the one mystery that is God.[10]

In Schulz's view, the Byzantine commentators "strike a balance" between these "diametrically opposed" possibilities.[11] Schulz also thinks that the authentic Early Christian tradition passed to Byzantium via Antioch and Theodore; in his view, "the predominately 'Aeropagitic' forces that shaped the history of the Byzantine liturgy"[12] are a bit of a detour. The commentary of Nicholas Cabasilas, however, "after an almost thousand-year-long, rambling history of symbolization and interpretation, once again gave the central liturgical event the authentic place it had in the early Christian and Antiochene heritage."[13] All these propositions need to be qualified, and the last one thrown out altogether. Here we can only suggest the outlines of a critique that would put the Byzantine commentaries into proper perspective.

First, there is a problem with treating liturgy like scripture. Scripture narrates the past and predicts the future. Liturgy is present activity directed toward the living God. Any description of the experience of this encounter will appear allegorical, yet it may be a literal reading of the liturgical rite. As Frances Young has shown through patristic evidence, the early Eucharist was experienced as a sacrifice of communion and a thank offering granting access to heaven. Only later, and at first only in the West, do we find the idea that it is also a participation in the earthly sacrifice of Christ.[14] The evidence of early Eucharistic prayers bears this out. The earliest Eastern prayers do not even mention the Last Supper. Instead, they have a thanksgiving offering of "reasonable sacrifice and bloodless worship,"[15] a phrase that in early Jewish-Christian tradition refers to the angelic service.[16] Even when the Eucharistic prayers obtain an anamnesis of the death and Resurrection of Christ,[17] it does not imply a reenactment of Christ's death, as Young observes, "but . . . simply draw[s] attention to the nature of the Eucharist as a thankful memorial of Christ's sacrificial death, and as the means of receiving the benefits of his Covenant-sacrifice."[18] Though it had an anamnesis, the Antiochene Eucharistic prayer also offered "reasonable and bloodless worship" and thanked God for having "raised us up to heaven and granted us the kingdom which is to come."[19] In this respect, Dionysius, who also interprets that Antiochene liturgy, is the more literal interpreter and Theodore the more allegorical. As Enrico Mazza observes, though Theodore avoids allegory in his exegesis:

> he makes extensive use of allegory in interpreting the liturgy; in fact, it is difficult to find anywhere else so marked an allegorical approach as in Theodore. This is another fact that bids us to avoid confusing the exegetical and sacramental vocabularies of these authors, even when the words are materially the same.[20]

What separates the commentaries of Theodore and Dionysius is not exegetical approach but theological presuppositions. Both deal with the same fundamental problem, "preserving the Nicene affirmation of the 'Godness of God.'"[21] "It is well known," says Theodore, "that the one who is eternal and the one whose existence has a beginning are greatly separated from each other, and the gulf found between them is unbridgeable" (Cat. Hom. 4.6).[22] A similar thought is at the center of Dionysius's apophatic theology. At its extreme, in Antiochene theology the gulf is never bridged, not even by the Incarnation. Theodore makes a distinction between the divine Word and the assumed man Jesus. For Theodore it is only in the Resurrection, after the assumed man has proved faithful even unto death, that the union between the man Jesus and the Word becomes in any way permanent. The persistence of this divide drives Theodore into historical imagery; only by imitation of the historical actions of Jesus can we hope to attain at our resurrection the closeness he now enjoys with the Word. Heaven for us becomes a future hope; it is no longer experienced as a present reality. The idea of participation in heavenly worship, which is still expressed in the liturgy, must be reimagined. The Eucharist becomes an imitation of Christ's historical Passion, which is in turn an image of his high-priestly offering in heaven (cf. Cat. Hom. 15.15). For Dionysius, in contrast, the gulf is bridged through the Incarnation and its converse, *theosis*, deification through union with God in Christ, which can be experienced even now and especially in the liturgy. In the end, the opposition is not between methods of exegesis, between historical and ahistorical approaches, or even between images of heavenly worship and life-of-Christ symbolism, but between salvation experienced as present reality or only as future expectation. The former is, in fact, the perspective of the Early Christians, and in this sense Dionysius is closer than Theodore to this tradition. At the same time, Theodore raises the question of the importance of Christ's earthly ministry for the Eucharist.

The Byzantine liturgy, in its texts, presents itself as participation in angelic worship, and there is no reason to believe that this was an innovation of the sixth century. This participation is evident already in our earliest witness to the liturgy, the eighth-century ***euchologion*** (a manuscript collecting the patriarchal prayers) *Barberini gr. 336*.[23] A few examples must suffice. First, the form that the Eucharistic anamnesis took in many of the Eastern anaphoras, including the Byzantine, already points beyond a focus on Christ's Passion: "remembering his saving Passion, his life giving cross, his three-day burial, his resurrection from the dead, his ascension into heaven, his session at your right hand, God and Father, and his glorious and fearful second coming" (anaphora of Saint Basil).[24] An emphasis on heavenly worship is also evident in the original prayer of entrance, which the bishop said as he and the congregation were about to enter the church at the beginning of the liturgy, and is now said at the Little Entrance with the Gospel book partway through the liturgy. This entrance prayer compares the present liturgy with the heavenly and asks that angels may also enter "to serve and glorify your goodness with us."[25]

The Cherubikon, a chant introduced in 573 that still accompanies the Great Entrance, the transfer of the Eucharistic gifts to the altar after the readings, presents an interesting case. In it the faithful are said to represent (*eikonizontes*) the angels who invisibly accompany the King of all, whom the faithful will receive in communion.[26] This hymn is similar to but contrasts with Theodore's idea that the deacons represent angels escort ing Christ to the tomb. Here, all the faithful, and not only the deacons, are images of the ranks (*taxesein*) of angels, a thoroughly Dionysian idea.[27] The contrast with Theodore may be intentional. His mystagogical homilies were known in Constantinople—they formed part of the dossier for his condemnation in 553—and it is not impossible that his symbolic interpretation of the liturgy had gained a measure of popularity, which might account for its reemergence in the eighth-century commentary of Patriarch Germanus of Constantinople.

The five Byzantine liturgical commentaries

Maximus the Confessor (580–660)

Of the Byzantine liturgical commentaries, only Maximus the Confessor's *Mystagogy* (*Myst.*, ca. 628–30) might be seen as intentionally balancing the positions of Theodore of Mopsuestia and (pseudo-)Dionysius the Areopagite.[28] All three belong to the era of Christological controversy over the relationship of Christ's humanity to his divinity. Theodore died in 428, the year the controversy erupted when Nestorius, archbishop of Constantinople, rejected the title of Theotokos, birth-giver of God, for Mary. Nestorius followed an Antiochene theology like Theodore's, which made a sharp distinction between the human Jesus and the divine Word. Only the man could be born, not God.[29] Nestorius was opposed by Saint Cyril of Alexandria and condemned by the Council of Ephesus in 431. Some of St. Cyril's followers later rejected the subsequent Christological definition of the Council of Chalcedon (451) that Christ was one person in two natures, preferring instead Cyril's formula of "one incarnate nature of God the Word." The next two centuries were taken up by unsuccessful efforts to overcome the schism Chalcedon engendered. It was in this context that the Dionysian writings were introduced, first by followers of Cyril in discussions that took place in Constantinople in the 530s. Partly as a result of these discussions, Theodore of Mopsuestia and his writings, including his catechetical homilies, were condemned as Nestorian by the Council of Constantinople in 553. And Maximus earned his title "confessor" for his opposition to monotheletism, the imperially imposed doctrine that Christ had only one will, which was another attempt to reconcile the Cyrillian non-Chalcedonians with the decisions of Chalcedon. The *Mystagogy*, though written before Maximus became embroiled in the monothelete controversy, applies a neo-Chalcedonian lens to the interpretation of the *synaxis*, the Eucharistic liturgy, explicitly taking Dionysius's commentary as its starting point.

Maximus begins with the same problem as Theodore and Dionysius, the unbridgeable divide between God and humans (*Myst.* prol.):

> [I]t is necessary that we understand correctly the difference between God and creatures. . . . For nothing whatsoever, whether being or nonbeing, is linked to [God] as a cause, no being or what is called being, no nonbeing or what is called nonbeing, is properly close [*engys*] to [God]. [God] is in fact a simple existence, unknowable and inaccessible to all and altogether beyond understanding which transcends all affirmation and negation.

Maximus addresses this divide in part through meditations on divisions in the church building, which we will examine below. Here we will note that Maximus's solution to this divide is the same as Dionysius's: it is through the Incarnation and *theosis*, realized through imitation of the heavenly liturgy and, ultimately, of God. The Christian who helps those in need obtains "nearness [*engytēta*] to God" and deification (*theōsis*) "through grace and participation" by taking on "in happy imitation the energy [*energeian*] and characteristic of God's own doing good" (*Myst.* 24). By imitating God's activities (energies) in the world, humans become participants in those activities, and thus become close to God and share in divine life.

Here, however, Maximus adds what might appear an Antiochene note: it is specifically imitation of Christ's Passion (*pathos*) that leads to deification. For Maximus, this does not involve a liturgical reenactment of Christ's death and burial. Rather, he understands the Passion to be an ongoing reality: in the Incarnation, God became "poor for us and t[oo]k upon himself by his own suffering [*sumpathōs*] the sufferings [*pathē*] of each one and 'until the end of time,' always suffering [*paschontos*] mystically out of goodness in proportion to each one's suffering [*pathou*]" (*Myst.* 24). Maximus asserts that "the one in need of having good done to him is God"—he cites Matthew 25:40—as is the one who meets that need: "All the more reason, then, will that one be God who by loving [people] in imitation of God heals by himself in divine fashion the hurts of those who suffer" (ibid.). Within the liturgy, the readings give instruction for Christian living, the Gospel in particular "proposing to the zealous some suffering [*kakopatheian*] on behalf of the Word" (*Myst.* 13).

Listed in the following table, are the liturgical moments Maximus comments on, in their order of appearance.

Maximus interprets the liturgy three times. He first shows what the liturgy means for the present worshippers (*Myst.* 9–13). Next, he assigns meanings in terms of the eschatological future (14–21). The third pass shows what happens in the present liturgy for the contemplative soul (22–23). Each time Maximus, like Dionysius, sees in the

Liturgy of the Catechumens
Entrance of bishop and people Ascent of bishop to throne Readings with responses and blessings Descent of bishop from throne Dismissal of catechumens Closing of doors
Liturgy of the Faithful
Entrance of mysteries (Great Entrance) Kiss of peace Creed "Holy, Holy, Holy" (Sanctus in the Eucharistic prayer) Our Father "One Is Holy" (Call to communion/communion)

liturgy an upward progression culminating in communion. The entrance of the worshippers into the church with the bishop symbolizes their conversion (9); they are instructed through scripture readings (10), closed off from the world by the closing of the doors, reconciled with each other through the kiss of peace, "set . . . in the number of the angels" through the Sanctus, come to know God as Father, and are "divinized by love . . . to the extent possible" through communion (13). In the next pass, the dismissal of the catechumens and the closing of the doors represents the last judgment after the Gospel has been preached to the whole world (14–15); then the faithful enter into "the nuptial chamber of Christ" (15; entrance of the mysteries), where they are fully united with all things through the Word (17; kiss of peace), praise the Trinity with the angels (19; Sanctus), are fully adopted as children of God (20; Our Father), and have "all of God entirely fill them and leave no part of them empty of his presence" (21; communion). Finally, the contemplative soul puts away all distraction as "the Word . . . leads it to knowledge of theology . . . granting it an understanding equal to the angels" and intimate knowledge of the Trinity (23).

Before embarking on these three levels of interpretation, however, in perhaps another nod to the Antiochene concern for history, Maximus places the entire liturgical celebration into a salvation-historical context through an interpretation of the bishop's entry into the church (*Myst.* 8). This entrance is "a figure and image" of Christ's first coming in the flesh, culminating in the ascension, represented by the bishop's ascent to his throne in the apse. By his coming, Jesus "freed human nature" from corruption and enslavement to the devil, "redeemed all its debt," and now, "in exchange for our destructive passions he gives us his life-giving Passion [*pathos*] as a salutary cure which saves the whole world." The saving work of Christ is thus presented as an event in the past that both precedes and makes possible the present liturgical experience, which in turn anticipates the final fulfillment.

Maximus begins his commentary with an extended meditation on the church building (*Myst.* 1–5). The church building had become a symbol for the Christian community in liturgical practice well before Maximus's time. Both initiatory and penitential rites depended on the contrast between being inside and outside the building. In initiation, the dismissal of the catechumens before the Eucharistic portion of the Divine Liturgy and the barring of the doors made a sharp distinction between the baptized and the nonbaptized and gave symbolic value to both the readings and the Eucharist. Similar symbolic meanings are apparent in the practice of canonical penance, in which the penitents were excluded from the building and gradually readmitted, first only to hear the readings, then to attend the Eucharistic prayers, and finally to partake of communion. Hagia Sophia may have been designed with this penitential practice in view, if the ninth-century *Diēgēsis* can be believed. According to this text, the bands of green marble stretching across the width of the floor parallel to the altar, called "rivers" here and elsewhere, indicated where penitents were to stand for the liturgy.[30]

Jean-Claude Larchet is right to say that "while Pseudo-Dionysius . . . underlines the difference at the heart of unity, Maximus underlines . . . the unity at the heart of difference."[31] He begins with the unity, represented by the church as both building and people. "Holy Church," says Maximus in *Mystagogy* 1, "bears the imprint and image of God since it has the same activity [*energeian*] as [God] does by imitation and in figure." Just as God, "who brought into existence all things . . ., contains, gathers and limits [*synechei kai synagei kai perigraphei*] them. . . . Maintaining about himself as cause, beginning, and end all beings which are by nature distant from one another," so the church does for the congregants. God accomplishes this gathering of all things into unity through Christ:

> who encloses [*perikleiōn*] in himself all beings. . . . As the center of straight lines that radiate from him he does not allow by his unique, simple, and single cause and power that the principles of beings become disjoined at the periphery but rather he circumscribes their extension in a circle [*kyklō perigraphōn*] and brings [*agōn*] back to himself the distinctive elements of being which he himself brought into existence.

So does the church, which bestows on its members through baptism "one divine form and designation, to be Christ's and to carry his name," so that, "converging with all the rest and joining together with them," they can "be and be seen [*horasthai*] as one body formed of many members . . . really worthy of Christ himself, our true head."

Larchet is not correct when he says that in this chapter on unity Maximus only has in view the church as people.[32] Maximus is describing the *synaxis*, the visible gathering of "the numerous and of almost infinite number" of

> men, women, and children who are distinct from one another and vastly different by birth and appearance, by nationality and language, by customs and age, by opinions and skills, by manners and habits, by pursuits and studies, and still again by reputation, fortune, characteristics, and connections (*Myst.* 1),

all pouring in through the doors from all sides of the church and converging under the central dome as the bishop makes his way around the ambo and down the solea at the first entrance of the liturgy (cf. *Myst.* 8, 9). The words he uses to describe the divine (hence, also the church's) activity in this ingathering—"enclose," "circumscribe . . . in a circle," terms he adds to the image of God as center point he borrows from Dionysius[33]—evoke the image of a centrally planned church such as Hagia Sophia containing the gathered faithful.[34] Maximus draws further attention to this vision of the building enclosing the *synaxis* of the faithful by his emphasis on the closing of the outer doors, after the catechumens have been dismissed, as the main moment of transition in the liturgy, which both separates the faithful from worldly distractions and introduces them "into the spiritual world, that is, into the nuptial chamber of Christ" (*Myst.* 15; cf. 13, 23). Once the outer doors are closed, the church is the image of this nuptial chamber of Christ, or rather, of Christ himself, drawing the faithful to himself and encircling them in his embrace.

In the remainder of the treatise Maximus explores further how this unity in Christ is achieved, again using the church as model, first as building, then as the faithful in action through the liturgy. "While it is one house in its construction, it admits of a certain diversity in the disposition of its plan by being divided"—by the chancel barrier around the altar—"into an area exclusively assigned to priests and ministers, which we call a sanctuary, and one accessible to all the faithful, which we call a nave" (*Myst.* 2). In successive chapters, this distinction in the construction of the church serves as an analogue for the distinction between intelligible and sensible within the created world (2), between heaven and earth in the sensible world (3), between the body and soul in the one human person (4), here he adds the distinction between sanctuary and altar table, soul and mind), and between active and contemplative parts of one soul (5). Unstated but implied through the use of Chalcedonian terminology and in the role Christ plays in uniting disparate elements is the distinction between the natures in the person of Christ, whose incarnation is the way the unbridgeable chasm between Creator and creatures is ultimately overcome. The church, which represents Christ:

> is one in hypostasis [*kata tēn hypostasin*] without being divided into its parts by reason of the differences between them, but rather by their relationship to the unity it frees these parts from the difference arising from their names. It shows to each other that they are both the same thing, and reveals that one is existing [*hyparchon*] for the other in turn what each one is being [*ōn*] for itself (*Myst.* 2).[35]

The meaning of these parts is manifested by the activity that takes place in them:

> Thus, the nave is the sanctuary in potency by the relationship of the liturgy [*mystagogias*] toward its end, and in turn the sanctuary is the nave in act by possessing the principle of the inseparable [*adiastatou*][36] liturgy; which remains one and the same in its two parts (*Myst.* 2).

The division in the plan of the church corresponds to the division of the liturgy into two parts by the closing of the doors. The liturgical focus of the first part, the hymns and readings, is the ambo in the center of the nave; the focus of the second part is the altar table. Yet, in spite of this division, the liturgy, like the church, remains one.

Patriarch Germanus of Constantinople (d. 733)

With Germanus of Constantinople's *Historia ekklesiastikē kai mystikē thēoria* (HE, ca. 725, *On the Divine Liturgy*),[37] one hundred years after Maximus's commentary, there reemerges a life-of-Christ symbolism resembling that of Theodore of Mopsuestia (HE 37). And here I would like to stress that with respect to liturgy, a recourse to historical symbolism is not literal but anagogic:

> By means of the procession of the deacons and the representation of the fans, which are in the likeness of the Seraphim, the Cherubic Hymn signifies the entrance of all the saints and righteous ahead of the cherubic powers and the angelic hosts, who run invisibly in advance of the great king, Christ, who is proceeding to the mystical sacrifice. . . . It is also in imitation of the burial of Christ. . . . The altar is an image of the holy tomb, and the divine table is the sepulchre in which, of course, the undefiled and all-holy body was placed.

Germanus's underlying theology is not Theodore's, however, but almost the opposite. Writing at the beginning of the imperial policy of Iconoclasm—he was deposed in 730 for his defense of the icons—Germanus uses his commentary on the liturgy to show that the divine Word can be depicted in Christ. Theodore might well have said that, just as the man Jesus could be born of a woman but not the divine Word (cf. Cat. Hom. 6.3–4), so only the man could be depicted. Germanus demonstrates that God the Word is depicted in the liturgy.

Especially significant in this regard is his interpretation of the reading of the Gospel (HE 31):

> The Gospel is the coming of God, when he was seen by us. He is no longer speaking to us as through a cloud and indistinctly, as He did to Moses. . . . Nor does he appear through dreams as to the prophets, but He appeared visibly as a true man. He was seen by us as a gentle and peaceful king who descended quietly like rain upon the fleece, and we have beheld his glory, as of the only begotten Son, full of truth and grace. Through him, the God and Father spoke to us face to face, and not through riddles. . . . We have heard and seen with our eyes that He is the wisdom and word of God, and we all cry "Glory to You, O Lord."

God himself was seen in the Son. One thinks especially of the Pantokrator icon that would begin to occupy the domes of churches in the immediate post-Iconoclasm period. This icon echoes the Eucharistic prayer, in which God is addressed as "Father Pantokrator" and "the Father of our Lord Jesus Christ . . . who is the image of your goodness, the identical seal, manifesting you the Father in himself" (anaphora of Saint Basil).[38]

Germanus places the Passion into a larger life-of-Christ narrative that extends from his birth to his ascension. The entrance of the Gospel represents Christ's first coming. This entrance used to be at the beginning of the liturgy. Now, however, the entrance takes place after the people have gathered in the church and during the singing of the third of three antiphons, which for Germanos represent the prophecies of Christ's coming (HE 23). Christ's first coming (24) is now identified specifically as his birth, so that the Trisagion (Thrice Holy Hymn) is like the three gifts of the Magi (25). While for Theodore the Resurrection takes place during the anaphora, Germanus sees an image of the Resurrection just before the opening dialogue of the anaphora. An angel (a deacon) rolls away the stone (raises the aer, the veil covering the gifts) and "proclaims the resurrection on the third day" in the diaconal exhortation before the dialogue: "'Let us stand aright'—behold, the first day!—'Let us stand in fear'—behold, the second day!—'Let us offer in peace'—behold, the third day!" (41). From here to the end of the liturgy, the imagery is of heavenly worship, in which "we" participate:

> Having come into "the unity of the faith and the communion of the Holy Spirit" [anaphora of Saint Basil] through the dispensation of the One who died for us and is sitting at the right hand of the Father, we are no longer on earth but standing by the royal throne of God in heaven, where Christ is. . . . (41).

Germanus, certainly following Maximus's lead, begins his commentary with an encomium on the church as a whole, understood as both building and people:

> The church is the temple of God, a holy place, a house of prayer, the assembly of the people, the body of Christ. It is called the bride of Christ. . . . The church is an earthly heaven in which the supercelestial God dwells and walks about. It represents the crucifixion, burial, and resurrection of Christ (HE 1).[39]

As with the commentary on the liturgical actions, the life-of-Christ imagery appears alongside that of heavenly worship. This is also evident in his depiction of the sanctuary, which corresponds to both the tomb of Christ and "the heavenly and spiritual altar [*thysiastērion*, sanctuary][40] where the earthly and material priests who always serve the Lord represent the spiritual, serving, and hierarchical powers of the immaterial and celestial Powers" (6). Germanus finds meanings for each part of the building and its furnishings, some aligned with his interpretation of the liturgy, some seemingly arbitrary. The apse, apparently because of its shape, is both the cave of Christ's birth and the cave of his tomb (3). The latter makes sense in the context of the liturgy since the altar table, on which "lies the true and heavenly bread, the mystical and unbloody sacrifice," represents the shelf on which Christ lay in his tomb (and also the throne of God on which Christ rested in heaven in the body, 4), but the former seemingly bears no relation to the liturgy. The ciborium over the table represents the place where Christ was crucified and also alludes to the Ark of the Covenant (5). The ambo is the stone rolled away from the tomb on which the angel sat, proclaiming the Resurrection, both because of its oval shape and because the deacon stands on it to read the Gospel (10). The chancel barriers have functional significance, separating the place of the people from "the Holy of Holies," "the place of prayer," which only the priests may enter, but also remind Germanus of the bronze barriers around the Holy Sepulchre in Jerusalem (9). The bema (synthronon) is Christ's throne, where he reigns with the apostles and on which he will come to judge the world (7).

Nicholas and Theodore of Andida (late eleventh century)

After Iconoclasm, Germanus's commentary attained quasi-official status. It was widely circulated and frequently updated to reflect changes in the liturgy. The name attached to it, however, was not Germanus but Basil the Great, putative author of the liturgy itself.[41] That is the name under which it was known by the authors of the next commentary, the late eleventh-century *Summary Meditation* [*Protheōria*] *on the Symbols and Mysteries Accomplished in the Divine Liturgy* (*Protheoria*, P, cited in *Patrologia graeca*, PG) by Nicholas, bishop of Andida, revised by his successor, Theodore.[42] Though written in the province of Pamphilia Secunda, it makes frequent reference to the practice of Hagia Sophia. By this time, the liturgy of the Great Church had spread throughout the regions under the control of the Patriarchate of Constantinople.[43]

In the *Protheoria*, the first commentary written after Iconoclasm, we get a full-blown life-of-Christ symbolism, stretching throughout the liturgy from the *prothesis*, the preparation of the gifts before the liturgy, to the removal of the gifts from the altar after communion, with almost every part of the liturgy being assigned some symbolic attachment to Christ's life. For the authors of the *Protheoria* the liturgy is an icon, just like the Gospels (P 2, in PG 140.420A–C), good sermons (they mention two by Gregory Nazianzen), and the cycle of icons in the church (P 3, in PG 140.420C–421B). The liturgy, therefore, is to be understood according to the theology of the icon. The life-of-Christ imagery does not offer a history or make the past present, but instead makes an encounter with Christ possible through portrayal or depiction. In order to do this, the liturgy, like the other examples mentioned, must present the whole life of Christ, an "exact image [*aparallaktos eikōn*] of the life-giving and divine body" missing none of its members (P 1, in PG 140.420A).

Sometimes the authors are quite imaginative in their assignment of symbolic meaning. To satisfy the requirements of their life-of-Christ imagery, for example, they turn the *prothesis* into a representation of Christ's birth. The *prothesis* in this text identifies both the preparation of the gifts before the liturgy and the place where the preparation took place, in Middle Byzantine architecture a chapel to the north of the sanctuary.[44] In the eighth and ninth centuries, in response to the burial symbolism attached to the transfer of the gifts at the Great Entrance, the rite of preparation of the gifts before the liturgy had become quite elaborate, with a developed sacrificial symbolism. Part of the rite involves cutting a "Lamb," the part of the bread that will be consecrated, out of a loaf. For Nicholas and Theodore, this cutting out of the Lamb is the removal of Christ from Mary's body by birth. The deacon who performs this rite is the angel of the Annunciation. The bread thus prepared remains hidden in the *prothesis*, just as Christ remained hidden until his appearance at his baptism (P 9–10, in PG 140.429B–432A). Since for the authors the crucifixion is represented by the elevation of the gifts before communion, the established symbolism of the Great Entrance as burial procession is problematic, so they make it the entry into Jerusalem (P 18, in PG 140.442B). They resort to more creativity to account for the Eucharistic prayer, which becomes an image of the betrayal and trial (P 19–24, in PG 140.444B–449B).

Like Germanus, the authors of the *Protheoria* assign meaning to parts of the church building to reinforce their interpretation of the liturgy. Discussion of the architectural symbolism is not gathered together at the beginning of the treatise, as in Maximus and Germanus, but occurs as occasioned by the order of service. The sanctuary (*thysiastērion*) and the altar table (*trapeza*), for example, receive a good deal of attention during the discussion of the transfer and deposition of the gifts. Since this transfer represents the entry into Jerusalem, the *prothesis* chamber is seen as Bethany; the sanctuary, Jerusalem; and the table, the upper room. The table is also the cross, the tomb, the Resurrection, and the place of the ascension. Located between four pillars supporting the ciborium, representing the four pillars of the earth, the table also stands for the center of the earth, where Christ has "wrought salvation" (Ps 73:12). To the objection that not every table has a ciborium, the authors suggest that the half dome over the apse can produce the required symbolism (P 18, in PG 140.441B–444A).

Sometimes the authors stretch traditional understandings to new meanings. For example, the marble strips parallel to the sanctuary in Hagia Sophia, traditionally called "rivers," are used to show that the entrance of the bishop into the sanctuary, during which he crosses these rivers, is an image of Christ's baptism (P 14, in PG 140.436C–D). Sometimes the authors present a traditional interpretation that they abandon in their interpretation of the liturgy. The *prothesis* chamber is superior to the Holy of Holies of the Jewish Temple, since the table on which the Lamb of God is sacrificed is greater than the table of the Holy of Holies (P 7, in PG 140.425B). Clearly, the rite of the *prothesis* had come to be understood as representing the sacrifice of Christ. But when our authors finally get to the rite, they interpret it as the birth of Christ, as we have seen.

The Andidans are the first Byzantine authors to mention a curtain closing off the sanctuary, though it appears that this was a monastic practice. John Chrysostom had mentioned a curtain at Antioch already in the fourth century.[45] In the liturgy known to the *Protheoria* there is a closing of the doors to the sanctuary right after the Great Entrance. These doors were not opened again until after the anaphora, which, along with the drawn curtain, at least in the monasteries, and the veiling of the gifts, provided the grounds for interpreting the anaphora as the night of Christ's betrayal and trial (P 21, in PG 140.445B–448A). By the time of the next commentary, the sanctuary doors will be opened before the creed. When this happens, notes Taft, the diaconal command "The doors! The doors!" no longer refers to the outer doors to the church but concerns instead the sanctuary doors. And the response desired is not to secure the doors but to open them.[46]

Nicholas Cabasilas (d. ca. 1396/7) and Symeon of Thessaloniki (d. 1429)

The final two commentators, Nicholas Cabasilas and Symeon of Thessaloniki, need to be discussed together. Writing within a few decades of each other, in the same city, under similar theological and political conditions, they nevertheless represent quite different approaches to the liturgy. The empire, though in the twilight of its history, was culturally near its zenith.[47] Beset by the Ottoman Turks from the East and the Venetians from the West, the Byzantine Empire was yet experiencing a theological renaissance, when the liturgy reached its full development. The Latin occupation of Constantinople from 1204 to 1261 was followed by a growing interest in Western theology along with attempts at reunion with the Western church. This was also the period of the full flowering of the *hesychast* movement in the person of Gregory Palamas, whose controversial ideas were vindicated at a council in Constantinople in 1351. These trends could come into conflict, but, as Marcus Plested has shown in a recent work on the reception of Thomas Aquinas in Byzantium, there were no easy alignments among the different possibilities. It was just as possible to be pro-Palamite, pro-Western, and antiunion as it was to be anti-Palamite, anti-Western, and prounion.[48] For Schulz, as we have seen, Cabasilas embodies the endpoint and, as a return to the ideas of Theodore of Mopsuestia, the pinnacle of the Byzantine liturgical commentaries, and therefore he treats him last. Perhaps this is because Cabasilas is very knowledgeable of and sympathetic toward Western theology, or because his Palamism, if it is present, is not pronounced. Symeon's commentary, however, is chronologically the last and to some degree can be seen as a reaction to Cabasilas.

Nicholas Cabasilas

Nicholas Cabasilas was from a prominent Byzantine family. His uncle, Nilus Cabasilas, was archbishop of Thessaloniki, the immediate successor to Gregory Palamas (d. 1359) in that see. Nicholas took his famous uncle's surname, but he himself was never ordained. Nonetheless, he wrote two significant works of liturgical commentary. Besides his *Commentary on the Divine Liturgy* (DL),[49] which will concern us here, he wrote the equally famous *The Life in Christ* (LC),[50] which discusses other sacraments but also includes communion as the conclusion of the rites of initiation and thus supplements the commentary on the liturgy. The detailed description he gives to the liturgical rites, even to those parts inaccessible to the laity such as the *prothesis* and the silent prayers, suggests that he was working from texts as well as experience.

In his *Commentary on the Divine Liturgy*, Cabasilas accepts the basic presupposition of the *Protheoria* that the whole life of Christ must be represented in the liturgy, using the same image: "The whole celebration of the mystery is like a unique portrayal of a single body, which is the work of the Savior" (DL 1.7). But he takes a different, some would say more rational, starting point. He begins with the affirmation that the main point of the Eucharist is the consecration of the bread and wine into the body and blood of Christ, that this consecration itself constitutes the sacrifice, and this sacrifice "commemorates the death, resurrection, and ascension of the Savior, since it transforms these precious gifts into the very body of the Lord, the Body which was the central figure in all these mysteries" (DL 1.6). Thus, everything that comes before the consecration has to represent moments in Christ's life before his death, and what comes after the consecration has to represent moments after the ascension, principally, the giving of the Holy Spirit to the Church, which is symbolized by the *zeon*, the warm water added to the chalice just before communion (DL 37.3–37.5). The *prothesis* before the liturgy, when the gifts are prepared, posed some problem for this approach, since this rite had developed strong sacrificial symbolism. Cabasilas treats it as prophetic action anticipating the consecration (DL 11.A.1–3).

In general, however, Cabasilas keeps "pictorial thinking" to a minimum.[51] The liturgy has to accomplish two things: the sacrifice, and the preparation of the people to receive the sacrifice. The sacrifice is accomplished in the consecration; the purpose of the rest of the liturgy is to dispose the worshippers to faith. Here we see Cabasilas's engagement with developments in Roman Catholic sacramental theology. We can compare what Cabasilas says with Thomas Aquinas's answer to the objection that, since the sacrament is completed by the words of Christ, no other words are needed. Thomas answers that "the celebration of this mystery is preceded by a certain preparation in order that we may perform worthily that which follows after" (*Summa Theologica* III.83.4). To the kind of preparations Thomas has in mind, through hymns and prayers, Cabasilas adds the distinctive Byzantine emphasis on the liturgy as icon of Christ:

> Even if one maintains that the readings and psalms serve another purpose—for they were introduced in order to dispose us to virtue and to cause God to look favourably on us—that does not mean that the same ceremonies cannot at once urge us to virtue and illustrate the scheme of Christ's redemptive work (DL 1.9).

It is no surprise, then, that Schulz finds that Cabasilas's treatment of these ideas "give[s] evidence of an unusual, almost scholastic, conceptual clarity."[52]

Cabasilas does not assign meaning to the building; in fact, he barely mentions the architecture except as it concerns a rite or gesture. The only symbolic meaning attached to the architecture in the commentary is by implication—for example, the transfer of gifts is like Christ's entry into Jerusalem (DL 24.3), therefore, one might infer that the sanctuary represents Jerusalem, as it does in the *Protheoria.* Perhaps because Cabasilas does not explain the building, and in order to bring his thinking into line with Maximus, Myrrha Lot-Borodine prefaces her seminal discussion of the commentary with a chapter on the consecration of the altar table from *The Life in Christ.*[53] An examination of Cabasilas's treatment of this rite shows us why he does not comment on the building or its parts. While the liturgy presents the life of Christ, the altar is an image neither from the life of Christ nor of the heavenly liturgy, but rather of Christ and the human person. In order to construct the altar, the bishop must follow the model of his own soul; he must "make the heart an altar . . . being altogether turned in on himself and bent down in body [=prostration before entering the sanctuary]. Thus, as far as he is able, he exhibits the altar in himself before he enters the sanctuary" (LC 5.3 [SC 361: 5.10]).[54] Similarly, the relics of the saints placed in the altar are images of Christ; in fact, they are the presence of Christ almost to the same degree as the "mysteries," the Eucharistic elements. "There is nothing more akin to the Mysteries than the martyrs. . . . These bones are a true temple of God and an altar, while that which is made with hands is the imitation of the real" (LC 5.6 [SC 361: 5.25–5.26]).

Cabasilas, however, reveals how the parts of the building were perceived in his day—at least, the nave and the sanctuary, which are clearly earth and heaven. When the priest steps out of the sanctuary at the end of the liturgy, he is descending from heaven to earth (DL 53):

> After the sacrifice is completed with its concluding doxology and the holy rites have been duly performed, one should note how the priest brings to an end, as it were, his communing with God, and gradually descends from these heights to converse with mankind. . . . [W]ithin the sanctuary he addresses himself to God and prays secretly on his own behalf. Then he leaves the sanctuary and standing in the midst of the congregation he says aloud, so that everyone can hear, the prayer of common supplication. . . . After the distribution of the bread and after the psalm, the priest says the last prayer over the people. This is not only said outside the sanctuary and in a manner which can be heard by all, but the words of the prayer are addressed directly to the congregation itself, thus showing the increasing extent to which the priest is now associating himself with the people.

The contrast Cabasilas draws, in keeping with his focus on the actions of the liturgy, is in the manner of the prayer. But it is clear that he also knows the inside–outside dichotomy with respect to the sanctuary. That a layman should present such a clerical perspective suggests that it was the common understanding.

Symeon of Thessaloniki

Symeon was a monk in Constantinople before becoming archbishop of Thessaloniki, the second largest city in the empire, about 1416. He also played an active role in the politics of his day, encouraging his city to resist submission to both the Latin Venetians and the Turks. The city submitted to the Venetians in 1423, but fell to the Turks one year after Symeon's death in 1429. Symeon wrote two commentaries on the Divine Liturgy. The first, the *Explanation of the Divine Temple* (E), was a stand-alone work addressed to the people of Crete, which was also under Venetian control. The second, *On the Sacred Liturgy* (SL), is part of a larger work that includes a refutation of heresies, followed by commentary on the sacraments. Both are available in a recent edition with English translation.[55] There is also a commentary on the consecration of a church, *On the Sacred Temple* (PG 155.305–61A), which follows *On the Sacred Liturgy* and is of some interest to the present topic, but which does not have a modern edition or translation.

Bornert says both that Symeon represents a return to the approach of Dionysius and Maximus and that historical symbolism predominates in his writings.[56] We can straighten out this apparent paradox somewhat if we recognize that Symeon's goal is not to be an original thinker but a collector and a presenter of tradition. He "borrows," he says, "whatever we have learned from those before us" (E 14). The multiplication of symbolisms is in fact an aspect of Symeon's method, and it sustains in part his understanding of the liturgy. He rejects efforts, such as Cabasilas's, to assign one meaning to each rite. Bornert says that Symeon reinterprets Dionysius by replacing the contemplation of the intelligible with "a fundamentally historical vision of salvation."[57] We might rather say that Symeon is using Maximus and Dionysius to counteract what had become a lopsidedly historical interpretation of the liturgy. At the same time, it should be acknowledged that Symeon uses Dionysius to bolster an increasingly clericalized view of the liturgy.

Though he does not mention his fellow Thessalonian and near contemporary by name, it is clear that Symeon's polemic is partly directed at Cabasilas. The latter, in keeping with his effort to rationalize the liturgical symbolism, argued forcefully against venerating the gifts in the Great Entrance procession because they had not been consecrated. "If any of those who prostrate themselves before the priest carrying the offerings adores [*proskynousi*] them as if they were the Body and Blood of Christ and prays to them as such, he is led into error" (DL 24.5). Symeon counters that those who think this way have rejected the findings of the Seventh Ecumenical Council. The gifts are true icons of Christ, having been prepared as such in the *prothesis*, and can be venerated [*proskynein*] (E 66). Here he shows that, while he recognizes the qualitative importance of consecration, he also understands the rest of the liturgy to be more than illustrative.

Symeon presents a subtler critique of Cabasilas's method in connection with the consecration. For Cabasilas, the consecration is the moment of the sacrifice, the high point in the liturgy and of the life-of-Christ symbolism. After the *epiclesis* of the Holy Spirit:

> the offerings are consecrated, the sacrifice is complete; the splendid Victim [*hiereion*], the most holy Body of the Lord, which really suffered the outrages, insults and blows, which was crucified and slain, which under Pontius Pilate bore such splendid witness; that Body which was mocked, scourged, spat upon, and which tasted gall. . . . (DL 27).

"When the sacrifice has thus been completed, the priest, seeing before him the pledged of God's love, the Lamb of God, uses him as his intercessor. . . ." (DL 33.1). This elaboration on the wounds of Christ suggests the kind of satisfaction soteriology more generally associated with post-Anselmian Western theology. Symeon seems to echo Cabasilas but takes a different approach. When the bishop invokes the spirit on himself and the gifts, he says:

> he immediately sees the living [*zōnta*] Jesus lying before him, He Himself being truly the bread and the chalice. The universal offering, and the common propitiation, the living [*zōsa*] delight, the infinite joy, the kingdom of heaven, and the only true good lies before all on the divine altar. Therefore the hierarch confidently prays for all (E 75).

Symeon uses the term *hiereion*, sacrificial victim, only once, in the commentary on the Sacred Liturgy. In accordance with his interpretation of the Great Entrance as Christ's second coming, he describes how Christ will appear to those in unity: "Jesus will be the most beautiful sacrificial victim in the midst of all His saints, being the peace and unity of all, priest and victim offered in the sacred-service, uniting all and being united to them" (SL 138). At his coming, Christ, "the living [*zōtikē*] and inexhaustible delight, the One who is perceived as slain through the Lamb, will be seen to be both living [*zōn*] and scourged" (SL 136). Symeon emphasizes that it is the living Jesus who is present to the worshippers. The true contemplation of the economy is of Christ himself in his sacrificed but resurrected body (E 64).

Why does Symeon feel the need to critique Cabasilas on this point? Perhaps, in part, it results from his general antipathy toward what might be perceived as Latinizing theology. But I suspect he feels that Cabasilas gives insufficient weight to the possibility of the present experience of deification, the chief concern of the *hesychast* revival, in which Symeon also participated. As John Meyendorff notes:

> Palamas and his disciples did not mean to promote an esoteric method of spirituality intended only for a limited number of elect distinguished thereby from the mass of the faithful, but simply wished to express the real intimacy established by the incarnation between God and all Christians.[58]

There has been much debate about the degree to which *hesychast* spirituality is present in Cabasilas's works, and one has to conclude that it is muted at best.[59] What we see in Symeon, then, may be a *hesychast* response to Cabasilas. This would account for Symeon's return to Dionysius and Maximus, from whose works the *hesychasts* drew heavily.

The influence of these two writers is especially strong in Symeon's treatment of the church building, with which, following the model of Maximus, he opens his *Explanation*. Like Maximus, he begins with the division of the church into sanctuary and nave, which he takes as a figure of Christ's two natures, the soul and body of the human being, invisible and visible reality, and heaven and earth. Symeon throws in the narthex to get an image of the Trinity as well as the ecclesiastical hierarchy, à la Dionysius, of bishop, priests, and deacons, or clergy, laity, and penitents. The narthex also represents earth, the nave, heaven, and the sanctuary, "that which is above the heavens" (E 16). Symeon mentions both a chancel screen with pillars and a curtain closing off the sanctuary (E 17).

The Dionysian hierarchy is mentioned again in connection with the synthronon; the throne represents Christ seated at the right hand of the Father, and the steps "the order and ascent of each of the clergy and the angels" (E 20). In Dionysian fashion, the priests in the sanctuary begin the liturgy, which is then taken up by the deacons, readers, and cantors, who form "the middle chorus of the heavenly beings," and then finally by the laypeople, who "join with the singers appealing for the mercy of God" (E 25). Symeon later asserts that this is why the sanctuary doors are closed after the Great Entrance, since only those "in the priestly action" are permitted to see the mysteries. Invoking Dionysius's hierarchies, he says that "the hierarch approaches the altar directly, the priests and ministers do so through him," and "through the priests and ministers, the people partake of . . . communion and the sacred hymns" (E 70). In *On the Sacred Liturgy* 137–40 and 151–56, he is more specific about the order of communion by rank and by where it is taken: bishops, priests, and deacons take it inside the sanctuary; subdeacons, readers, monks, and ordinary laity, in that order, outside.

The symbolic significance of elements of the building is partly determined by the location of the clergy, as representing the "otherworldly powers," and what they are doing. When they pray in the narthex as they prepare to enter the church, the nave represents heaven. Symeon mentions that monasteries have curtains separating narthex from church so there is something to pass through (E 22). Similarly, when they bring relics to a church being consecrated and are about to enter it, the whole church is heaven (*On the Sacred Temple*, PG 155.321C–D):

> But when we do prayers inside . . . in front of all the people, we regard the sanctuary as another heaven, and the nave signifies the things of earth, for within it everyone comes and gathers together, and together they send up prayers to God (E 22).

New with Symeon is the symbolic use of the narthex. Also new is an interpretation of the bishop's exit from the sanctuary to go to the back of the church to begin the liturgy. While the entrance of the bishop into the church at the beginning of the liturgy (Maximus) or into the sanctuary during the antiphons (Germanus) had been interpreted as Christ's appearance in the world at his birth, Symeon goes one step further: This "descent to the West" represents the descent of the Word from heaven all the way to Hades (E 23, SL 24).

Conclusion

The liturgical commentaries offer us good evidence for the development of the liturgical forms. But do they give us evidence of how the liturgy was experienced? Here we must be cautious. As we have seen, none of our commentaries is a simple record of agreed-upon meanings; all reflect the theological tendencies of their time and all have a theological purpose toward which they direct their interpretations. Nevertheless, the fact that they can pursue theological discourse through an appeal to the liturgy shows that liturgy was an accepted source of theological meaning. The authors must presuppose some common understandings of the liturgy even as they suggest new ones. Thus, Germanus's argument that the use of life-of-Christ symbolism in the liturgy demonstrates that Christ can be depicted must mean that this was an accepted way of viewing the liturgy, at least in his clerical circles; we must always keep in mind the position within the liturgy of an author and his intended audience. All the commentaries after Germanus presuppose the iconic character of the liturgy, that it provides a true image of Christ. Not all the proposed symbolisms were universally or even generally accepted, of course.

Some of the symbolism was accepted, however, and influenced the development of the liturgy. It is hard to see how the *prothesis* could have developed the way it did without the understanding of the Great Entrance as Christ's burial procession. Other developments related to this basic interpretation of the liturgy include the recitation of hymns borrowed from Holy Week services in connection with the deposition of the gifts and the conversion of the aer into a shroud. Given the persistence of this symbolism in the commentaries after Germanus, it is amazing how slowly these changes, apart from the *prothesis*, took place. The hymns did not begin to appear in the texts until the fourteenth century, the same century that the aer began to be embroidered with the crucified Christ.[60] Looked at another way, we may suspect some resistance to these developments.

These developments affect mainly the clergy, not only because they were the ones performing the actions but also because they happened outside of the laity's hearing and sometimes sight. One thing the commentaries clearly show is a growing clericalism. This was not primarily the result of the way the liturgy was being interpreted; it may have been in large part inadvertent. Already by Maximus's time, the prayers of the celebrant were mostly inaudible. When the office of antiphons was tacked on to the beginning of the liturgy, sometime between Maximus and Germanus, the original entrance prayer with its request for angelic accompaniment, when bishop and people entered together, now became the prayer of the entry into the sanctuary of the bishop alone. The clergy alone now came to represent the angels, even during the Cherubikon (cf. Germanus, HE 37). The liturgy became the province of the clergy alone; they alone joined the heavenly worship. The *Protheoria* hypothesizes that the purpose of the prayer behind the ambo at the end of the liturgy was to provide a summary of the whole liturgy so that those standing outside the sanctuary might have some sense of what happened in the liturgy (P 38). Even a layman such as Nicholas Cabasilas describes this prayer as the descent of the priest from heaven to converse with ordinary humanity. Symeon's return to Dionysius's hierarchies only served to reinforce this dichotomy.

Developments in the liturgy and its understanding influenced developments in church architecture and decoration.[61] Over time, the sanctuary became more closed off from the nave. After Iconoclasm, the space between the parapet and architrave of the chancel barrier began to be filled in with icons, at first attached to the architrave or the supporting columns. By the thirteenth century the permanent iconostasis appears.[62] The other liturgical center, the ambo, also moved toward the sanctuary. Schulz has correctly and insightfully coordinated the development of the iconographic program in Byzantine churches with the post-Iconoclastic liturgical commentaries.[63] These changes can be charted through archaeology, of course, but the commentaries reveal the underlying attitudes toward the liturgy that inspired them or made them possible. One might think that the separation of sanctuary from nave would be a sufficient cause for the later developments, but Maximus's *Mystagogy*, with its emphasis on unity in diversity, shows us that this is not the case.

Notes

1 On these fourth-century mystagogies, see Edward Yarnold, *The Awe-Inspiring Rites of Initiation: Baptismal Homilies of the Fourth Century* (Slough, UK: St. Paul Publications, 1972), and the newer edition, recast to mirror the Roman Catholic Rites of Christian Initiation of Adults (Edinburgh: T&T Clark, 1994); and Enrico Mazza, *Mystagogy: A Theology of Liturgy in the Patristic Age*, trans. Matthew J. O'Connell (New York: Pueblo, 1989).

2 René Bornert, *Les commentaires byzantins de la Divine Liturgie du VIIe au XVe siècle* (Paris: Institut Français d'Études Byzantines, 1966).

3 Hans-Joackim Schulz, *Die byzantinische Liturgie: Vom Werden ihre Symbolgestalt* (Freiburg: Lambertus-Verlag, 1964).

4 Hans-Joachim Schulz, *Die byzantinische Liturgie: Glaubenzeugnis und Symbolgestalt*, 2nd ed. (Trier: Paulinus, 1980); translated by Matthew J. O'Connell as *The Byzantine Liturgy: Symbolic Structure and Faith Expression* (New York: Pueblo, 1986). References to Schulz's work will be to this English translation.

5 Hans-Joachim Schulz, *Die byzantinische Liturgie: Glaubenzeugnis und Symbolgestalt*, 3rd ed. (Trier: Paulinus, 2000).

6 Hugh Wybrew, *The Orthodox Liturgy: The Development of the Eucharistic Liturgy in the Byzantine Rite* (Crestwood, NY: St. Vladimir's Seminary Press, 1990).

7 A transcript of the sole manuscript containing Theodore's *Catechetical Homilies* (MS Mingana syr. 561) with an English translation was published in two separate volumes by Alphonse Mingana, *Commentary of Theodore of Mopsuestia on the Nicene Creed*, Woodbrooke Studies 5 (Cambridge: W. Heffer, 1932) and *Commentary of Theodore of Mopsuestia on the Lord's Prayer and on the Sacraments of Baptism and the Eucharist*, Woodbrooke Studies 6 (Cambridge: W. Heffer, 1933). A facsimile edition of the text with French translation was published by Raymond Tonneau and Robert Devreese, *Les homélies catéchétiques: Reproduction phototypique du ms. Mingana Syr. 561 (Selly Oak Colleges' Library, Birmingham)* (Vatican City: Biblioteca Apostolica Vaticana, 1949). Quotations will be from Mingana's translation, while citations to the homilies will use Tonneau and Devreese's numbering.

8 Texts of both *The Celestial Hierarchy* and *The Ecclesiastical Hierarchy* by (pseudo-)Dionysius the Areopagite are in Günther Heil and Adolf Martin Ritter, Eds., *Corpus Dionysiacum II: De coelesti hierarchia, De ecclesiastica hierarchia, De mystica theologia, Epistulae,* Patristische Texte und Studien 67, 2nd rev. ed. (Berlin: De Gruyter, 2012); translations are in Colm Luibheid, trans., *Pseudo-Dionysius: The Complete Works, Classics in Western Spirituality* (New York: Paulist Press, 1987). Alexander Golitzen, *Mystagogy: A Monastic Reading of Dionysius Aeropagita,* Cistercian Studies Series 250 (Collegeville, MN: Liturgical Press, 2013), xxxvi, calls the Ecclesiasical Hierarchy "the core and pivot of the Dionysian system."

9 See Bornert, *Commentaires byzantins*, chap. 1; the precise terms come later, 248–49, where they are used to contrast Nicholas Cabasilas and Symeon of Thessaloniki.

10 Robert F. Taft, "The Liturgy of the Great Church: An Initial Synthesis of Structure and Interpretation on the Eve of Iconoclasm," *Dumbarton Oaks Papers* 34–35 (1980–81): 45–75, esp. 61.

11 Schulz, Byzantine Liturgy, 29.

12 Schulz, *Byzantine Liturgy*, 134.

13 Schulz, *Byzantine Liturgy*, 193. This particular sentence does not reappear in the third German edition, but one can find similar statements there, and Schulz's position is clear from the structure of the book, which makes Dionysius a representative of liturgical understanding in the age of Justinian.

14 Frances Young, *The Use of Sacrificial Ideas in Greek Christian Writers from the New Testament to John Chrysostom*, Patristic Monograph Series 5 (Cambridge, MA: Philadelphia Patristic Foundation,1979; 2nd ed., Eugene, OR: Wipf & Stock, 2004), 270, 279–81.

15 Thus the earliest version of the Alexandrian anaphora of Saint Mark, found in Strasbourg Papyrus gr. 254 in the Bibliothèque Nationale et Universitaire de Strasbourg: "giving thanks . . ., we offer the reasonable sacrifice and this bloodless service." Text in Anton Hänggi and Irmgard Pahl, *Prex eucharistica: Textus e variis liturgiis antiquioribus selecti* (Fribourg: Éditions Universitaires, 1968), 116–18; trans. R. C. D. Jasper and G. J. Cuming, in *Prayers of the Eucharist: Early and Reformed*, 3rd rev. ed. (Collegeville, MN: Liturgical Press, 1990), 52–54.

16 In *Testament of Levi* 3:5–7, Levi has a vision in which he sees "the angels of the presence, who . . . offer the Lord a pleasing odor, a reasonable and bloodless offering." Text in M. de Jonge, *The Testaments of the Twelve Patriarchs: A Critical Edition of the Greek Texts* (Leiden: Brill, 1978); trans. H. C. Kee in *The Old Testament Pseudepigrapha I: Apocalyptic Literature and Testaments*, ed. James H. Charlesworth (Garden City, NY: Doubleday, 1983), 788–95.

17 On the development of the Eucharistic prayers in the fourth and fifth centuries, see Paul F. Bradshaw and Maxwell E. Johnson, *The Eucharistic Liturgies: Their Evolution and Interpretation* (Collegeville, MN: Liturgical Press, 2012), chap. 4.

18 Young, *Sacrificial Ideas*, 278–79. Young critiques the idea, popular since Gregory Dix's *The Shape of the Liturgy* (London: Dacre Press, 1945) and still common in liturgical theology, to which Schulz also subscribes, that anamnesis means "re-presentation" of the thing remembered.

"Adequate grounds for this assertion have yet to be produced," she says in ibid., 278–79n112. For a more recent critique of Dix, see Bryan D. Spinks, *Do This in Remembrance of Me: The Eucharist from the Early Church to the Present Day*, SCM Studies in Worship and Liturgy (London: SCM Press, 2013), 1–2.

19 Anaphora of Saint John Chrysostom. Attempts by John Fenwick, *The Missing Oblation: The Contents of the Early Antiochene Anaphora* (Bramcote, UK: Grove Books, 1989) and Robert Taft, "St. John Chrysostom and the Byzantine Anaphora That Bears His Name," in *Essays in Early Eastern Eucharistic Prayers*, ed. Paul F. Bradshaw (Collegeville, MN: Liturgical Press, 1997), 195–226, and elsewhere, to argue from the Syriac anaphora of the Twelve Apostles that the original Antiochene anaphora had no offering are not convincing. The Syriac anaphora may be in part a translation of the Antiochene anaphora, but it was likely translated for a tradition that did not have an offering in the anamnesis. Most of the Syriac anaphoras in Hänggi and Pahl's *Prex eucharistica* lack such an offering, including Syriac Basil. The doubling of the offering in Chrysostom's anaphora can be accounted for by its adaptation to the Byzantine liturgy, which already had an offering, of "your own from your own," connected with a popular refrain. This leaves "reasonable and bloodless worship" as the original content of the Antiochene offering.

20 Mazza, *Mystagogy*, 181–82n12.

21 Frances Young, *Biblical Exegesis and the Formation of Christian Culture* (Cambridge: Cambridge University Press, 1997), 168. For Theodore in particular, see Rowan A. Greer, *Theodore of Mopsuestia: Exegete and Theologian* (London: Faith Press, 1961), esp. chap. 3, "Theodore's Christology," on which the following discussion depends in part.

22 Tonneau and Deevresse, *Les homélies catéchétiques*, translate the last portion as "il y a une telle distance entre eux qu'il n'est pas possible que les deux se trouvent ensemble."

23 Stefano Parenti and Elena Velkovska, Eds., *L'Eucologio Barberini gr. 336*, 2nd rev. ed. (Rome: Edizioni Liturgiche, 2000).

24 This section of the anaphora is missing from Barb. gr. 336; the ninth- or tenth-century Monastery of Grotteferrata, MS Crypt. Gr. γ β vii is used to fill the lacuna; translation from Jasper and Cuming, *Prayers of the Eucharist*, 119.

25 My translation.

26 On the melodic structure of the Cherubikon and how it mirrors the imagined heavenly worship by repeating patterns of the Sanctus and the Trisagion, see Laura Steenberge's essay in this volume, "*We Who Musically Represent*: Text-Setting in the Medieval Cheroubika.".

27 Thus, the Cherubikon is not an image from Theodore sneaking into the liturgy in Dionysian garb, as Schulz suggests, *Byzantine Liturgy*, 36–37.

28 Maximus the Confessor, *Mystagogy*, text: Christian Boudignon, Ed., *Maximi Confessoris Mystagogia: Una cum Latina interpretatione Anastasii Bibliothecarii*, Corpus Christianorum, Series Graeca 69 (Turnhout: Brepols, 2011). Translations: George C. Berthold, *Maximus Confessor: Selected Writings*, Classics of Western Spirituality (New York: Paulist Press, 1985), 181–225; and Julian Stead, *The Church, the Liturgy and the Soul of Man: The "Mystagogia" of St. Maximus the Confessor* (Still River, MA: St. Bede's Publications, 1982). Secondary literature: Bornert, *Commentaires byzantins*, 83–124; Schulz, *Byzantine Liturgy*, 43–49, 173–77 (3rd German ed., 117–24); and Wybrew, *Orthodox Liturgy*, 94–101. Unless otherwise stated, quotations will be from the Berthold translation. Robert F. Taft, "Is the Liturgy Described in the Mystagogia of Maximus Confessor Byzantine, Palestinian, or Neither?" *Bollettino della Badia Greca di Grottaferrata* 8 (2011): 233–70, has recently reaffirmed that Maximus comments on the Byzantine liturgy and not that of his native Palestine, as was suggested by Joseph Patrich, "The Transfer of Gifts in the Early Christian Churches of Palestine: Archaeological and Literary Evidence for the Evolution of the 'Great Entrance,'" in *Pèlerinages et lieux saints dans l'Antiquité et le Moyen Âge: Mélanges offerts à Pierre Maraval*, ed. Béatrice Caseau-Chevallier, Jean-Claude Cheynet, and V. Déroche, Centre de Recherche d'Histoire et Civilisation de Byzance Monographies 23 (Paris: Association des Amis du Centre d'Histoire et Civilisation de Byzance, 2006), 347–50.

29 Theodore, Cat. Hom. 6.3, says, for example, "our blessed Fathers . . . do not teach that the Divine nature of the Only Begotten was born of a woman . . . but they followed the Sacred Books which speak differently of natures while referring (them) to one prosopon . . ."

30 *Diēgēsis peri tēs oikodomēs tou naou tēs megalēs tou Theou ekklēsias tēs eponoazomenēs Hagias Sophias*, in Theodore Preger, Ed., *Scriptores originum constantinopolitanarum* (Leipzig: B. G.

Teubner, 1901): 102–03. Cyril Mango, "Byzantine Writers on the Fabric of Hagia Sophia," in *Hagia Sophia from the Age of Justinian to the Present*, ed. Robert Mark and Ahmet Ş. Çakmak (Cambridge: Cambridge University Press, 1992), 45–50, shows that, though this text is in part fictitious, it also appears in part to rely on sources. There was an order of penitents in Constantinople, still mentioned in the ninth century, who were permitted to attend the liturgy without communing, and it is entirely possible that they were assigned to a specific place in the church. On evidence for continued penitential practice, see Robert F. Taft, *A History of the Liturgy of St. John Chrysostom*, vol. 5, *The Precommunion Rites*, Orientalia Christiana Aanalecta 261 (Rome: Pontifical Oriental Institute, 2000), 181–82.

31 Jean-Claude Larchet, "La symbolique spirituelle de l'église selon la *Mystagogie* de Saint Maxime le Confesseur," in *L'espace liturgique: Ses éléments constitutifs et leur sens*, ed. C. Braga, Conférences Saint-Serge, LIIe Semaines d'Études Liturgiques (Rome: Edizione Liturgiche, 2006), 51.

32 Larchet, "La symbolique spirituelle de l'église," 50 and n4.

33 Dionysius Areopagita, *The Divine Names* 5.2. Text in Beate Regina Suchla, *Corpus Dionysiacum I: De divinis nomibus*, Patristische Texte und Studien 33 (Berlin: De Gruyter, 1990); trans. in Luibheid, *Pseudo-Dionysius*. This circularity in the Byzantine perception of divinity can be connected to the circumambulation ceremonies in the Armenian rite; see Christina Maranci's essay in this volume. Similarly, the actions of enclosing and circumscribing can be linked to the *periēgēsis* (tour around) of the Byzantine ekphraseis of sacred space and the role the dome plays in structuring this experience. On these topics, see Ruth Webb's and Ravinder Binning's essays in this volume; for the psychosomatic effect of the dome, see also Bissera V. Pentcheva, *Hagia Sophia: Sound, Space and Spirit in Byzantium* (University Park, PA: Pennsylvania State University Press, 2017), chap. 1.

34 Larchet, "La symbolique spirituelle de l'église," thinks he can show that Maximus has a simple basilica, without dome, in view in the rest of the treatise. But Maximus refers only to the parts of the church he needs for the image he wants to create. It is likely that, since the liturgy he describes is that of the Great Church, the building he has in view is Hagia Sophia, the home of that liturgy, with which he would have been familiar from his service in the imperial court and when he was a monk near the capital. Where else might he have observed the large, multifarious crowd he describes, if not the imperial capital? Whatever church he has in mind, it is not a monastic church. The image of an encircling church building would have been strengthened by the curved colonnades in front of the semicircular exedra at each corner of Hagia Sophia. For Maximus's biographical details, see Taft, "The Liturgy Described in the Mystagogia," 227–29.

35 I have modified Berthold's translation to show Maximus's use of technical language used elsewhere of persons and natures for both Trinitarian theology and Christology: *hypostasis*, mode of existence (*tropos hyparxeōs*), principle of being (*logos ousias*); see, for example, Lars Thunberg, *Man and the Cosmos: The Vision of St Maximus the Confessor* (Crestwood, NY: St. Vladimir's Seminary Press, 1985), 37–38, 116–17, 119–20.

36 Accepting Boudignon's reconstruction of this difficult phrase, which in the majority tradition reads *idias autou*; cf. Boudignon, *Maximi Confessoris Mystagogia*, clxxviii–clxxix.

37 Germanus of Constantinople, *Historia ekklesiastikē kai mystikē thēoria*, text and translation: Paul Meyendorff, *St Germanus of Constantinople: On the Divine Liturgy* (Crestwood, NY: St. Vladimir Seminary Press, 1984). Secondary literature: Bornert, *Commentaires byzantins*, 83–124; Schulz, *Byzantine Liturgy*, 67–76, 184–88 (3rd German ed., 155–65); Wybrew, *Orthodox Liturgy*, 123–28; and Taft, "The Liturgy of the Great Church." The text presented by Meyendorff is that established by Nilo Borgia, *Il commentario liturgico di S. Germano Patriarca Constantinopolitano e la versione latina di Anastasio Bibliotecario*, Studi Liturgici 1 (Grottaferrata: n.p., 1912).

38 Noting its relationship with the anaphora of Basil, Charles Barber says that the Pantokrator "is the icon of the Father-in-the-Son. . . . That the Pantokrator can be seen confirms, perhaps more than any other icon, the possibility of Christian representation"; Barber, "From Transformation to Desire: Art and Worship after Byzantine Iconoclasm," *Art Bulletin* 75 (1993): 8–16, esp. 14. On Christ in the dome, see Ravender Binning's essay in this volume.

39 Vasileios Marinis, "The *Historia Ekklesiastike kai Mystike Theoria*: A Symbolic Understanding of the Byzantine Church Building," *Byzantinische Zeitschrift* 108 (2015): 753–70, unpacks the complex symbolism of the full passage, noting especially the importance of the rite of church consecration for its understanding.

40 Germanus and subsequent commentators distinguish between the altar table (*trapeza*) and the sanctuary (*thysiastērion*), the area around the altar table reserved to the clergy.

41 It was also occasionally ascribed to Cyril of Jerusalem. Bornert, *Commentaires byzantins*, 125–60, sorts through the manuscript evidence to affirm Germanus's authorship and the text established by Borgia.

42 Nicholas of Andida and Theodore of Andida, *Summary Meditation* [*Protheōria*] *on the Symbols and Mysteries Accomplished in the Divine Liturgy*, text: J.-P. Migne, *Patrologia graeca* (PG), 161 vols. (Paris: J.-P. Migne, 1857–66), 140:417–68. Secondary literature: Bornert, *Commentaires byzantins*, 181–206; Schulz, *Byzantine Liturgy*, 89–98, 188–90 (3rd German ed., 182–93); and Wybrew, *Orthodox Liturgy*, 139–44. There is no critical edition or, to my knowledge, published translation of the *Protheoria* in a modern language. Roland Betancourt is preparing an English translation for publication. Betancourt has also translated a didactic poem based on the *Protheoria* attributed to Michael Psellus, "A Byzantine Liturgical Commentary in Verse: Introduction and Translation," *Orientalia Christiana Periodica* 81 (2015): 433–72). I am grateful to him for discussing these projects with me; e-mail messages with author, January 23, January 27, May 29, 2015.

43 On the spread of the Byzantine rite, see Taft, "The Liturgy Described in the Mystagogia," 265–67.

44 At some point the place for preparation of the gifts at Hagia Sophia moved from the *skeuophylakion*, a structure outside the main church to the north, to the interior. How soon this occurred is a matter of dispute. Germanus, HE 36, says that the preparation, which he calls the *proskomidē*, took place in the *skeuophylakion*. For an argument from architectural evidence that the gifts were prepared in the northeastern section of Hagia Sophia, see Neil K. Moran, "The Musical 'Gestaltung' of the Great Entrance Ceremony in the 12th Century in Accordance with the Rite of Hagia Sophia," *Jahrbuch der Österreichischen Byzantinistik* 28 (1979): 178–81; and Moran, "The Skeuophylakion of the Hagia Sophia," *Cahiers Archéologiques* 34 (1986): 29–32. Moran's arguments for the early period founder on the textual evidence.

45 Frans van de Paverd, *Zur Geschichte der Messliturgie in Antiocheia und Konstantinopel gegen Ende des vierten Jahrhunderts: Analyse der quellen bei Johannes Chrysostomos, Orientalia Christiana Analecta 187* (Rome: Pontifical Oriental Institute, 1970), 41–47.

46 Robert F. Taft, *A History of the Liturgy of St. John Chrysostom*, vol. 2, *The Great Entrance*, Orientalia Christiana Analecta 200 (Rome: Pontifical Oriental Institute, 1975), 408–11.

47 See the exhibition volume Helen C. Evans, Ed., *Byzantium: Faith and Power (1261–1557)* (New York: Metropolitan Museum of Art, 2004).

48 Marcus Plested, *Orthodox Readings of Aquinas: Changing Paradigms in Historical and Systematic Theology* (Oxford: Oxford University Press, 2012), 58–60.

49 Nicholas Cabasilas, *Commentary on the Divine Liturgy*, text: René Bornert, Jean Gouillard, and Pierre Périchon, Eds., *Nicholas Cabasilas: Explication de la Divine Liturgie*, trans. Sévérien Salaville, Sources Chrétiennes 4bis (Paris: Éditions du Cerf, 1967). Translation: J. M. Hussey and P. A. McNulty, trans., *Nicholas Cabasilas: A Commentary on the Divine Liturgy* (London: SPCK, 1960). Secondary literature: Bornert, *Commentaires byzantins*, 215–44; Schulz, *Byzantine Liturgy*, 124–32, 190–96 (3rd German ed., 232–44); Wybrew, *Orthodox Liturgy*, 158–64; Myrrha Lot-Borodine, *Un maître de la spiritualité byzantine au XIVe siècle: Nicolas Cabasilas* (Paris: Éditions de l'Orante, 1958); and Constantine N. Tsirpanlis, *The Liturgical and Mystical Theology of Nicholas Cabasilas* (New York: n.p., 1979). Quotations will use the Hussey and NcNulty translation but give the additional paragraph numbers from the Sources Chrétiennes edition.

50 Nicholas Cabasilas, *The Life in Christ*, text: M.-H. Congordeau, Ed. and trans., *Nicolas Cabasilas: La Vie en Christ*, 2 vols., Sources Chrétiennes 355 (books I–IV) and 361 (books V–VII) (Paris: Éditions du Cerf, 1989–90). Translation: Nicholas Cabasila, *The Life in Christ*, trans. Carmino J. deCatanzaro (Crestwood, NY: Saint Vladimir's Seminary Press, 1974). For a recent treatment, see Nicholas Denysenko, "*The Life in Christ* by Nicholas Cabasilas: A Mystagogical Work," *Studia Liturgica* 38 (2009): 242–60.

51 The phrase is Schulz's, *Byzantine Liturgy*, 190.

52 Schulz, *Byzantine Liturgy*, 125.

53 Lot-Borodine, *Un maître de la spiritualité byzantine*, 13–22, discussion of Maximus, 17–19.

54 It is possible to see here a reference to *hesychast* practices; see Congourdeau's note on this passage, Sources Chrétiennes 361, 18n10. I give the SC paragraph number in brackets. Golitzen,

Mystagogy, 16–17, 20–24, argues that this idea of the soul as true temple is central to Dionysius; he traces it to Syrian ascetic circles.

55 Symeon of Thessaloniki, *Explanation of the Divine Temple* and *On the Sacred Liturgy*, text and translation: Steven Hawkes-Teeples, *St. Symeon of Thessalonika: The Liturgical Commentaries*, Studies and Texts 168 (Rome: Pontifical Institute of Medieval Studies, 2011). Secondary literature: Bornert, *Commentaires byzantins*, 245–63; Schulz, *Byzantine Liturgy*, 114–24 (3rd German ed., 216–31); and Wybrew, *Orthodox Liturgy*, 164–71.

56 Bornert, *Commentaires byzantins*, 248, 250.

57 Bornert, *Commentaires byzantins*, 251.

58 John Meyendorff, *A Study of Gregory Palamas*, trans. George Lawrence (Crestwood, NY: SVS Press, 1998), 38.

59 See Denysenko, "*The Life in Christ*," 257–58. In a recent article, Ioannis Polemis argues that Cabasilas is even anti-Palamite; Polemis, "Nikolaos Kabasilas's *De Vito in Christo* and Its Context," *Byzantinische Zeitschrift* 106 (2013): 101–32.

60 On this, see Taft, *The Great Entrance*, 216–17 (on the aer-shroud) and 244–50 (on the hymns).

61 On these developments, see Robert Ousterhout, "The Holy Space: Architecture and Liturgy," in *Heaven on Earth: Art and the Church in Byzantium*, ed. Linda Safran (University Park, PA: Pennsylvania State University Press, 1998), 81–120; Vasileios Marinis, *Architecture and Ritual in the Churches of Constantinople, Ninth to Fifteenth Centuries* (Cambridge: Cambridge University Press, 2013).

62 Annemarie Weyl Carr, "Images: Expressions of Faith and Power," in Evans, *Byzantium: Faith and Power*, 144; Sharon Gerstel, *Beholding the Sacred Mysteries. Programs of the Byzantine Sanctuary* (Washington, D. C.: Published by College Art Association in association with University of Washington Press, 1999), 9–11.

63 Schulz, *Byzantine Liturgy*, *passim*. Some details need to be corrected. Images from the life of Christ appeared much earlier than he supposes, for example, already in the ninth century in Constantinople; see the extracts from Constantinus Rhodius, *Description of the Church of the Holy Apostles*, in *The Art of the Byzantine Empire 312–1453*, by Cyril Mango, Sources and Documents in the History of Art (Englewood Cliffs, NJ: Prentice-Hall, 1972), 199–201.

5 Christ's all-seeing eye in the dome

Ravinder S. Binning

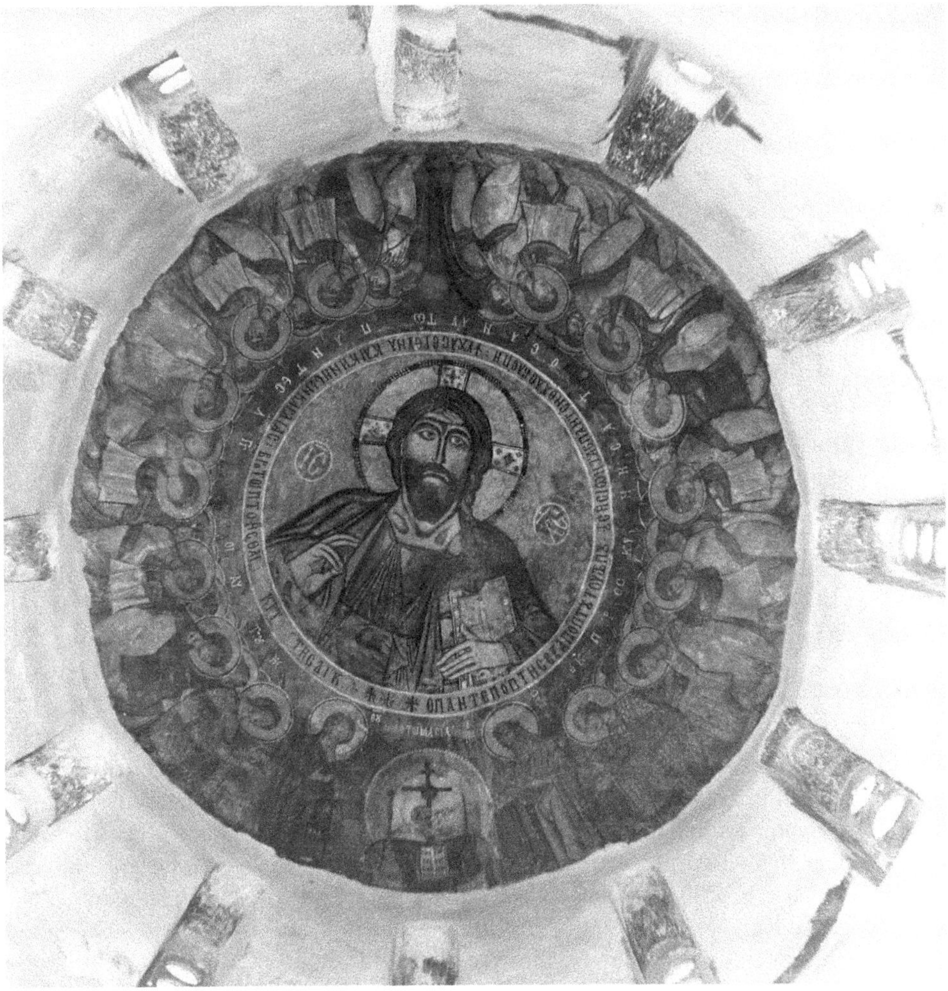

Figure 5.1 Christ with Deesis, hetoimasia, and angels in cupola, fresco, 1105–06, Panagia Theotokos Church, Trikomo, Cyprus.

The Byzantine Institute and Dumbarton Oaks Fieldwork Records and Papers, MS.BZ.004-H.69.22, Image Collections and Fieldwork Archives, Dumbarton Oaks, Trustees for Harvard University, Washington, DC. Photograph: Courtesy of Image Collections and Fieldwork Archives, Dumbarton Oaks, Trustees for Harvard University, Washington, DC.

In the small village of Trikomo, in northwest Cyprus, in the single-aisled Panagia Theotokos Church (ca. 1130), a colossal portrait of Christ presides over the dome's apex (Figure 5.1).[1] Set against a hazy red background, the leonine figure, draped in a purple himation, holds a closed gemmed codex in one hand while offering a blessing with his other. He is surrounded by an angelic procession set against a celestial blue background. This cosmic backdrop places the scene in a different space and time: here, Christ is the apocalyptic God-Man, yet to be seen but promised to come, depicted with fierceness and wrath.[2] A chilling message, written around the dome, dramatizes the image of Christ and confronts the viewer:

> He who sees all from the distant place
> Sees all those who enter here.
> He examines their souls and the movements of their hearts.
> Mortals, tremble (with fear) before the Judge [at Judgment!].[3]

The poetic inscription is imperative in its exhortation: "tremble with fear" (*ptoeisthe*). This fear (*ptoa*) is not diffuse, quotidian angst but the abject horror fit for events of divine appearance. The same root (*eptoiēthein*) appears in antiquity. In book twenty-two of *The Odyssey*, it describes how aghast Penelope's suitors were when, on Odysseus's homecoming, Athena appears: "Then Athena, held up her aegis, the bane of mortals, from on high from the roof, and the minds of the suitors were horror-stricken."[4] In the Byzantine context, a form of this word also appears in a strophe in Romanos the Melodist's sixth-century sung sermon, the "*Kontakion* on the Second Coming," wherein the first-person narrator declares at the thought of the Last Judgment, "Reflecting fearfully on your fearsome judgment seat and the day of judgment, Lord Excellent above-all, I shiver and am terrified [*ptooūmai*], convicted by my conscience."[5]

Powerful, too, is the title given to Christ: "All-seeing" (*pantepoptēs*). In the eleventh century, Anna Dalassene, mother of the ruling emperor Alexios I Komnenos (1081–1118), built a nunnery to Christ *Pantepoptēs* (now Eski Imaret Mosque) on Constantinople's fourth hill, which remained a significant Byzantine foundation for centuries. But besides the one at Trikomo, the only other inscription using this word is in the Hagios Nikolaos Church (1337–38) in the Greek village of Platsa, near Mani.[6] There, Christ is invoked as "all-seeing," begged to admire the renovations done to the church, and to grant pardon for the sins of the builders.[7] In the Septuagint, "all-seeing" appears only in the book of Maccabees to describe the Lord of the Israelites who, in retaliation against Antiochus IV Epiphanes, casts an unseen pestilence: "But God All-Seeing [*panepoptēs*], the God of Israel, smote him with an incurable and invisible plague" (2 Macc. 9:5).[8] It follows, then, that for Early Christian theologians, "all-seeing" carried associations with the Old Testament Lord's infallible justice and punishment, which could destroy entire populations instantly. Clement of Alexandria (ca. 150–215) in his *Paedagogus*, used it to describe God's ceaseless watch over the impieties of Sodom and Gomorra:[9] "the All-seeing Word, whose notice those who commit impieties cannot escape, cast His eye on them."[10]

Wrought with paradox, the language of Trikomo's inscription is labyrinthine in its prosody and syntax. This is fitting given its subject matter: the fearsome mystery of Christ's invisible presence. It manipulates the root and meanings of "visible" (*optos*), first to describe Christ's omnivision (*pantepoptēs*), then his invisible status as "out of sight" (*ex apotopou*). Divine invisibility is thus contrasted with the image's visibility (*optos*), and the interaction of word and image constructs a dynamic of concealment and limitation. Further, divine omnivision contrasts with the viewers's gaze, which fixes itself on a given point during the process of reading. The syntax contains a chiastic construction: "he who sees all . . . sees all who enter [*O pantepoptēs . . . pant(as) enthade blepei*]," entwining the viewer's gaze with Christ's invisible one. When these words are read aloud, however, one is reminded of the connection between the sound of the root—*optos* in several words and the word for both "face" and "voice" (*ops*).[11] When considered vis-à-vis the image, the prosodic connections between visible (*optos*) and face (*ops*) create another dimension to this riddle. It leads the mind to two temporalities: the present act of looking at Christ's face in the image, then to the future, when the true face (*ops*) of the Judge will become revealed at the Last Judgment.

All of these aspects compel a reconsideration of this familiar icon of Christ, long called "Pantokrator." In the history of art, the austere, middle-aged, and purple-robed Christ of the Second Coming is considered among Byzantium's tutelary images. Yet, several scholars have attempted to explain the portrait type by ascribing it single titles (such as "Pantokrator") only to find that the Byzantines themselves rarely labeled the image as such.[12] Charles Barber once wrote the portrait type was "a problematic icon . . . one that many authorities have despaired of defining."[13] The problem lies in the fact that, unlike certain Marian icons such as the Hodegetria or Eleousa, "Pantokrator" derives from neither a toponymic nor fixed poetic or hymnographic origin.[14] In labeling the image both "Judge" and "All-seeing," the Trikomo example shows that this face of Christ could assume several titles at once. However, it is impossible to affix a static definition to such an image, even if that image asserts the aspirations to construct a timeless and penultimate vision of the Divine. However celebrated as objects of transcendental tradition, religious images are always works in progress, with meanings that shift according to the desires, practices, and emotions of people.

I argue that this instability between title and image is productive; it reveals the diverse, creative modes through which sources from Middle Byzantium imagined and anthropomorphized a supernatural, all-seeing energy. On the one hand, taking the Trikomo composition and its inscription as a case study, this essay offers new approaches to the problem of the "Pantokrator" image by considering first how, paradoxically, the two-dimensional portrait was constructed and perceived in three dimensions.[15] It investigates how this face of Christ reflects what I shall call a Byzantine poetics of fear and how its spatial framing inventively dramatizes judgment as an immersive spatial, aural, and affective experience.

On the other hand, a case study of this fresco opens up questions about how the history of art may be put in dialogue with the history of emotions.[16] It invites an analysis of how this image fulfilled desires to rehearse what many sources concurred was fearful, and how the visual, spatial, verbal, and temporal manipulation of it mediated this central emotion. In locating, historically, fear in relation to art, I do not claim that the contemporary historian can retroactively project fixed horizons of response onto long-gone viewers. I do, however, claim that the relation between fear and art may be charted by reconstructing how this fresco's material and spatial conditions immersed viewers, physically, in a scenario widely considered in the Middle Ages to be the most terrifying event imaginable.

"Condemned by their own judgment": encountering Christ in the dome

The famous late twelfth-century ekphrasis of the no longer extant Holy Apostles Church in Constantinople by Nikolaos Mesarites (b. 1164) offers a point of departure for analyzing the personalized fear that sustained engagements with such a Christ portrait could induce. His formal description of both the portrait and its space in the dome matches the figure and its spatial arrangement at Trikomo (Figure 5.1). We assume, however, that, given its location in Constantinople, the mosaic at the Holy Apostles was of grander scale and execution than the fresco at the small church at Trikomo. Nevertheless, Mesarites begins his section on the image by anthropomorphizing Christ as "the God-Man" (*theanthrōpon*). He then dramatizes the portrait's spatial presentation, claiming it makes Christ appear as "leaning and peering out as though from the rim of heaven, at the point where the dome begins."[17] This line reveals that, like our Trikomo example, Christ's figure assumed the zenith of the space, an aerial point whose core formed the center out of which the spherical shape radiated. Mesarites imagines the immersive space under the dome in a microcosmic sense: both "the arches of heaven" (*ouranias antygas*) and "globe" (*sphaira*) conjure astronomical associations.[18]

From his ekphrasis, it is unclear where Mesarites would have stood to perceive the portrait "at the point where the dome begins." Considering the details he provides about Christ's face, he could have been standing either directly under the center or just to the right of it, the area to which Christ's eyes on the Trikomo ceiling are directed. From his vantage point, Christ's face produced a penetrating effect:

> His eyes, to those having procured a clean understanding, are merry and accessible and instill the joy of contrition in the souls of the immaculate of heart and of the poor in spirit. For the eyes of the Lord are upon the righteous, as the psalmist says. His look is gentle and wholly soft, turning neither to the left nor to the right, but universally toward all and toward each individually, simultaneously. Such are those eyes to those who have a clean understanding; to those, however, who are condemned by their own judgment, the eyes are irascible, impenetrable, boding of ill; the face, having been made savage, is full of fear, boldness, and hardness. For the face of the Lord is this way for those doing evil.[19]

The face was open to interpretation since Christ's gaze opened a dialectical space into which the viewer's conscience could project possibilities. For those with a clear conscience—perhaps an impossibility given the precondition of original sin on all mortals—Christ's face projected welcoming assurance. For sinners, Christ was imagined in fearsome human terms as subject to the vicissitudes of passion, indeed, as possessing "wreckless abandon" (*itamon*). Mesarites's words each have different shades of meaning. Christ's eyes are "boding of ill" (*dusantētoi*) and "irascible, furious" (*orgiloi*) (see Figure 5.4). His face has become "savage" (*itamon*), which is also translatable as "charging headlong," and "fearsome" (*ekphoboun*), as well as "hardened" or "rugged" (*sklērias*).

Mesarites also remarks how the space of the dome dramatized the illusion of Christ's panoptic power. The portrait's "eyes" (*blemmata*), he claims, turn "neither to the left nor to the right, but universally toward all and toward each individually [*merikōs*], simultaneously [*ama*]." Though its language is hyperbolic and characteristic of ekphrasis, this phrase dramatizes the mosaicist's level of execution. Beyond producing a realism that confronted the viewer, the mosaic may have been installed in such an orientation that placed Christ's eyes on a spatial plane that met viewers from all angles. Further, the phrase is identical to the inscription at Trikomo that imagines divine sight transversing and perceiving part and whole, simultaneously. Additionally, both claim that this gaze occurs, supernaturally, in the present—the Trikomo inscription uses the present tense of "seeing" (*blepei*), and Mesarites makes reference to a psalm: "For the eyes of the Lord are upon the righteous" (Ps. 34[35]:15). This all-seeing eye was central to Byzantine conceptions of divinely sanctioned justice.[20] Christ's gaze was believed to be ceaseless, even when evil forces operated unpunished.[21] To this point, Theodore the Stoudite (759–826) once averred: "Thoughtful servants [of God] do not say 'Until when?' . . . For the Sleepless Eye knows what is good in each case."[22] Theodore's word "sleepless" (*akoimētos*) here gains further significance when we notice that the Trikomo image, for example, creates the illusion of Christ's eyes as always open.

Confronting the image of Christ, according to Mesarites, created a moment of intense self-reflexivity: "to those, however, who are condemned by their own judgment [*autokatakriton*], the eyes are irascible, impenetrable, boding of ill." The experience was absorptive, forcing an inward meditation on guilt. The face was chameleonic, contingent on the viewer's psyche. One sees the same bifurcation described by Mesarites on the famous sixth-century icon of Christ at the Holy Monastery of St. Catherine in Sinai (Figure 5.2).[23] Gazing at the face, the viewer becomes lost in the dialectic of the eyes, in the field of possibility ranging from harsh reproach to calm assurance. The sort of fear surrounding this image thus emerged from condemnation by one's "own judgment," a participation in its suspense (Figure 5.3). Romanos the Melodist, in the strophe quoted earlier, also emphasizes in another passage how the conscience becomes indicted when imagining the Last Judgment: "Reflecting fearfully on your trial and the day of judgment, Lord excellent above all, I shiver and am terrified, convicted by my conscience [*suneidēseōs*]."[24]

Mesarites was not the only Byzantine observer struck by this image's confrontational power. Another source details a miraculous encounter with the same mosaic at the Holy Apostles. The tale comes from Paul, the tenth-century bishop of Monemvasia, a fortified Byzantine town in the Peloponnese, but focuses on encountering the very same image in the dome of the Holy Apostles Church in Constantinople.[25] Paul composed a series of moralizing stories modeling pious behavior. One story, "The Man Who Confessed to an Image of Our Lord Jesus Christ," urged belief in the transformative power of confession to the same image of Christ, seen in the dome of the Holy Apostles:

> There was a man of considerable prominence who had great faith in the holy high priest John Chrysostom. Happening to fall ill, this man ordered his domestic servants to take him up and to bring him into the venerable temple of the Holy Apostles where the healing tomb of the divinely inspired father, Chrysostom, is located. He came into the venerable temple and prayed with tears in his eyes. Then he lay down on the bed, which had been prepared for him on the ground. Lying there, racked with pain, he called upon Christ's holy apostles and our holy father Chrysostom to be merciful to him in his critical condition. Then there came to his mind the improper deeds he had done. As he thought over these sins, he lamented and said "Woe is me, the wretched impenitent! How can I set out on that journey from which there is no return? How will I withstand the reproach of the fearful judge who is no respecter of persons? How to endure those eternal and unbearable

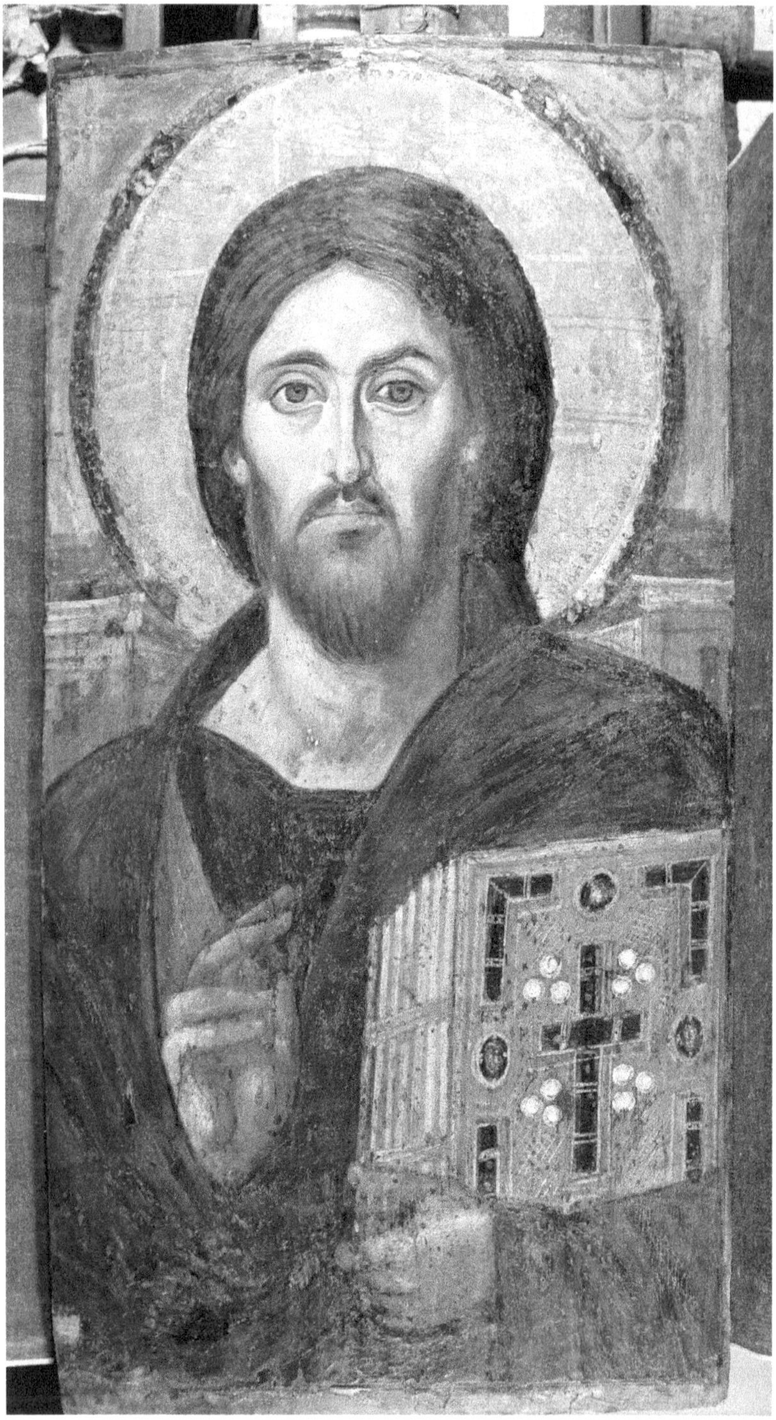

Figure 5.2 Christ Pantokrator, bust, sixth century, tempera on wood panel, The Holy Monastery of St. Catherine, Mount Sinai, Egypt.

Photo: Reproduced through the courtesy of the Michigan-Princeton-Alexandria Expedition to Mount Sinai.

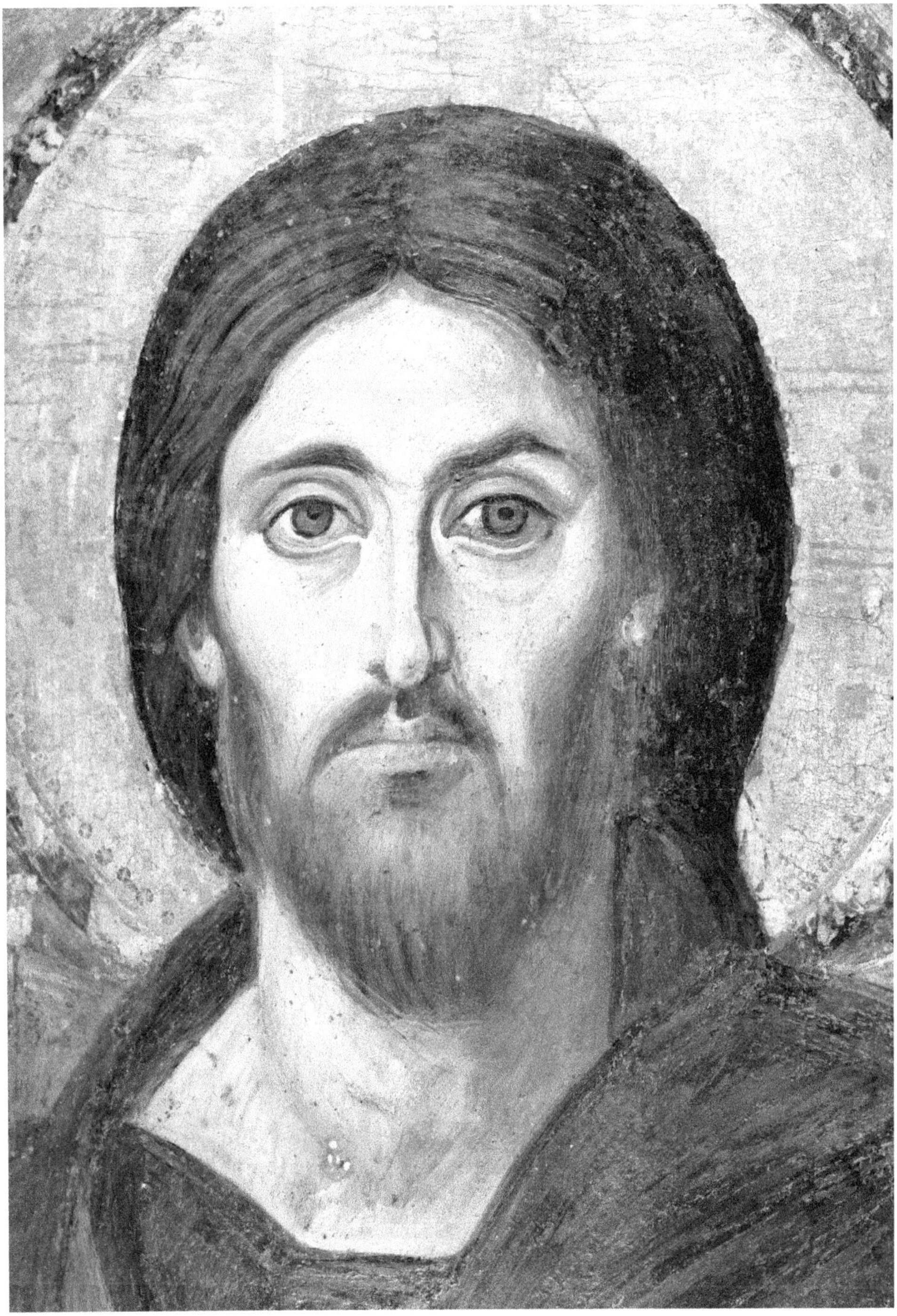

Figure 5.3 Detail of Figure 5.2, Christ Pantokrator. Detail showing Christ's eyes.
Photo: Reproduced through the courtesy of the Michigan-Princeton-Alexandria Expedition to Mount Sinai.

> torments?" Such, and more beside, were the things he said in his wailing. About the sixth hour, when everybody else had gone out of that holy temple, he was left alone. He gazed intently into the dome of that holy temple and saw there the icon of our Lord Jesus Christ, to which he began to address words such as: "Lord, if I shiver and dread the sight of your icon, made by human hands, how will I look upon you, the fearful judge—wretch that I am!—when you come to pass judgment on all creation? Oh Lord, I am a sinner. Forgive me, for I have not kept your commandments." Then he began openly to confess his sins to the icon. Yet, having said all, he then recalled another, yet one more grievous sin, which had been committed. At that he began to say: "One other sin I have committed, Lord, but I dare not name it aloud. I dare not, lover of men; I dare not, oh merciful one." While he was saying this there came to him a voice out of the holy icon: "Speak!" When he had named this last sin, there came another utterance: "Your sins are forgiven." This fearsome voice having resounded, the man rose up in good health. . . .[26]

Woven into Paul's sermons, the tale may have reached both lay and monastic audiences. From his informal description of the image, we gather that the listeners in provincial Monemvasia were familiar with both the portrait type and its domical installation, as well as the Holy Apostles Church (*Apostoleion*) and the saving grace of John Chrysostom's tomb there. Paul calls the image "the icon of our Lord Jesus Christ," then later refers to the one portrayed as "the fearful judge," in line with the words of Trikomo's inscription. The audience would have been familiar with the tradition of domical Christ images. He also refers to Christ as "no respecter of persons" (*aprosōpolēptos*)—a title that Romanos the Melodist uses to describe Christ in his "*Kontakion* on the Second Coming."[27]

The tale uses the event of a miracle to exhort faith in the image of Christ, by claiming that it mediates an engagement with the unseen but all-seeing God. It even reveals the presence of skepticism in accepting the divine presence contained within an image "made by human hands." Countering such skepticism may have been the cause for the story. Paul's argument is essentially the same as the message in the inscription at Trikomo. The image, as representation, does not defer or predict the presence of Christ but visually figures his invisible watch. The visible asserts the invisible. In turn, the viewer's fear of the image becomes a physical, sensed assurance of his faith in the invisible.

The tale transmits this message indirectly: first, our protagonist, though taken by fear, sees the mosaic only as artifice: "if I shiver and dread the sight of your icon, made by human hands, how will I look upon you, the fearful judge—wretch that I am!—when you come to pass judgment on all creation?"[28] The fearsome divine voice (*phoberas phōnēs*) corrects him, asserting that the image is not a simulation but an actualization of real-time judgment, just as the Trikomo inscription claims.

The physical and psychological nuances in Paul's words reveal how the Byzantines conceptualized their experiences of religious fear. While the semantics offers a period-specific insight, Byzantine theories of this emotion deserve further scholarly investigation.[29] The fear in this tale seems similar to the phenomenon evoked by Trikomo's inscription: an abject, bodily terror—"I shiver and dread" (*phrittō kai dedoikai*). One of the words for fear, *phrittō*, relates to hair bristling or standing on end. This fear is sensed as a somatic possession, a frigid arousal of the flesh. Paul's parallel word for "dread" (*dedoika*) describes a more diffuse sense of angst, a state of mind rather than of body. This fear is a totalizing, altered state. We gather that this specific sort of psychosomatic, reverential fear was perceived as mental and tactile, even a self-disciplinary experience and exercise. One psalm verse describes such fear of God as a desired, mystical state, a psychosomatic mortification related to crucifixion or flagellation: "Nail onto my flesh, the fear of you. For I am afraid of your judgments" (Ps. 118[119]:120).[30] Fear was the inscription of penitence on the body—and it was the image that served as the instrument of discipline.[31]

Judgment without verdict at Trikomo

How was fear under the dome constructed as an immersive, physical experience? The Trikomo composition (Figure 5.1) largely preserves the same effects described by Mesarites and Paul of Monemvasia. There, the inscription, composition, and dome combine to create a microcosmic experience of Christ's circular power: *pan* (all), root of *pantepoptēs* (all-seeing), carries a spatial sense, in line with scripture's description: "he looks to the ends of the earth, and sees under the whole heaven" (Job 28:24). The dome, as a spherical enclosure overhead, furthers the immersive effect of containment. The pictorial composition drives the sense of torque as the entire scene radiates out from the central medallion in registers. This overwhelming circularity is manifested as the composition impels the eye to move in a circle while reading the inscription. The neck, too, must bend back to perceive the ceiling, furthering the disorientation produced by the spiraling of the composition.

It is possible that Trikomo's composition, with Christ surrounded by the angelic procession, was a common choice for domical decoration, for a nearly identical scheme in fresco adorns the dome at the Holy Apostles Church at Pera Chorio of about 1180, located 74 kilometers west of Trikomo (Figures 5.4, 5.5).[32] There, Christ, in a medallion, also occupies the apex. Though little of the Pantokrator's face survives through the ruination, we may nevertheless discern some of its characteristic severity, especially in the eyes after its cleaning (Figure 5.5). At both Trikomo and Pera Chorio, a large group of attendant cherubim—some of them wearing the Byzantine court *loros*—crouch under Christ's force. Pictured in a procession, each figure offers supplication. These are no longer the graceful, delicate cherubim that travel between heaven and earth; instead, each is a lugubrious member of a grave assembly. At both Trikomo and Pera Chorio a verse from Hebrews 1:6, "And let all the angels of God worship him," is inscribed. At Trikomo, the passage forms another circular band around Christ; at Pera Chorio, the words are between the angels' heads.

At Pera Chorio, the pictorial space under Christ is lost. At Trikomo, however, the angelic procession culminates with humanity's intercessors at either end: the Virgin and John the Baptist. With hands out, they motion, pleading on humanity's behalf. An empty throne, labeled "hetoimasia," stands between them, reflecting Psalm 9:7, "The Lord shall endure forever: he has prepared his throne for judgment." The prepared throne symbolizes the prophecy, "we must all appear before the judgment seat of Christ, so that each one may be recompensed for his deeds in the body, according to what he has done, whether good or bad" (2 Cor. 5:10). Noteworthy is the fact that the word *pantepoptēs* ("all-seeing") appears directly above the empty throne (Figure 5.6). This combination of word and image furthers the prophetic effect by inserting an open space between invisible authority (signaled by the word) and absence (signaled by the open seat awaiting occupation.)

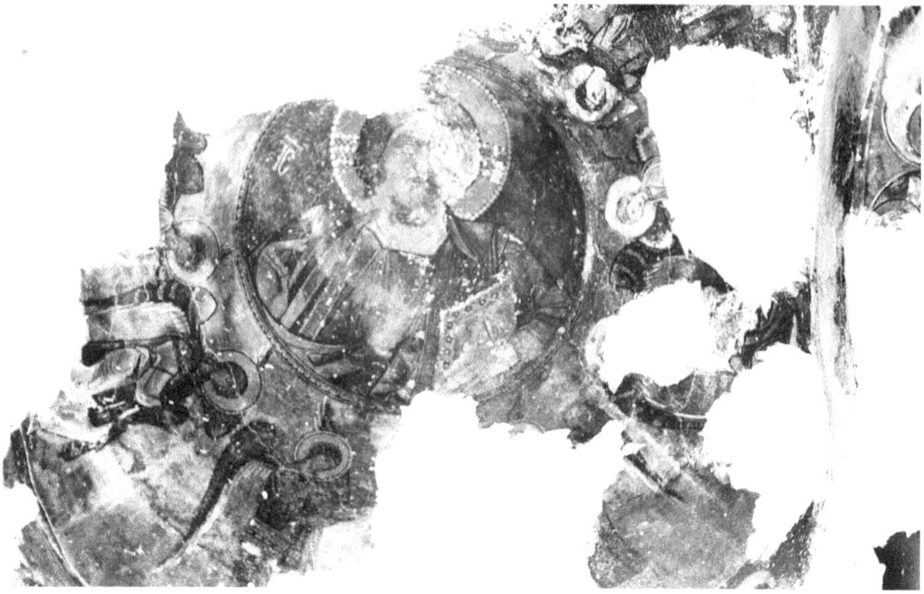

Figure 5.4 Pantokrator before cleaning, fresco, twelfth century, Holy Apostles Church, Pera Chorio, Cyprus, from Ernst Hawkins and Arthur Megaw, "The Church of the Holy Apostles at Perachorio and Its Frescoes," *Dumbarton Oaks Papers* 16 (1962): 277–350, fig. 3.

Nicosia Department of Antiquities B9576, with permission of Image Collections and Fieldwork Archives, Dumbarton Oaks Trustees for Harvard University, Washington, DC.

The *hetoimasia* gives the composition ominous tension. For the viewer, it forces and coordinates an awareness of present and past in relation to the event promised in Matthew 25:

> [31–34] When the Son of man shall come in his glory, and all the holy angels with him, then shall he sit upon the throne of his glory: And before him shall be gathered all nations: and he shall separate them one from another, as a shepherd divides his sheep from the goats: And he shall set the sheep on his right hand, but the goats on the left. Then shall the King say unto them on his right hand, Come, ye blessed of my Father, inherit the kingdom prepared for you from the foundation of the world. . . . [41] Then shall he say also unto them on the left hand, Depart from me, O cursed, into everlasting fire, prepared for the devil and his angels.

In the fresco, while the scene of judgment is set, the final verdicts remain uncertain. In signaling but not actually representing the event prophesied by Matthew, the composition simulates judgment without an outcome. The unfolding effect is one of going before Christ himself. A fearful, interstitial space takes form in the viewer's experience, a suspenseful moment that resists discursive meaning. The interstice leaves the viewer in a space between prophecy and its fulfillment. He stands before the image, indicted. Mesarites describes this state of reflexivity in his ekphrasis, with the phrase "condemned by their own judgment" (*autokatakriton*).

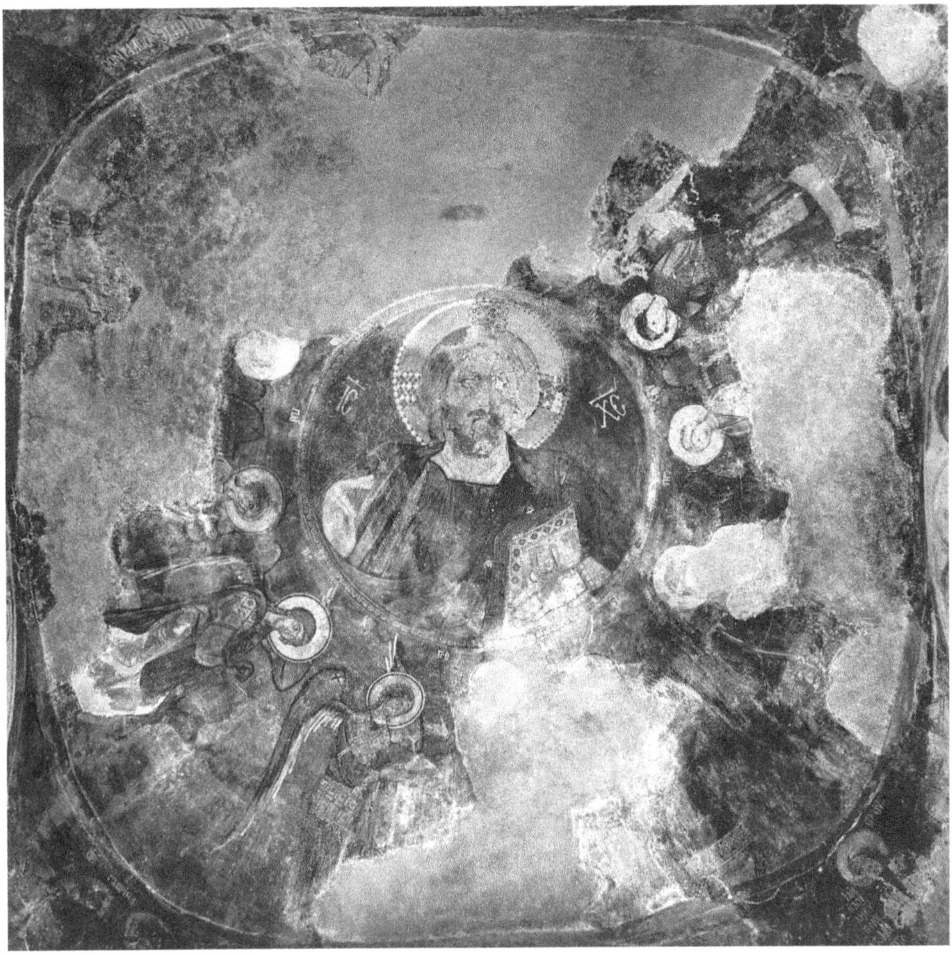

Figure 5.5 Pantokrator after cleaning, from Hawkins and Megaw, *Dumbarton Oaks Papers* 16 (1962): Fig. 3.

Nicosia Department of Antiquities B9531, with permission of Image Collections and Fieldwork Archives, Dumbarton Oaks Trustees for Harvard University, Washington, DC.

Judgment becomes more vivid as the inscription around Christ states, "He examines their souls and the movements of their hearts."[33] The inscription warns that the divine watch penetrates the viewer's private space of thought, memory, and intention, reflecting scripture: "Neither is there any creature that is not manifest in his sight: but all things are naked and opened unto the eyes of him with whom we have to do" (Heb. 4:13). Text, image, and space work in orchestration to simulate the oppressive experience of being the object of divine surveillance. Yet how does one go beyond uncovering how the building and images framed this simulation and render this in words, the phenomenon of architecture provoking a binding, submissive spatial experience? Here, we may look to a modern source for a more precise description of this sensation of being watched. The Surrealist author Roger Caillois, in his influential essay on camouflage, imagines this sensation of being overwhelmed, even abducted, by space itself:

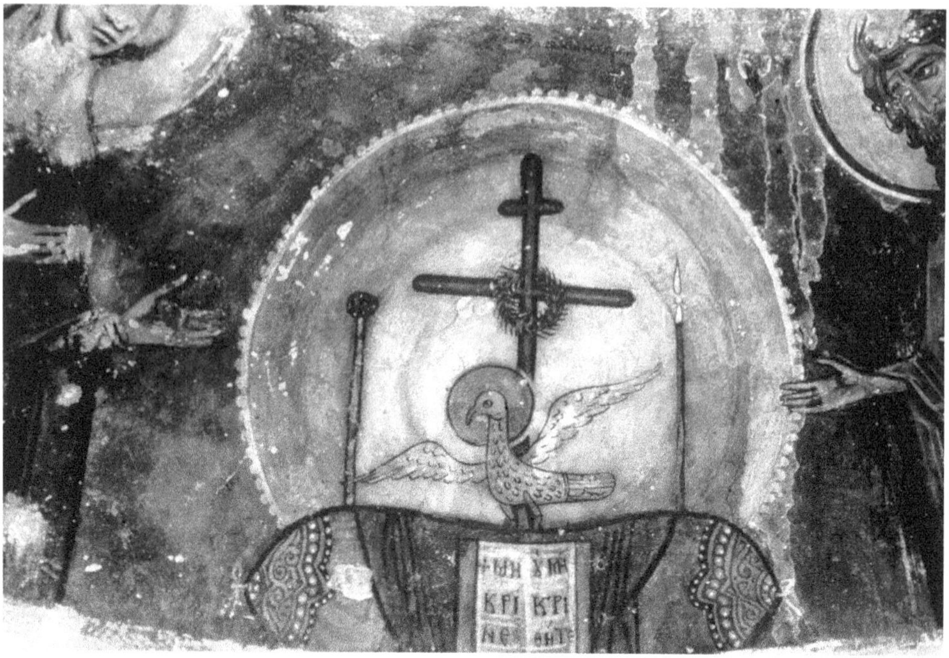

Figure 5.6 Hetoimasia, fresco, 1105–06, Panagia Theotokos Church, Trikomo, Cyprus.
The Byzantine Institute and Dumbarton Oaks Fieldwork Records and Papers, MS.BZ.004-BF.S.1979.2938D, Image Collections and Fieldwork Archives, Dumbarton Oaks, Trustees for Harvard University, Washington, DC. Photograph: Courtesy of Image Collections and Fieldwork Archives, Dumbarton Oaks, Trustees for Harvard University, Washington, DC.

> *I know where I am, but I do not feel as though I'm at the spot where I find myself.* To these dispossessed souls, space seems to be a devouring force. Space pursues them, encircles them . . . it ends by replacing them. Then the body separates itself. . . . He tries to look at himself from any point whatever in space. . . . And he invents spaces of which he is the "convulsive possession."[34]

With recourse to an unknown "he," Caillois places the reader in another's body to describe the feeling of being contained. If we replace the divine gaze with Caillois's conception of "space," we may be able to reconstruct the original sensations of fear under the "eye" of the dome: an engulfing space—itself conceived by Caillois as a mystical force—is like the effect of being the object of Christ's surveillance. Self-awareness under the dome's expanse is akin to the circular sensation of being watched "from any point whatever in space." Indeed, if one identifies with the composition's simulation of binding judgment, he absconds possession of himself: his "soul" and the "movements" of his "heart" are rendered transparent. Placed in Christ's simulated theater of justice, the faithful viewer, "condemned by his own conscience," à la Mesarites's description, becomes Christ's "convulsive possession," to borrow Caillois's phrase.

Architecture constructs the judgment scenario. The spatial effects of the domical presentation are far different from those elicited by seeing images of the Last Judgment

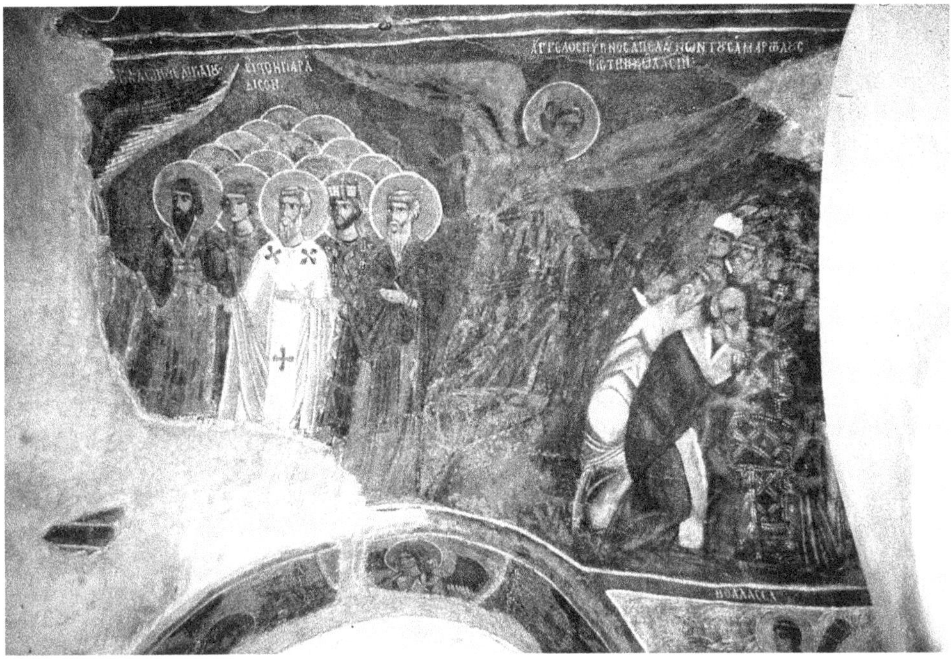

Figure 5.7 Last Judgment in Narthex, fresco, first third of the twelfth century, Hagios Nikolaos tis Stegis Church, Kakopetria, Cyprus.

The Byzantine Institute and Dumbarton Oaks Fieldwork Records and Papers, MS.BZ 004 BF.S.1979.1556D, Image Collections and Fieldwork Archives, Dumbarton Oaks, Trustees for Harvard University, Washington, DC. Photograph: Courtesy of Image Collections and Fieldwork Archives, Dumbarton Oaks, Trustees for Harvard University, Washington, DC.

in other architectural installations. The Trikomo ceiling fresco implicates the viewer far differently than, for example, the narrative fresco of the Last Judgment fresco in the early twelfth-century monastic church of Hagios Nikolaos tis Stegis at Kakopetria, in the Tröodos Mountains 117 kilometers west of Trikomo (Figure 5.7).[35] Gazing at the fresco at Kakopetria, in the lunette over the north door, the beholder is now an observer removed, a degree, from divine judgment. The result is a distinct empathetic response. The body itself is not placed in a zone of participation. Although others are portrayed undergoing torment, the image does not force the same sense of self-reflexivity and interstitial tension. Judgment is depicted, not predicted and enforced.

Imagining Christ's terrifying voice

The suspenseful interstice effect opened by Trikomo's composition comes, in part, from perceiving Christ on the verge of administering his verdict. The "fearsome voice" (*phoberā phōnē*) that resounded "out of the holy icon" (*ek tēs hagias eikonas*) was central to Paul of Monemvasia's tale. Yet did Byzantine makers aspire to construct Christ in the image as though speaking to the viewer—or at least, to evoke the effect of his terrifying voice?[36] Evidence suggests so. At Trikomo, the inscription's placement on the ceiling, along with its imperative tone, renders it as though it were a message from above. The same effect is magnified at the Timios Stauros Church at Pelendri,

Cyprus (ca. 1171–72), west of Trikomo.[37] There, an inscription also frames Christ's medallion in the dome, stating: "I am the Judge and God of All. Leaning down from on high before the Judgment I advise [you] to obey my laws, whoever wishes to escape the tortures."[38] Because it affixes the first-person pronoun to the image (*ego*), the inscription becomes a direct command from Christ himself. It speaks on the image's behalf, literally giving it a voice. The text continues to assert the first person, by qualifying divine laws (*nomous*) with the first-person possessive (*emous*). As at Trikomo, too, Christ's strident adjuration to the viewer has no direct scriptural parallel; the inscription calls attention to how the image in the dome mediates the paradox of Christ's presence as an approach (*pelō*) from on high (*hypsōthen*).

In all of these examples, Christ's raised right hand evokes gestures of speech. Emperor Leo VI (r. 866–912), in "Sermon 28," an ekphrasis of the church in the Monastery of Kauleas in Constantinople, composed between 886 and 912, remarks that the mosaic's illusionism was a result of Christ's being presented with an immediacy that exuded a near-rhetorical address to the viewer:

> Now the [structure] which is above the beautiful pavement and forms the roof is raised in the shape of a half-sphere. In the midst of it is represented an image of Him to whom the craftsman has dedicated the church. You might think you were beholding not a work of art, but the Overseer and Governor of the universe Himself who appeared in human form, as if He had just ceased preaching and stilled His lips.[39]

Here, Leo inversely dramatized Christ's vocal presence, as having just made a proclamation. Christ in the half-sphere (*hēmikyklion*) is called not Judge but "Overseer and Governor of the universe" (*epoptēn kai kybernētēn*). The "Overseer" (*epoptēn*) again uses the same root, *opos*, as Trikomo's inscription to describe the divine omnivision.

Architectural space facilitates the fantasy of Christ's voice, dramatized in combination with his image that, following Leo's passage, is presented as though on the brink of adjudication. In Paul's tale, the space of the dome is the conduit for communication from the divine voice—its spatiality places the image as though in a tunnel, on the other side of a dramatic dialogue. Indeed, there exists in other Middle Byzantine churches inventive modes through which the space of the dome mediates the illusion of Christ the Judge's booming voice. At the Church of the Koimesis at Daphni Monastery, built in the early twelfth century, the looming, domical Christ mosaic is given voice by the scrolls of the surrounding Old Testament prophets, Moses, David, Solomon, Elijah, Isaiah, Daniel, Joel, Jonah, Habbakuk, Zephaniah, and Malachi (Figure 5.8). Pure gold space separates the centralized Pantokrator image from the prophet cycle. The aniconic distance between the two registers represents the time stretch between prophecy on one side, and its fulfillment, in the appearance of Christ's face in the center, on the other. That space also structures two sides of a dialogue: Jonah, for example, bears a scroll with his own verses: "I called to the Lord my God from my affliction and he heard me" (Jon. 2:2). Habakkuk, meanwhile, proclaims: "Lord, I have heard your speech and was afraid" (Hab. 3:2). Here, another interstitial space emerges in the distance between the prophets and Christ: one in which Christ's voice is implied, and desired. Like Paul's tale, in these prophetic scenarios, the voice confirms the Divine's invisible presence.

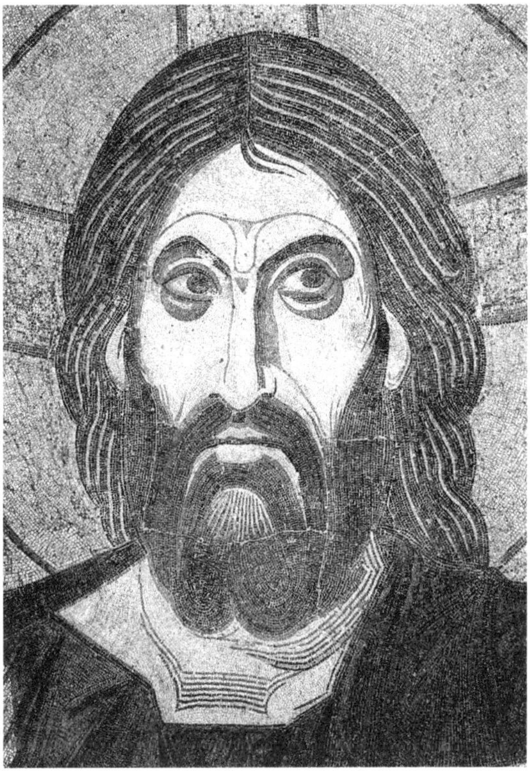

Figure 5.8 Pantokrator with prophets in cupola, mosaic, ca. 1100, Church of the Koimesis, Daphni Monastery, Daphni, Greece, from Ekkehard Ritter, Corpus for Wall Mosaics in the North Adriatic Area, ca. 1974–1990s.

The Byzantine Institute and Dumbarton Oaks Fieldwork Records and Papers, MS.BZ.002 BF.T.Da.014B, Image Collections and Fieldwork Archives, Dumbarton Oaks, Trustees for Harvard University, Washington, DC. Photograph: Courtesy of Image Collections and Fieldwork Archives, Dumbarton Oaks, Trustees for Harvard University, Washington, DC.

We cannot be certain if such a prophet cycle framed the Pantokrator image at Trikomo, although a similar scheme exists at the late-Komnenian Church of the Panagia Chryseleousa at Strovolos, Cyprus.[40] There, Habakkuk appears holding a scroll with the exact same verse as seen at Daphni. Ezekiel and Zephaniah, with head raised toward Christ, are present along with Jonah and Micah, each bearing a solemn expression. At Strovolos, however, Christ is a full figure, with head haloed and feet stretching out of the circular frame. Habakkuk holds the same verse as seen in the Daphni example, which declares God's fearsome voice; this repetition cannot be merely coincidence, especially considering the great distance between Strovolos on Cyprus and Daphni in mainland Greece.

The Old Testament is the source for the notion of a divine speech asserting, invisibly, God's surveillance over mortal events. Beginning in Genesis, the Lord's voice punishes Adam who, like Paul's sick man, thinks he can conceal his transgressions:

> they heard the voice of the Lord God walking in the garden in the cool of the day: and Adam and his wife hid themselves from the presence of the Lord amongst

> the trees of the garden. . . . And the Lord God called unto Adam, and said unto him, "Where art thou?" And he said, "I heard thy voice in the garden, and I was afraid, because I was naked; and I hid myself" (Gen. 3:9–10).

In addition, other Old Testament accounts of the divine voice highlight its terrifying tone. In Deuteronomy, Moses says, "Out of the heavens He let you hear His voice to discipline you; and on earth He let you see His great fire, and you heard His words from the midst of the fire" (Deut. 4:36).[41]

A canon for that terrible day

The interstice of divine judgment, the inspiration to penitential confession, and the pleas for salvation from unendurable tortures found their dramatic rehearsal in the space and time of a liturgical event. Aspects about the domical image that we have been exploring, from its spatial presentation in the apex of the cupola to its evocation of a terrifying divine voice, reflect the ritualized fantasies about the horrible possibilities of Christ's Second Coming. A striking parallel exists between the language of Byzantine liturgical works about the Last Judgment and the image of Christ the Judge and its poetics at Trikomo. The actual lines of transmission, copying, and reproduction of this Christ icon, from Constantinople to Cyprus, for example, remain unclear. Yet, given the language of fear surrounding it, liturgical developments may have enhanced desires to envision it in pictorial and spatial form. Liturgical sources surely conditioned both Mesarites's ekphrasis and Paul of Monemvasia's story about encountering the image of Christ at Holy Apostles Church in Constantinople. As I shall argue, the closest ritual source with which we may connect the fresco lies in the Triodion, a hymnal for the Lenten-Paschal season.

However, first it is necessary to chart the origins of how the liturgy intensified the imagination about Christ's Second Coming, and how later, the Triodion's hymns became standard in Byzantine spirituality. The earliest documented Byzantine ritual centered on Christ's Second Coming is the tradition of the Sunday of the Last Judgment in the weeks before Lent, which may have originated in Constantinople. At least in the ninth or tenth century, the cathedral liturgy of Hagia Sophia specified for that Sunday the reading of Matthew 25:31–46, the Gospel passage wherein the events of the Second Coming are foretold.[42] The earliest preserved Byzantine hymn about the Last Judgment is Romanos's mid-sixth-century *kontakion*, "On the Second Coming." Narrative in thrust, Romanos's *kontakion* lists both the numerous historical predictions and then describes, vividly, the cosmic upheaval at the End of Time. He damns the Jews, doomed to retribution when they "see Him whom they pierced."[43] Then, John the Baptist, Elijah, Matthew, and Enoch are acknowledged for their visionary powers.[44] The poem narrates apocalyptic catastrophes: strophes 7–11, for example, predict the reign of the Antichrist ("Then the lawless one will build a temple and lead astray / The proper assembly of Hebrews and others").[45] The piece vividly recounts what to expect at the heavenly tribunal and offers a humble prayer for intercession and mercy: "And through the intercessions of the Virgin, Mother of God, spare me / And do not tear me from thy Sight, O most just Judge."[46]

The Middle Byzantine works under investigation, however, are closer in time to a different hymn, though one probably influenced by Romanos's *kontakion*.[47] Although we have no surviving documents attesting to the rituals of twelfth-century Cyprus, the language of trembling in Trikomo's inscription, even the mode through which it implicates the body, resembles the tenebrous poetics of the Triodion hymn. The Triodion (literally, "three odes") is a service book with hymns for the ten weeks before Pascha, of which three were in preparation for the six weeks of Lent and Holy Week.[48] Though some have argued it is impossible to trace the Triodion to a single original document, as a compilation of liturgical poetry, it can be divided into three developmental strata.[49] The first stratum emerged from a Palestinian context, especially the Monastery of Mar Saba. Its hymns were composed between the fifth and ninth centuries and attributed to Andrew of Crete, John of Damascus, Cosmas of Maiouma, Theophanes Graptos, and Stephan the Sabaite.[50] In the early ninth century during Iconoclasm, however, a second stage of the Triodion developed at the Monastery of St. John the Forerunner at Stoudios in Constantinople. It was here that its famous iconophile abbot (*hegoumenos*) Theodore (759–826) and his brother Joseph (762–832) composed new poetry for the pre-Lenten season and compiled the works of the earlier Palestinian hymnographers.[51]

With such reform came a new hymnography for the Sunday of the Last Judgment, or Meatfare Sunday (*Apokreas*), the eighth Sunday before Easter. In the earliest surviving copy of the Triodion, in the Monastery of St. Catherine in Sinai, Sinai graecus 734, a canon titled "On the Second Coming" and lacking authorial attribution is assigned for Meatfare Sunday.[52] Another early Triodion copied at the Grottaferrata Monastery in Italy in the eleventh century and now at the Vatican Library (Vatican City, Biblioteca Apostolica Vaticana, MS Gr. 771), largely preserves the same hymn but claims its author to be Theodore the Stoudite himself.[53] It is not entirely certain whether Theodore actually composed it, although Derek Krueger has argued that its fiery language resembles the *catecheses* that Theodore delivered at the Stoudios Monastery about the Last Judgment.[54] The Triodion itself was reworked until 1522, when, in Venice, a printed edition appeared; that edition forms the 1872 Vatican publication, which constitutes the received tradition of the Triodion today and maintains that Theodore the Stoudite composed the canon "On the Second Coming."[55]

Cypriot manuscripts of the Triodion do not survive from this period, and it is therefore unclear as to whether its "On the Second Coming" was ever sung in the Panagia Theotokos Church at Trikomo in the twelfth century. Nevertheless, the canon offers a glimpse into the emotions surrounding the coming appearance of Christ. Considering the fact that the reform of the Triodion occurred during the period of Iconoclasm, its effect on the liturgy influenced the anthropomorphic portrayal of Christ after the controversy and, thus, the images of the Judge in the dome.[56] The poetry also marks a new phase in the development of a penitential spirituality in Byzantium, first for monastic contexts. Krueger has charted the depth of the Stoudite Triodion's liturgical influence in Byzantium in shaping the range of spiritual emotions, especially regarding the Last Judgment.[57] The Triodion's canon places the reader in the scenario of judgment as it is read in the first person.[58] It thus structures a more personalized declaration of fear in comparison with Romanos's earlier *kontakion*. The reader's conscience is placed on trial, as he is unified with all of humanity in the binding future judgment. I provide only the first canon out of eight, since the entire piece is quite long.

> I shiver contemplating and foresee dreading the horrible day of your unspeakable coming, when you will sit to judge the living and the dead, O my God, All-powerful!

> At the time when you will come, O God, with thousands and ten thousands of celestial ruling angels! May you deem me in my misery worthy to meet you in the clouds, O Christ!
>
> Come, my soul, seize with the mind that hour and day when God will stand against you visibly! Wail and lament to be found clear of guilt in the hour of the Trial!
>
> It terrifies and puts me out of my wits! The inextinguishable fires of Gehenna, the odious worm, the grinding of sharp teeth! But free me, forgive me, Christ, and stand me among the elect!
>
> May I, though wretched, also hear Your desired voice that calls your holy to joy and may I take the inexpressible blessing of the Kingdom of Heaven.
>
> May you not enter in judgment against me, carrying my deeds, examining my words, and judging my sins. But in your compassion, neglecting my wrongs, save me, O All-powerful one![59]

Materializing this scenario as a distant but impending reality, Trikomo's ceiling fresco reflects all of the poem's tensions. Both the poem and the fresco implicate one in the same theater of judgment, salvation, and punishment. Staring at Christ overhead, the viewer, just like the narrator, is immersed in different interstices: between intimacy and distance, desire and denial, forgiveness and punishment, paradise and torture.

Certain words of the hymn match those found in Paul of Monemvasia's tale: *phrittō* for bodily fear and *dedoikōs* for rampant dread. Could Paul have been paraphrasing Theodore Stoudites's canon "On the Second Coming"—"I shiver contemplating and foresee dreading" (*phrittō ennoōn, dedoikōs proorō*)—when he tells the protagonist, "I shiver and dread the sight of your icon" (*phrittō kai dedoika*)? Further, one cannot ignore the similarities between Paul's character calling to mind the "unbearable torments" and the first-person narrator of the canon imagining the terrors of Gehenna; or, the very constructions in both texts revolving around anticipation and revelation, whether in the mind's eye—"the Canon of the Second Coming"—or as mediated by an image in the story of self-indictment narrated by Paul of Monemvasia.

The Triodion canon marks a different mechanics of penitence. For unlike Romanos's piece, the Triodion's "On the Second Coming" commands the reader to inscribe emotional states on himself: the fear one feels in reciting the poem leads to an altered state. Lurid in its language of sensation, the poem describes the mind battling to comprehend the magnitude of the coming event. Physical trembling emerges in the process of "foreseeing" (*proorō*). Further, the use of the word *labe*, from *lambanō*, fuses cognition with bodily sensation. I render *labe*, "seize," though the word itself assumes many meanings, often with physical, even violent nuances.[60] "Come, my soul, seize with the mind that hour and day": apprehending the event is to enter a state of mental turbulence. Its language engages with the scenario as if it were a spiritual exercise, one based on self-identification and in-situ projection. The narrator tells himself to "Wail and lament in order to be found clear of guilt in the hour of the Trial!" As in Paul of Monemvasia's tale, undergoing the physical pangs of fear leads to spiritual absolution and relief.

Directly addressing Christ, the hymn calls the Savior by numerous titles, each in line with the anthropomorphic depiction. Yet not once is he called "Pantokrator," the title always applied in art history—though at the end of the first and last lines, he receives essentially the same title of "all-powerful" (*pantodyname*). This title contrasts with Romanos's earlier *kontakion*, as Romanos ends each strophe with "O Judge." In the Triodion canon, Christ occupies the "Kingdom of Heaven" (*basileia ouranōn*) and is imagined to have a voice (*phōnē sou*). The poem also makes vivid the figural, human impression of Christ through its description of him as "seated," showing the same emotions as a despotic judge, and scrutinizing every word, deed, and sin. Above all, the poem pictures Christ in the mind's eye as "standing over" (*epistē*) the living and the dead, "visibly" (*emphanōs*).

Per the canon's language, the architectural staging of the image positions Christ as spatially "standing against" (*exista*), even "imposing," vis-à-vis the body of the viewer. Both Romanos and the author of the Triodion canon use the same vivid term for the state of "being laid bare" or "exposed to view" (*tetrachēlismenoi*) before Christ at the Last Judgment.[61] This word, argues Marjorie Carpenter, is hard to define, though it comes from the vocabulary associated with wrestling competitions.[62] While it means to overpower or lay bare, the physical sense of the term comes from the pulling back of one's neck to expose the throat (to be cut, in some cases). It follows that, to see the image of Christ in the dome, one must perform the same motion of bending the head far back and exposing the throat. The very act of looking at the aerial portrait places one in a position of physical vulnerability and, thus, submission.

The ceiling fresco, shaped by the dome's coil, pictures Christ "in the clouds" just as the Triodion canon describes. And, in line with the canon, Christ in the image appears with a cherubic assembly. Although the canon dramatizes their number as "thousands and ten thousands," did the fresco painter attempt to render such a large assembly at Trikomo? It is possible that he, in conceiving of the viewer's position below, sought to create the illusion of angels receding into an endless background space, dramatized by the ceiling and the dome's shape. Considering the canon, we may take this play between the heavenly space of the image and the physical space of the viewer further. For in addition to fear, the canon's language yearns: "May you deem me in my misery worthy to meet you in the clouds, O Christ!" and "Stand me among the elect!" Space emerges between the planes on which the viewer stands and the one that Christ occupies. Following the canon, it would seem that very space would assert the sensation of physical distance from Christ and, thus, intensify desires "to meet" him.

Preparing for the hour of judgment

This same language of yearning manifests itself elsewhere in the Panagia Theotokos at Trikomo: this time, in an epigram in the apse. I will conclude with this final poetic coordination between Last Judgment and Trikomo's monumental program, as it ties the building and its frescoes to the memory of a lost individual. Written in the unknown donor's voice, the short poem relates the building itself (and, by extension, its fresco program) to spiritual redemption. Underneath a painting of the Virgin Theotokos (to whom the church was dedicated) reads the penitential plea (Figure 5.9):

> O Queen of All and Superior of All
> Chaste Lady and Mother of the Lord
> see the desire of my suffering soul
> and become my intercessor in the hour of Judgment
> that I may be spared the lot of [hell].[63]

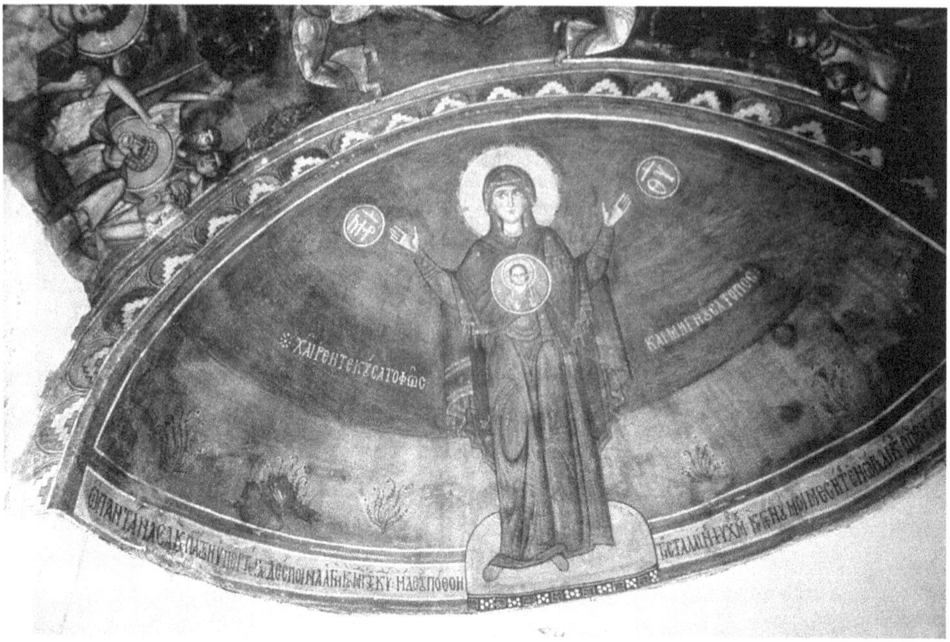

Figure 5.9 Theotokos in apse with epigram, fresco, 1105–06, Panagia Theotokos Church, Trikomo, Cyprus.

The Byzantine Institute and Dumbarton Oaks Fieldwork Records and Papers, MS.BZ.004 BF.S.1979.2938D, Image Collections and Fieldwork Archives, Dumbarton Oaks, Trustees for Harvard University, Washington, DC. Photograph: Courtesy of Image Collections and Fieldwork Archives, Dumbarton Oaks, Trustees for Harvard University, Washington, DC.

The language parallels that of the Triodion canon by sharing in the first-person narrator's woeful tone and uses the same word (*talainēs*) to describe the soul as "suffering." The text foresees and, thus, makes reparations for the "hour of Judgment." Other inscriptions from Byzantine Cyprus, especially those at the Panagia Phorbiotissa Church at Asinou (1105), 105 kilometers west of Trikomo, share even more closely in the language of the Triodion canon. The apse at Asinou virtually reproduces the same poem as at Trikomo, which survives with the donor's name. However, in the south wall of the central bay of Asniou's naos, another inscription offers the entire building itself in exchange for protection "on the terrible Day of Judgment"—those very words that open the Triodion canon.[64] According to these apsidal inscriptions, these churches were donations to the Virgin. As permanent structures, buildings such as the Panagia Theotokos at Trikomo materialized "desire" (*pothos*) for redemption and became the medium through which an individual could memorialize pleas long after his death.[65] A currency of salvation, dictated by the divine Judge himself, endowed them with their value.

The different architectural forms orchestrate the votive mechanics. While Trikomo's ceiling inscription locks the viewer in a chiastic relation with Christ's sight ("He who sees all . . . Sees all those who enter"), anticipating judgment, the apsidal inscription does the opposite: it begs the Virgin to "see the desire" of "a miserable soul."[66] On the one hand, this apsidal inscription was placed for mortal eyes. Each time the poem was read, another prayer would have been entered on behalf of the donor's soul.[67] On the other hand, a far stranger effect is inherent to its placement and formula: in asking the Virgin to "see," the painted words were executed with the desire to meet her gaze above. This address to the eyes above recalls the hymnographic tradition wherein Romanos concludes his *kontakion* by requesting that the just Judge "not tear" him "from Thy sight."[68] Indeed, it was the all-seeing Judge and the Virgin intercessor, evoked throughout the program, who were the most important viewers for Trikomo's building and fresco cycle. These invocations asserted and memorialized the belief that judgment of the living was a continuous, invisible process. Painting and architecture asserted this and created a zone of participation that rehearsed "the hour of the Trial."

Notes

1 All translations are by the author, unless otherwise noted. The date of the frescoes is contested. Considering the technical similarity to the decoration of the Panagia Phorbiotissa Church at Asinou (1105), it is likely the pupils of the master there carried out Trikomo's program—a point first argued by David Winfield, "Hagios Chrysostomos, Trikomo, Asinou: Byzantine Painters at Work," in *Praktika tou protou diethnous Kyprologikou synedriou, Leukosia, 1969* (Nicosia: Hetaireia Kypriakōn Spoudon, 1972), 285–91; and Annemarie Weyl Carr and Laurence J. Morrocco, *A Byzantine Masterpiece Recovered: The Thirteenth-Century Murals of Lysi, Cyprus* (Austin, TX: University of Texas Press, 1991), 53. Those dates are different from those proposed in Andreas Stylianou and Judith Stylianou, *The Painted Churches of Cyprus: Treasures of Byzantine Art* (London: Trigraph for the A. G. Leventis Foundation, 1985), 486. The fresco is also mentioned in Henry Maguire, *The Icons of Their Bodies: Saints and Their Images in Byzantium* (Princeton, NJ: Princeton University Press, 1996), 116–17. For history and reconstruction of the monument, see Athanasios Papageorghiou, *Christian Art in the Turkish-Occupied Part of Cyprus* (Nicosia: Holy Archbishopric of Cyprus, 2010), 429–41.

2 Rev. 19:15, New International Version: "Coming out of his mouth is a sharp sword with which to strike down the nations. He will rule them with an iron scepter. He treads the winepress of the fury of the wrath of God Almighty."

3 Normalized by Andreas Rhoby, *Byzantinische Epigramme auf Fresken und Mosaiken* (Vienna: Verlag der Österreichischen Akademie der Wissenschaften, 2009), 376–77: "'Ὁ παντεπόπτης ἐξ ἀπ όπ του τοῦ τόπ ου / τοὺς εἰσιόντας π άντ(ας) ἐνθαδε βλέπ ει, / ψυχὰς ἐρευνᾷ καὶ κίνησιν καρδίας. / Βροτὸι, π τοεῖσθε [τὸν] κρ<ι>τ[ὴν τ(ὸν) τῆς δίκ[ης]."

4 A. T. Murray, trans., *The Odyssey with an English Translation* (London: William Heinemann, 1919), 22.297–99: "δὴ τότ' Ἀθηναίη φθισίμβροτον αἰγίδ' ἀνέσχενὺψόθεν ἐξ ὀροφῆς: τῶν δὲ φρένες ἐπ τοίηθεν."

5 Romanos the Melodist, *Sancti Romani melodi cantica: Cantica genuina*, ed. Paul Maas and C. A. Trypanis (Oxford: Clarendon Press, 1963), 266: "φρίττω καὶ πτοοῦμαι ὑπο τῆς συνειδήσεως." Translation from Marjorie Carpenter, *Kontakia of Romanos, Byzantine Melodist* (Columbia, MO: University of Missouri Press, 1972), 372.

6 Rhoby, *Byzantinische Epigramme*, 229–31, no. 135.

7 Rhoby, *Byzantinische Epigramme*, 230.

8 *The Septuagint with Apocrypha*, ed. Lancelot Brenton (Peabody, MA: Hendrickson Publishers, 1986), accessed online, www.ellopos.net/elpenor/greek-texts/septuagint/default.asp: "ὁ δὲ π ανεπ όπ της Κύριος ὁ Θεὸς τοῦ Ἰσραὴλ ἐπ άταξεν αὐτὸν ἀνιάτῳ καὶ ἀοράτῳ πληγῇ."

9 *Clément d'Alexandri: Le pedagogue*, trans. Marguerite Harl and Henri Marrou, Sources chrétiennes, no. 70 (Paris: Éditions du Cerf, 1960), bk. 3, chap. 8. sec. 44, paragraph 1, l. 4: "Οἱ Σοδομῖται ὑπ ὸπ ολλῆς τρυφῆς ἐξοκείλαντες εἰς ἀσέλγειαν, μοιχεύοντες μὲν ἀδεῶς, π ερὶ <δὲ> τὰ παιδικὰ ἐκμανῶς ἐπ τοημένοι, ἐπ εῖδεν αὐτοὺς ὁ παντεπόπτης λόγος, ὃν οὐκ ἔστι λαθεῖν ἀνόσια δρῶντας, οὐδὲ ἐπ ηρέμησεν τῇ ἀσελγείᾳ αὐτῶν ὁ ἄγρυπ νος τῆς ἀνθρωπότητος φύλαξ."

10 William Wilson, *Ante-Nicene Fathers, Vol. 2*, ed. Alexander Roberts, James Donaldson, and A. Cleveland Coxe (Buffalo, NY: Christian Literature Publishing Co., 1885), accessed online, www.newadvent.org/fathers/02093.htm.

11 Bissera Pentcheva has recently explored the complex shades of meaning between *ōpon* (face), *ōps* (eye), and *ops* (voice), especially in relation to divine revelation and Christ's face on the apse mosaic at the Holy Monastery of St. Catherine in Sinai; Pentcheva, "The Aesthetics of Landscape and Icon at Sinai," *Res* 65–66 (2015): 192–209, esp. 198–201.

12 See, for example, the earliest identification of this sort of Christ portrait on an encaustic icon at Sinai in Kurt Weitzmann and John Galey, *The Monastery of Saint Catherine at Mount Sinai, the Icons* (Princeton, NJ: Princeton University Press, 1976), 13–15. The icon at Sinai called "Pantokrator" in fact has no accompanying title other than the *nomen sacrum* "XC IC." Previous studies devoted to this image and the title "Pantokrator" include Jane Timken-Matthews, "Pantocrator: Title and Image" (PhD diss., New York University, 1980); Thomas Mathews, "The Transformation Symbolism in Byzantine Architecture and the Meaning of the Pantokrator in the Dome," in *Church and People in Byzantium: Society for the Promotion of Byzantine Studies, Twentieth Spring Symposium of Byzantine Studies, Manchester, 1986*, ed. Rosemary Morris (Birmingham: Centre for Byzantine, Ottoman and Modern Greek Studies, University of Birmingham, 1990), 200–20; and Salvatore Barbagallo, *Iconografia liturgica del Pantokrator* (Rome: Studia Anselmiana, 1996).

13 Charles Barber, "From Transformation to Desire: Art and Worship after Byzantine Iconoclasm," *Art Bulletin* 75, no. 1 (1993): 7–16, esp. 13.

14 For the distinction as applied to Marian icons and a summary of this tradition of naming within the historiography of Byzantine art, see most recently introduction to Bissera Pentcheva, *Icons and Power: The Mother of God in Byzantium* (University Park, PA: Pennsylvania State University Press, 2006), 3, 75–80, 174–80. Important earlier studies include Gordana Babic, "Il modello e la replica nell'arte bizantina delle icone," *Arte Cristiana* 76 (1988): 61–78; and André Grabar, "Remarques sur l'iconographie byzantine de la Vierge," *Cahiers Archéologiques 26 (1977): 169–78*

15 The view of Byzantine monumental art as spatially immersive, uniquely dimensional, and perspectival begins with Otto Demus, *Byzantine Mosaic Decoration: Aspects of Monumental Art in Byzantium* (Boston, MA: Boston Book & Art Shop, 1955). For further studies of Byzantine works and their phenomenal dynamism, see Rico Franses, "When All That Is Gold Does Not Glitter: On the Strange History of Looking at Byzantine Art," in *Icon and Word: The Power of Images in Byzantium: Studies Presented in Honor of Robin Cormack*, ed. Anthony Eastmond and Liz James (Aldershot: Ashgate, 2003), 13–24. Bissera Pentcheva, "Hagia Sophia and Multisensory Aesthetics," *Gesta* 50, no. 2 (2011): 93–111; and Pentcheva, *The Sensual Icon: Space, Ritual, and the Senses in Byzantium* (University Park, PA: Pennsylvania State University Press, 2010), 121–54.

16 Two important collections have emerged regarding the place of fear in the history of emotions, now the subject of complex cross-disciplinary scholarly inquiry. Michael Laffan and Max Weiss, eds., *Facing Fear: The History of an Emotion in Global Perspective* (Princeton, NJ: Princeton University Press, 2012); and Jan Plamper and Benjamin Lazier, *Fear across the Disciplines* (Pittsburgh, PA: Pittsburgh University Press, 2012).

17 Glanville Downey, "Nikolaos Mesarites: Description of the Church of the Holy Apostles at Constantinople," *Transactions of the American Philosophical Society* 47, no. 6 (1957): 855–924, esp. 901, sect. XIV: "Αὔτη ἡ σφαῖρα ὡς ἐξ οὐρανίας ἄντυγος τῆς ταύτης ἀρχῆς π ρὸς τὸ τοῦ ναοῦ ἔδαφος καὶ π άντα τὰ ἐν αὐτῷ π αρακύπ τοντα τὸν θεάνθρωπ ον ἡμῖν εἴκονικῶς ὑπ οδεικνύει Χριστόν."

18 Downey, "Nikolaos Mesarites," 870. This tradition of associating domes with the cosmos is vast. See also Karl Lehmann, "The Dome of Heaven," *Art Bulletin* 27 (1945): 1–27; and Thomas F. Mathews, "Cracks in Lehmann's 'Dome of Heaven,'" *Source: Notes in the History of Art* 1, no. 3 (1982): 12–16.

19 Downey, "Nikolaos Mesarites," 901, sect. XIV, with modifications: "οἱ ὀφθαλμοὶ τοῖς ἀκατάγνωστον κεκτημένοις τὸ συνειδὸς ἵλαροὶ καὶ εὐπρόσιτοι καὶ γλυκασμὸν κατανύξεως ἐνσταλάττοντες ταῖς ψυχαῖς τῶν καθαρῶν τῇ καρδίᾳ καὶ π τωχευόντων τῷ π νεύματι. ὀφθαλμοὶ γὰρ κυρίου ἐπὶ δικαίους κατὰ τὸν ψάλλοντα. π ρᾶόν τε τὸ βλέμμα καὶ τὸ ὃλον προσήνεια, οὐκ ἐπ αρίστερα βλέπ ον οὐδ᾽ὑπ ερδέξια, ὁλικῶς δε ἅμα π ρὸς ἅπ αντας καὶ π ρὸς τὸν καθέκαστον μερικῶς. οἷς μὲν οὖν τὸ συνειδός ἐστι ἀκατάγνωστον, τοιοῦτοι οἱ ὀφθαλμοί, οἷς δ᾽αὐτοκατάκριτον τὸ οἰκεῖον κριτήριον, ὀργίλοι δυσπρόσιτοί τινες καὶ δυσάντητοι, ἀπηγριωμένον τὸ πρόσωπον, ἐκφοβοῦν, ἰταμὸν καὶ σκληρίας ὑπ όπ λεω. πρόσωπον γὰρ κυρίου τοῦτον ἓχον τὸν τρόπον ἐπὶ ποιοῦντας κακά."

20 For a discussion of the Pantokrator in relation to concepts of divine justice, see Maguire, *The Icons of Their Bodies*, 116–17.

21 See also Henry Maguire, "From the Evil Eye to the Eye of Justice," in *Law and Society in Byzantium: Ninth–Twelfth Centuries*, ed. Angeliki E. Laiou and Dieter Simon (Cambridge, MA: Harvard University Press, 1994), 217–40.

22 Theodore the Stoudite, *Theodorae Studitae epistulae*, ed. G. Fatouros (Berlin: De Gruyter, 1992), 2:544: "ἕως π ότε;' . . . οἶδεν ὁ ἀκοίμητος ὀφθαλμὸς τὸ ἀρκοῦν ἐπ 'ἀμφοτέροις."

23 Manolis Chatzidakis and Gerry Walters, "An Encaustic Icon of Christ at Sinai," *Art Bulletin* 49, no. 3 (1967): 197–208; Weitzmann and Galey, *The Monastery of Saint Catherine at Mount Sinai: The Icons*, 13–15; and Hans Belting, *Likeness and Presence: A History of the Image before the Era of Art*, trans. Edmund Jephcott (Chicago, IL: University of Chicago Press, 1994), 133.

24 Romanos the Melodist, *Sancti Romani melodi cantica*, 266: "φρίττω καὶ π τοοῦμαι ὑπ ὸ τῆς συνειδήσεως." See also Derek Krueger, "Romanos the Melodist and the Formation of the Christian Self in Early Byzantium," in *Proceedings of the 21st International Congress of Byzantine Studies, London, 2006: Plenary Papers*, ed. Elizabeth Jeffreys (Aldershot: Ashgate, 2006), 1:247–66.

25 That Paul was writing in the second half of the tenth century is a secure fact: excavations at Corinth have uncovered a tenth-century seal with the inscription "Lord, help your servant / Paul by the mercy of God Bishop of Monembasia." Further, Paul is mentioned in the *Life of Saint Paul the Lesser of Latros*, a hagiography written about 969. Elsewhere, Paul mentions the imperial reign contemporary to his writings. Paul opens "The Three Women Who Were Discovered in the Time of Constantine" by stating it takes place "in the reign of the Emperor Constantine, son of Leo and of Zoê, son-in-law of Romanus the Elder." Therefore, it is correct to place the tales roughly within the reign of Constantine VII Porphyrogennetos, and two centuries before the construction of the Panagia Theotokos at Trikomo. John Wortley, introduction to *The Spiritually Edifying Tales of Paul, Bishop of Monembasia* (Collegeville, MN: Cistercian Publications, 1996), 17–60.

26 John Wortley, trans., *The Spiritually Edifying Tales of Paul, Bishop of Monembasia* (Collegeville, MN: Cistercian Publications, 1996), 101–02, with my modifications. For the original Greek, see the critical edition, John Wortley, *Les récits édifiants de Paul, évêque de Monembasie, et d'autres auteurs* (Paris: Éditions du Centre National de la Recherche Scientifique, 1996), 92–95: "Ἀνήρ τις τῶν ἐμφανῶν εἶχε π ίστιν π ολλὴν εἰς τὸν ἅγιον τοῦ Χριστοῦ ἱεράρχην Ἰωάννην τὸν Χρυσόστομον. καὶ ἀσθενείᾳ π εριπ εσὼν π ροσέταξε τοὺς αὐτοῦ οἰκέτας βαστάσαι αὐτὸν καὶ ἀπ ενέγκαι ἐν τῷ σεβασμίῳ ναῷ τῶν ἁγίων Ἀπ οστόλων ἔνθα ἐστὶ καὶ ὁ ἰαματοφόρος τάφος τοῦ θεοφόρου π ατρὸς Χρυσοστόμου. ἐλθόντος οὖν αὐτοῦ εἰς τὸν τοιοῦτον σεβάσμιον ναὸν καὶ εὐξαμένου μετὰ δακρύων, ἀνεκλίθη ἐν τῇ ἑτοιμασθείσῃ αὐτῷ χαμαιστρώτῳ εὐνῇ. ἀνακειμένου οὖν αὐτοῦ καὶ ὀδυνωμένου καὶ ἐπ ικαλουμένου τοὺς ἁγίους τοῦ Χριστοῦ ἀπ οστόλους καὶ τὸν ἅγιον π ατέρα ἡμῶν Ἰωάννην τὸν Χρυσόστομον ἐλεῆσαι αὐτὸν κινδυνεύοντα, ἦλθεν εἰς ἔννοιαν τῶν π επ ραγμένων αὐτῷ ἀτόπ ων π ράξεων. ὡς οὖν ταύτας διελογίζετο, θρηνῶν ἔλεγεν. «οἴμοι τῷ ἀθλίῳ καὶ ἀμετανοήτῳ. πῶς πορεύσομαι ὁδὸν ἐξ ἧς οὐχ ὑπ οστρέψω, καὶ π ῶς ὑπ ενέγκω τὴν ἀπ ειλὴν τοῦ φοβεροῦ κριτοῦ καὶ τὰς αἰωνιζούσας καὶ ἀφορήτους βασάνους;» ταῦτα αὐτοῦ λέγοντος καὶ πλείονα τούτων καὶ ἀπ οδυπ ομένου, ὡς π ερὶ ὥραν ἕκτην, πάντων ὑπ οχωρησάντων ἐκ τοῦ αὐτοῦ σεβασμίου οἴκου, κατελείφθη μόνος καὶ ἀτενίσας π ρὸς τὸν τροῦλον τοῦ αὐτοῦ ἁγίου ναοῦ καὶ θεασάμενος τὴν τοῦ Κυρίου ἡμῶν Ἰησοῦ Χριστοῦ εἰκόνα, ἤρξατο π ρὸς αὐτὴν λέγειν τοιαῦτα· «εἰ τὴν σὴν εἰκόνα, δέσπ οτα, τὴν διὰ χειρῶν ἀνθρώπ ων γεγονυῖαν θεωρῶν, φρίττω καὶ δέδοικα, π ῶς σε τὸν φοβερὸν κριτὴν ὁ τάλας θεάσομαι, ὅταν ἔλθῃς κρῖναι τὰ σύμπ αντα; ἥμαρτον, δέσπ οτα. συγχώρησόν με τὸν μὴ φυλάξαντα τὰ σὰ π ροστάγματα.» ταῦτα αὐτοῦ λέγοντος, ἤρξατο ἐξαγγέλλειν τὰς ἑαυτοῦ ἁμαρτίας π ρὸς τὴν ἁγίαν εἰκόνα. καὶ ἐξειπ ὼν πάντας, ἐνθυμηθεὶς καὶ ἄλλην χαλεπ ὴν ἁμαρτίαν ἣν ἐπ οίησα, ἤρξατο λέγειν: «καὶ ἄλλην ἁμαρτίαν ἐπ οίησα δέσπ οτα, ἀλλ'οὐ τολμῶ ταύτην ἐξειπ εῖν. οὐ τολμῶ φιλάνθρωπε, οὐ τολμῶ ἐλεήμων.» ταῦτα οὖν αὐτοῦ λέγοντος, ἦλθεν αὐτῷ φωνὴ ἐκ τῆς ἁγίας εἰκόνος: «εἰπ έ» ὡς οὖν ταύτην ἐξεῖπ εν, ἦλθεν ἑτέρα φωνή: «ἀφέωνταί σου αἱ ἁμαρτίαι» ταύτης οὖν τῆς φοβερᾶς φωνῆς ἐνεχθείσης, ἀνέστη ὑγιὴς καὶ π ροσπ εσὼν ἐπὶ πρόσωπον, μεγάλῃ τῇ φωνῇ εὐχαριστηρίους ὕμνους ἀνέπ εμπ ε. . . ."

27 Maas and Trypanis, Romanos the Melodist, *Sancti Romani melodi cantica*, 273: "δικαία ἡ κρίσις σου καὶ ἀπροσωπόληπτος, κριτὰ δικαιότατε."

28 "εἰ τὴν σὴν εἰκόνα, δέσπ οτα, τὴν διὰ χειρῶν ἀνθρώπων γεγονυῖαν θεωρῶν, φρίττω καὶ δέδοικα, πῶς σε τὸν φοβερὸν κριτὴν ὁ τάλας θεάσομαι, ὅταν ἔλθῃς κρῖναι τὰ σύμπ αντα."

29 The relationship between Byzantine art and emotion was first charted by Henry Maguire in his landmark study *Art and Eloquence in Byzantium* (Princeton, NJ: Princeton University Press, 1981). Other studies of fear in the broader Middle Ages include Anne Scott and Cynthia Kosso, eds., *Fear and Its Representations in the Middle Ages and the Renaissance* (Turnhout: Brepols Publishers, 2002). For a philosophical exposition on the nature of fear as a passion, see bk. 2, chap. 15 in John of Damascus's *Expositio fidei*, trans. Rev. S. D. F. Salomond, www.voskrese.info/spl/Exact.html.

30 "καθήλωσον ἐκ τοῦ φόβου σου τὰς σάρκας μου· ἀπ ὁ γὰρ τῶν κριμάτων σου ἐφοβήθην."

31 The penitential act, as a mystical state of arousal in Byzantium, deserves more investigation, especially as it relates to images. A productive source for studying Western medieval penitential practice as both a performance and a phenomenological, even erotic, engagement with one's body and God may be found in Niklaus Largier, *In Praise of the Whip: A Cultural History of Arousal* (New York: Zone Books, 2007).

32 For thorough analysis of this monument, see Arthur H. S. Megaw and Ernest J. W. Hawkins, "The Church of the Holy Apostles at Perachorio, Cyprus, and Its Frescoes," *Dumbarton Oaks Papers* 16 (1962): 277–348.

33 "ψυχὰς ἐρευνᾷ καὶ κίνησιν καρδίας."

34 Roger Caillois, "Mimicry and Legendary Psychasthenia," trans. John Shepley, *October* 31 (Winter 1984): 16–32, at 30 (emphasis added).

35 I thank Annemarie Weyl Carr for generously sharing her forthcoming article, "Hell in the Sweet Land: Hell's Place in in the Last Judgments of Byzantine and Medieval Cyprus," in *The Damned in Hell in the Frescoes of Venetian-dominated Crete (13th–17th Centuries)*, ed. Angeliki Lymberopoulou and Vasiliki Tsamakda (forthcoming), wherein she collects and examines, meticulously, all twenty-four surviving Cypriot examples of the Last Judgment in painting. She locates the understudied narthex scenes at Hagios Nikolaos tis Stegis as the earliest surviving example of the Last Judgment as a scene.

36 See recently the discussion of sight and hearing in synesthetic experiences, especially of the terrifying sound of Moses, in Pentcheva, "The Aesthetics of Landscape and Icon at Sinai," 198.

37 On this understudied church, see Ionna Christoforaki, "Cyprus between Byzantium and the Levant: Eclecticism and Interchange in the Cycle of the Life of the Virgin in the Church of the Holy Cross at Pelendri," in *Epeteris Kentrou Epistimonikon Erevnon*, Annual of the Cyprus Research Centre, Nicosia: Cyprus Research Centre 22 (Nicosia: Cyprus Research Centre, 1996), 215–45; Tania Velmans, "Quelques programmes iconographiques de coupoles chypriotes du XII au XVe siècle," *Cahiers Archéologiques 32 (1984):* 140–42; and Athanasios Papageorghiou, "The Paintings in the Dome of the Church of the Panagia Chryseleousa, Strovolos," in *Medieval Cyprus: Studies in Art, Architecture and History in Memory of Doula Mouriki*, ed. C. Moss and Nancy P. Ševčenko (Princeton, NJ: Princeton U Press, 1999), 147–61, esp. 150–51.

38 Normalized by Velmans, 140, trans. in Papageorghiou, *Christian Art in the Turkish-Occupied Part of Cyprus*, 150: "Ἐγὼ κριτὴς τε καὶ Θ[εὸς] π άντων π έλω. . . . οχοίψας ὑψώθεν π ρὸ τῆς δίκης π αρεγκιῶμαι τοὺς ἐμους τυρεῖν νόμους ὡς τις θέλει τῶν ἐκφυγεῖν τὰς βασσάνους."

39 See Leo VI, "*Ekphrasis* of the Church in the Monastery of Kauleas," in Cyril Mango, *The Art of the Byzantine Empire: Sources and Documents, 312–1453* (Toronto: University of Toronto Press, 1986), 202–03: "τῷ δ'ἄνωθεν ἐπ ηρμένῳ εἰς ἡμικύκλιον σχῆμα, ὃ καὶ ὀροφὴν πληροῖ τοῦ καλοῦ ἐδάφους, εἰκὼν κατὰ μέσης τετύπ ωται, ᾧ τὸν οἶκον ἀφιέρωσεν ὁ τεχνίτης· εἴπ οις ἂν οὐ τέχνης ἔργον ὁρᾶν, ἀλλ'αὐτὸν ἐκεῖνον τὸν ἐν φύσει τῇ βροτῶν ὀφθέντα τοῦ π αντὸς ἐπ όπ την καὶ κυβερνήτην, ἄρτι π επ αυμένον τοῦ π αραινεῖν, καὶ τοῖς χείλεσιν ἐπ ιτεθεικότα σιγήν." Greek in *Leontos tou Sophou: Panygērikoi logoi*, ed. Archimandrite Akakios (Athens: Nikolaou Rousopoulos, 1868), 245.

40 Papageorghiou, "The Paintings in the Dome of the Church of the Panagia Chryseleousa, Strovolos," 147–61. The site was rearranged in 1817, then again in 1952; it stands presently as a vaulted basilica. In 1991 the Department of Antiquities conserved the wall paintings, uncovering the domical image.

41 See also Pentcheva, "The Aesthetics of Landscape and Icon at Sinai," 197–98.

42 Juan Mateos, ed., *Le Typicon de la Grande Église: Ms. Saint-Croix n. 40, Xe siècle; Introduction, texte critique, traduction et notes*, 2 vols. (Rome: Pont. Institutum Orientalium Studiorum, 1963), 1:2–3. Archimandrite Job Getcha, *The Typikon Decoded: An Explanation of Byzantine Liturgical Practice*, trans. Paul Meyendorff (Yonkers, NY: St. Vladimir's Seminary Press, 2012), 150; and

Derek Krueger, *Liturgical Subjects: Christian Ritual, Biblical Narrative and the Formation of the Self in Byzantium* (Philadelphia, PA: University of Pennsylvania Press, 2014), 181.

43 Marjorie Carpenter, trans., *Kontakia of Romanos, Byzantine Melodist* (Columbia, MO: University of Missouri Press, 1972), 373.

44 Carpenter, *Kontakia of Romanos*, 373.

45 Carpenter, *Kontakia of Romanos*, 374–75.

46 Carpenter, *Kontakia of Romanos*, 380.

47 Carpenter, introduction to Carpenter, *Kontakia of Romanos*, xiii–xxxix.

48 Getcha, *The Typikon Decoded*, 35.

49 Getcha, *The Typikon Decoded*, 37.

50 Getcha, *The Typikon Decoded*, 36.

51 My deep gratitude goes to Derek Krueger for his generous conversation regarding the relationship between the Triodion and this Christ image. He clarifies the production and cultural context of the Stoudite reform of the Triodion and offers a deeply insightful reading of their texts, with special sensitivity to poetics, affect, and the formulation of subjectivity. Present in his analysis, too, are several excellent points regarding the image of Christ in the dome; Krueger, *Liturgical Subjects*, 164–97, 215–22.

52 Krueger, *Liturgical Subjects*, 180.

53 Krueger, *Liturgical Subjects*, 181.

54 Krueger, *Liturgical Subjects*, 179.

55 Krueger, *Liturgical Subjects*, 172.

56 Little work has been done considering the relation between the Stoudite liturgical reform and Iconoclasm. For Byzantine Iconoclasm, see Leslie Brubaker and John Haldon, *Byzantium in the Iconoclast Era, c. 680–850: A History* (Cambridge: Cambridge University Press, 2011). For Theodore the Stoudite's life, writing, and impact on Byzantine spirituality, see Roman Cholij, *Theodore the Stoudite: The Ordering of Holiness*, Oxford Theological Monographs (Oxford: Oxford University Press, 2002).

57 Krueger, *Liturgical Subjects*, 180–86.

58 See Krueger, *Liturgical Subjects*, 180–86.

59 *Triōdion katanyktikon: Periechon apasan tēn anēkousan autō akolouthian tēs hagias kai megalēs tessarakostēs* (Rome: n.p., 1879), 34: "Τὴν ἡμέραν τὴν φρικτήν, τῆς π αναρρήτου σου π αρουσίας φρίττω ἐννοῶν, δεδοικὼς προορῶ, ἐν ᾗ προκαθίσεις κρῖναι ζῶντας καὶ νεκρούς, Θεέ μου Παντοδύναμε. / Ὅτε ἥξεις, ὁ Θεὸς, ἐν μυριάσι καὶ χιλιάσι, τῶν ἀγγελικῶν, οὐρανίων Ἀρχῶν, κάμὲ ἐν νεφέλαις, ὑπ αντῆσαί σοι Χριστὲ, τὸν ἄθλιον ἀξίωσον. / Δεῦρο λάβε μοι ψυχή, αὐτὴν τὴν ὥραν καὶ τήν ἡμέραν, ὅτε ὁ Θεὸς ἐμφανῶς ἐπ ιστῇ καὶ θρήνησον, κλαῦσον, εὑρεθῆναι καθαρά, ἐν ὥρᾳ τῆς ἐτάσεως. / Ἐξιστᾷ με καὶ φοβεῖ, τὸ π ῦρ τὸ ἄσβεστον τῆς γεέννης, σκώληξ ὁ π ικρός, τῶν ὀδόντων βρυγμός, ἀλλ ἄνες μοι ἄφες, καὶ τῇ στάσει με Χριστέ, τῶν ἐκλεκτῶν σου σύνταξον. / Τῆς εὐκταίας σου φωνῆς, τῆς τοὺς Ἁγίους σου προσκαλούσης, ἐπ ὶ τὴν χαράν, εἰσάκούσω καγὼ, ὁ τάλας καὶ εὕρω, βασιλείας οὐρανῶν, τὴν ἄρρητον ἀπ όλαυσιν.
Μὴ εἰσέλθῃς μέτ' ἐμοῦ, εἰς κρίσεν, φέρων μου τὰ π ρακτέα, λόγους ἐκζητῶν, καὶ εὐθύνων ὁρμάς· ἀλλ' ἐν οἰκτιρμοῖς σου, π αρορῶν μου τὰ δεινά, σῶσόν με, παντοδύναμε."

60 Mother Mary and Kallistos Diokleia translate it "call to mind." Mother Mary and Kallistos Ware, trans., *The Lenten Triodion* (London: Faber & Faber, 1984), 153.

61 The word appears in the second canticle of the Triodion's "On the Second Coming": "φοβερὸν γὰρ τὸ κριτήριον, ἐν ᾧ π άντες, τετραχηλισμένοι στησόμεθα." In Romanos's *kontakion*: "τὸν ἀγῶνα τὸν ἴδιον ἕκαστος δίκαιος ἐπ ιδείξει γηθόμενος,/ ὅτε τὰ ἔργα γεγυμνωμένα καὶ τεταχηλισμένα/ φανεροῦνται ἐναντίον τοῦ κριτοῦ καὶ βασιλέως." (Each righteous man, rejoicing, will conduct his own trial, When his deeds have been bared and exposed to view, in the presence of the Judge and King); Carpenter, *Kontakia of Romanos*, 378.

62 Carpenter, *Kontakia of Romanos*, 378.

63 "ὦ παντάνασσα καὶ π ά[ντων ὑπ ερτέρα: / δέσπ οινα ἁγνὴ καὶ μῆτερ τοῦ κυρίου: / ἴδε τὸν π όθον τῆς ταλαίν[η]ς ψυχῆς μου: / καὶ γενοῦ μοι μεσῖτις ἐν ὥρα δίκης / ὅπ ως ἐκφύγω μέρος τῆς . . . [ᾼδη]." Here, I modify Papageorghiou's translation in *Christian Art in the Turkish-Occupied Part of Cyprus*, 431. He places "κολάσεως" (punishment) as the last word; however, I argue that it should be Hades, "ᾼδη," whose two syllables preserve the meter. Nancy Ševčenko used this epigram to normalize the nearly exact same poem that appears on the apse at Asinou. She suggests "hell" as well; Nancy Ševčenko, "The Metrical Inscriptions in the Murals of the Panagia

Phorbiotissa," in *Asinou across Time: Studies in the Architecture and Murals of the Panagia Phorbiotissa Cyprus*, Dumbarton Oaks Studies XLIII, ed. Annemarie Weyl Carr and Andreas Nicolaïdès (Washington, DC: Dumbarton Oaks, 2012), 69–92, esp. 77.

64 Ševčenko ("The Metrical Inscriptions," 79) uses the Trikomo inscription to normalize the fragments in Asinou's apse. Here, I provide a slightly modified version of Ševčenko's reconstruction of Nikephoros Ischyrios's epigram: "Πολλοῖς τεθηλὼς ἀγαθοῖς ἐν τῷ Βίῳ / ὧνπ ερ χορηγὸς ὡράθης σὺ, π αρθένε / Νικηφόρος μάγιστρος οἰκτρὸς ἱκέτης / ἤγειρα τόνδε τὸν ναὸν μετὰ πόθου / ἀνθ᾽οὗπεραιτῶ προστάτιν εὑρηκέναι / ἐν ἡμέρᾳ σε τῇ φρικώδει τῆς δίκης."

65 This sense of a building's ontology as an offering in exchange for salvation has been explored somewhat in Natalia Teteriatnikov, "Private Salvation Programs and Their Effect on Byzantine Church Decoration," *Arte Medievale* 7 (1993): 47–63; and Ann Wharton Epstein, "Formulas for Salvation: A Comparison of Two Byzantine Monasteries and Their Founders," *Church History* 50, no. 4 (1981): 385–400.

66 "Ἴδε τὸν π όθον τῆς ταλαίν[η]ς ψυχῆς μου."

67 Bissera Pentcheva, "Epigrams on Icons," in *Art and Text in Byzantine Culture*, ed. Liz James (Cambridge: Cambridge University Press, 2007), 120–38.

68 Maas and Trypanis, Romanos the Melodist, *Sancti Romani melodi cantica*, 380: "καὶ μὴ ἀπ ορρίψῃς με ἀπ ὸ τοῦ π ροσώπ ου σου, κριτὰ δικαιότατε."

6 Transfigured

Mosaic and liturgy at Nea Moni

Lora Webb

The Transfiguration mosaic in the southwest corner of the eleventh-century *katholikon*, or main church, of the monastery of Nea Moni, on Chios, Greece, invites its viewer to Mount Tabor (Figure 6.1). In an area of striking blue, the figure of Christ curves in and up toward the apex of the space, following the shape of the conch in which the tesserae were set. His divinity is made visible to his apostles, though the damage of centuries has clouded his face to us. To either side of Christ we see the Old Testament prophets Elijah and Moses, while at his feet Peter, John, and James gesture toward

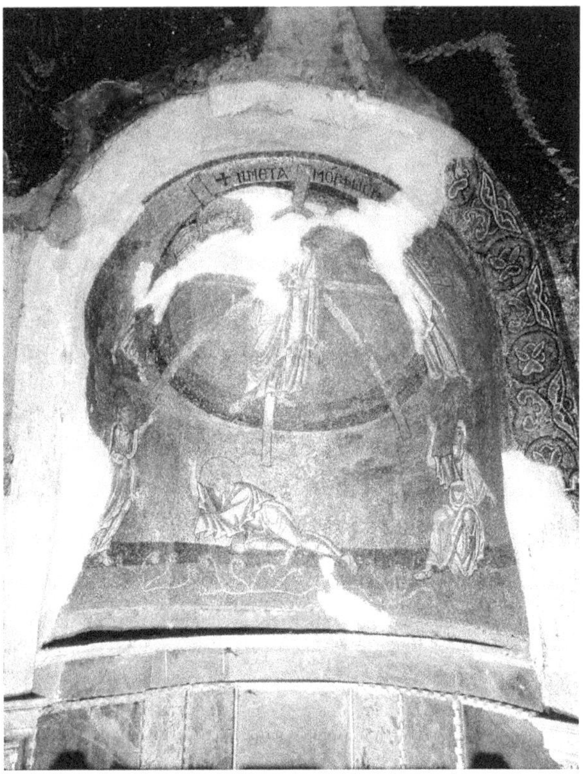

Figure 6.1 Transfiguration, Nea Moni, Chios, eleventh century.

From Doula Mouriki, *The Mosaics of Nea Moni on Chios*, 2 Vols. (Athens: Commercial Bank of Greece, 1985, 2: pl. 24) © Photographic Archive of Alpha Bank Publication.

the miracle taking place above their heads. The forms of the three apostles follow the outline of the blue circle, which reads like a void in the golden field of mosaic, adding greater depth to the already curved image. We could read the large blue mandorla as a receding space—like a portal opening up to the heavens—and it appears that Christ is pulled upward through this opening. However, at the same time as the mosaic opens up and away, the golden beams crisscrossing the blue catch Christ and push him forward. We might, then, see Christ in front of the blue and coming toward us. The layering of the pictorial elements makes it unclear whether Christ is receding or advancing. At one moment he seems to be floating up into the void and at another he appears to come out and into the space of the church; the mind toggles back and forth between the two. Thanks to the arrangement of the shimmering mosaic, Christ looks as he must have to the apostles on the mountain—at once stunningly, unspeakably divine but also close to the earth and nearly tangible.

The walls of Nea Moni were richly decorated with mosaic scenes of Christ's life, saints, and angels. The Transfiguration comes about halfway through the life of Christ cycle that adorns the eight conches of the naos.[1] These conches are the lowest level of mosaic decoration, placed just above the marble-covered walls where the architecture starts to round, turning inward and upward toward the now-reconstructed dome. Though the mosaics are studied in their own right as dazzling and rare examples of eleventh-century Byzantine monumental art, they were once only one part of a vibrant monastic community. In the main body of the church, the mosaics were originally experienced as a part of the daily offices. They are the sole tangible survivors of the ephemeral ritual life of the monastery. Just how the Transfiguration and the liturgy worked together is the subject of this paper.

In an attempt to grasp the vital experience of Nea Moni, I focus on one mosaic—the Transfiguration—and one liturgical day—August 6, the Feast of the Transfiguration. In doing so, I recognize that I am ignoring both the entirety of the decorative scheme and the whole of the liturgical year, but by limiting my focus, I am able to begin to fill in the sketch of the past left in the liturgical stage directions and the church building. Further, beginning with one image and one day exposes how the liturgy could highlight different parts of the church's decorative scheme throughout the year.[2] This paper, then, oscillates between two gravitational centers: one based in what is visible within the image, and one that addresses the rituals that enlivened its space. I argue that both require the subjective participation of the faithful, encouraging them—through eye and ear—to witness the event of the Transfiguration.

Closely reading both text and mosaic grants us a better idea of how the Transfiguration reached into the monks' world. Indeed, the *katholikon* of Nea Moni has already been rich ground for this type of comparison. William Tronzo and Charles Barber have written about the relationship among liturgical texts, ritual, and image, focusing on the Washing of the Feet mosaic in the inner narthex (Figure 6.2).[3] Tronzo's essay connects mosaic and liturgy, arguing that the ritual action of the hegumen washing the feet of the monks on Holy Thursday gave rise to the choice of subject matter of the mosaic. More specifically, Tronzo shows how the upper register uses Gospel quotations and depicts their message in representing Christ taking the towel and girding himself with it, and then standing before the basin of water. The same Gospel quotations are repeated in the liturgical texts specifying how the hegumen should perform this ritual mimesis of reenacting Christ's washing the feet of his disciples.[4] While Tronzo uses the liturgical texts to establish the origins of the iconography selected, he also uncovers how the mosaic extends beyond the pictorial frame into the rituals celebrated in the physical space in front of the image.[5] Barber has challenged Tronzo's claim.[6] For Barber, liturgical mimesis can blur the border between representation and prototype, making both share the same ontology. He quotes a poetic inscription surrounding the image of Christ in the tympanum of the narthex of the same Nea Moni, written by Theodore the Stoudite (759–826), that denies the right of the image to share essence with the prototype. By placing this ninth-century text next to the eleventh-century mosaic, Barber argues that the Byzantine viewers of the image of Christ were asked to venerate it relatively, that is, to recognize that the image does not partake of the essence of the prototype.[7] Yet, does Theodore's epigram offer us a key to the representational strategies of all mosaics within this church and Byzantium in general? The poem frames a nonnarrative image of Christ, whose connection to a specific liturgical action is therefore indirect. By contrast, the narrative representation of the Washing of the Feet and the close link between the Gospel passages quoted in the mosaic and their use in the liturgical mimesis of the ritual argue for a much stronger connection linking pictorial representation to its temporal expansion into the real space of ritual action. It is to this synergy between image and liturgical space that this paper turns, focusing on the Transfiguration and its liturgical celebration.

While the Cheroubikon hymn and Byzantine mystagogical texts give further evidence of this liturgical mimesis when they state that the priests and clergy processing along

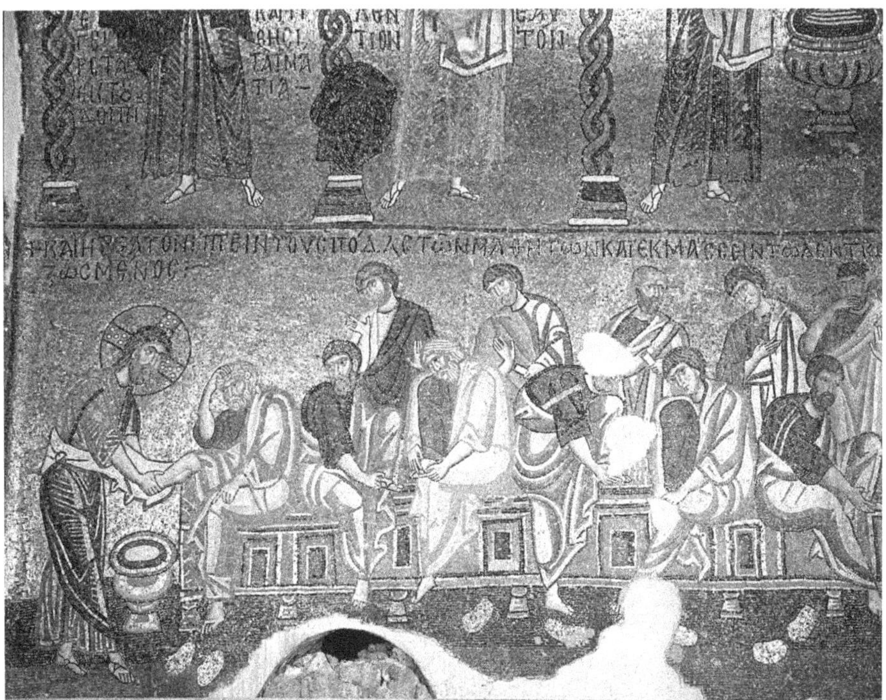

Figure 6.2 The Washing of the Feet, Nea Moni, Chios, eleventh century.

From Doula Mouriki, *The Mosaics of Nea Moni on Chios*, 2 vols. (Athens: Commercial Bank of Greece, 1985), 2: pl. 94 © Photographic Archive of Alpha Bank Publications.

the nave during the Great Entrance "image" in their movements the angels officiating around the throne of God in heaven, it is more difficult to determine exactly how the Byzantines perceived this ritual imitation by means of which terrestrial action enfigured its celestial models.[8] However, how the Transfiguration mosaic took part in the liturgical drama is even harder to pin down. In fact, it is impossible to know exactly how the monks of Nea Moni would have conceptualized image and ritual as a whole. But tracing the course of the Feast of the Transfiguration as it is described in a typikon—a document outlining the offices for a particular foundation—helps us to discover what was sung, read, and chanted in the company of the mosaic. Because no typikon survives from Nea Moni, I turn to the **Synaxarion** of Theotokos Evergetis, one of the most influential monastic rites of the eleventh century, to discover what was likely read and performed during the Feast of the Transfiguration.[9]

Before working through the liturgical events of August 6, however, it is crucial to understand more fully how the image of the Transfiguration plays with a viewer's conception of space. The mosaic form of Christ toggles between ascent and descent as he curves around and over the area directly in front of him. Christ's advance into the space of the real stands in stark contrast to the effect of illusionistic space constructed with perspective and foreshortening. In the case of the Transfiguration mosaic, terms such as "external coherence" and "negative (or reverse) perspective" are apt descriptors.

"External coherence" was coined by Alois Riegl to describe the effect of northern Renaissance portrait groups. In a painting such as Rembrandt's *Sampling Officials of*

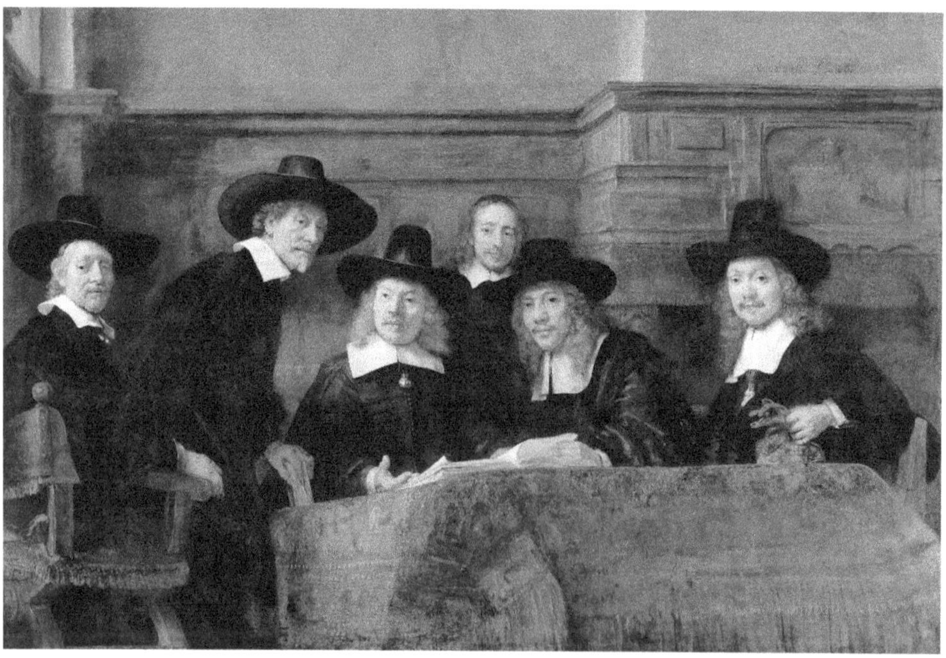

Figure 6.3 Rembrandt, *Sampling Officials of the Draper's Guild*, 1662, Rijksmuseum, Amsterdam.

the Draper's Guild (Figure 6.3), which Riegl considered to be exemplary in its handling of the concerns of group portraiture, the sitters remain individuals while also being united by their common business venture. Their gaze out at the viewer fuses our space in front of the canvas with theirs. Riegl called this interaction between the pictorial space and the real "external coherence," signifying how the pictorial projects into the immanent space of the beholder. The viewer further reads the sitter as conscious of an outside viewer, investing the painted figures with a liveliness or presence.[10] Riegl's external coherence is established mainly through the gazes of the individuals within the painting, and it is a subjective and psychological reaction to the regard of the subjects, which draws them closer to the viewer.

The Transfiguration functions slightly differently from the figures of the *Draper's Guild.* It has only one figure, who looks out toward the viewer—Christ—and the remaining figures attend to this central vision (Figure 6.1). Christ's face has been damaged, but from other versions of this event, such as the Transfiguration at the Monastery of St. Catherine, Mount Sinai, discussed below (Figure 6.4), we can imagine that he originally stared out at the space in front of him. Christ here is the visual and psychological touchpoint for the viewer, but the pictorial elements of the mosaic also unite the space of the viewer with that of the image. External coherence as defined by Riegl accounts for part but not all of the effect.

Otto Demus also discusses how figures in works of art engage the viewer with their gaze, but his subject was Middle Byzantine mosaics.[11] Like the group portrait, icons confront the beholder who stands in the physical space in front of them. The gazes of the saints are not directed only at the viewer, however, and Demus further notes that figures in mosaic programs look across domes and around pendentives at each other, enfolding the physical space of the church in the pictorial space.[12] More recently, Robert Nelson has taken up the idea of external coherence, arguing that the Byzantine beholder took part in an active and subjective viewing process through which Byzantine mosaics "lean through that '[pictorial] window'" and become a "part of the religious drama enacted below."[13]

External coherence establishes a sense of presence within the image, but I would argue that the mosaic constantly flips between Christ coming into the world and the viewer being pulled up into the celestial orbit. Where Riegl claims that the subjective effect of the painting is brought about only through the gaze of the sitter looking out, Demus extends the effect on the viewer to what he terms "negative perspective." "Byzantine 'perspective,'" he explains, "might be described as 'negative' perspective. It takes into account the space which surrounds, and is enclosed by, the image and which intervenes between the image and the beholder; and it aims at eliminating the perspective effects on the beholder's vision."[14] In other words, "negative perspective" does not develop recession and depth of pictorial space but, rather, expands into and engages the real, physical space in front of the image. The mosaics have a further uplifting effect in that the figures are arranged to look proportionally correct in the eyes of the viewer standing below:

> The beholder of Byzantine decorations, who sees the image undistorted in spite of the great height at which it appears, feels lifted up to its level, high above the ground. . . . In this way, not only spiritually but optically also, the Byzantine beholder was received into the heavenly sphere of the holiest icons; he participated in the sacred events.[15]

Though the icons, like Riegl's sitters, look out at the viewer, Demus points to the role optical effects can play in creating a unity between the viewer and the image; external coherence allows for the figure to seem to recognize the space in front of it, but Demus's "negative perspective" allows for the viewer to enter the space of the depicted figure.

The Transfiguration utilizes both external coherence and negative perspective at once. The rays within the blue mandorla behind Christ alternate between seeming to protrude and recede. Like toggling between options on a computer screen, the eye switches back and forth between the two. The ambiguous sense of space around the figure of Christ would be crucial in a church environment, where the Holy Spirit was called to descend into the Eucharistic bread and wine and thus enable those who partake in Christ to be lifted up, however fleetingly, to heavenly heights.

The crossing rays behind Christ simultaneously encroach on the space of the viewer and draw him up into the mosaic, but the landscape of the image also invites the viewer into the scene. Unexpectedly, this mosaic rendition of the Transfiguration, which is said to have taken place on a mountain peak, lacks this hilly terrain. Only a thin strip of green and a few flowering plants fill the bottom of the conch. Far from being detrimental to the monks' reading of this image, I argue that this absence joins the physical space of the church with the pictorial. The lack of the mountain in the Transfiguration scene is not unique to the Nea Moni mosaic. The episode is also rendered without the mountain peak in the central apse of the sixth-century Monastery of St. Catherine on Mount Sinai (Figure 6.4).

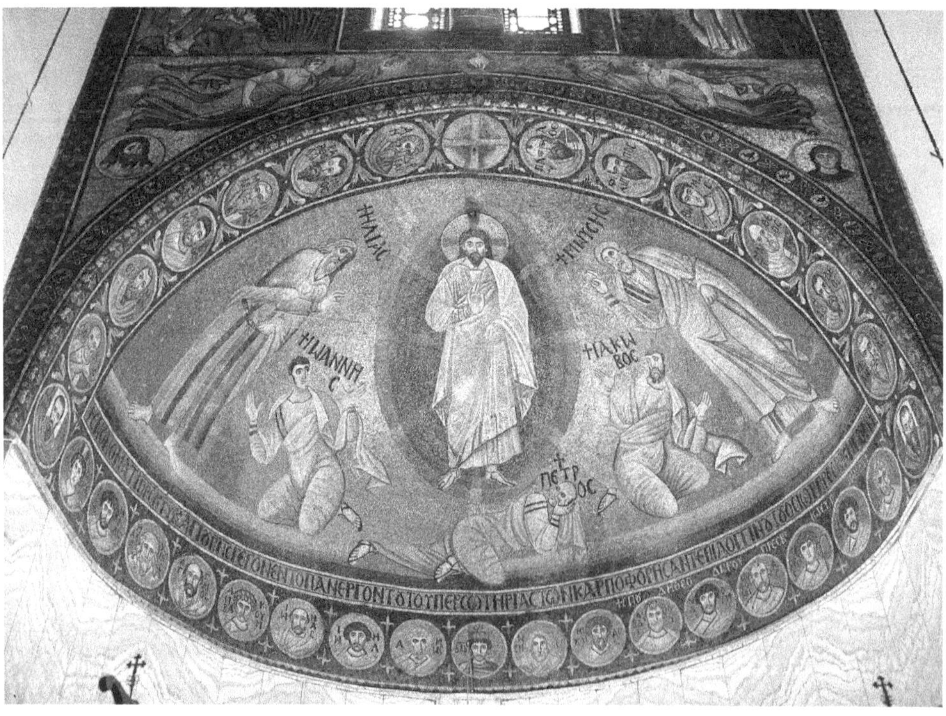

Figure 6.4 Transfiguration, Saint Catherine's Monastery, Mount Sinai, sixth century.
Published through the courtesy of the Michigan-Princeton-Alexandria Expeditions to Mount Sinai.

The composition in the apse at Sinai is similar to that at Nea Moni. Christ, resplendent in an ovoid mandorla, is surrounded by the blessing Old Testament prophets and startled apostles. Just above the apse are two scenes depicting Moses: to the left he approaches the burning bush and to the right he receives the Tablets of the Law. For Jaś Elsner, the entire program of the apse instructs the viewer on spiritual ascent as mirrored in the two scenes of Moses and in Christ's Transfiguration. Elsner argues that Moses and the Burning Bush mosaic can be read as a call for prophetic ascent. The next scene, in which he receives the Law, is the summit of mystical vision, where he is allowed to see God's back. Christianity, this mosaic argues, makes the divine face visually available in the way the apse mosaic showing the Transfiguration depicts Christ confronting en face both the apostles in the image and the viewer standing in the physical space of the church.[16] By eliminating the mountain, the mosaic makes this face-to-face encounter more immediate. In Elsner's view, the mosaic both activates the immediate experience of climbing Mount Sinai and conflates it with the imagined historical Transfiguration of Christ at Mount Tabor.[17] The mosaic in Nea Moni clearly carries the same message: the interior of the church should be thought of as a spiritual summit.

The history and legends of Nea Moni equally channel Sinai, though instead of the mountain, they emphasize the burning yet unconsumed bush. The impetus for the building of the foundation comes from the story of the miraculous discovery of an icon of the Virgin. According to legend, after setting the woods ablaze in order to discover the source of a mysterious light, three monks found the icon in a myrtle bush that had not burned. They took it back to their cave with them but awoke the next day to discover it was gone. Going in search of it, they found the icon had returned to the myrtle bush, and the monks once more brought it back to their cave. After the icon miraculously made its way back to the bush several times, the monks understood that it wanted to be at the site of the shrub, and they erected a building there.[18] The story of the burning bush draws its inspiration from the sacredness of Sinai and Moses's encounter there with a similar unconsumed bush. The fact that the two Transfiguration mosaics, one at Nea Moni and another at Sinai, function similarly should come as no surprise.

The Sinai mosaic leans into the physical space of the viewer. It, too, employs both external coherence and negative perspective. When looking at the Transfiguration, the viewer momentarily perceives the earth below the monks' feet as Mount Tabor, binding Sinai—an Old Testament site—the holy past of the Transfiguration, together with the earthly present. Though the site of a miraculous happening, Nea Moni does not rest on such "God-trodden" ground as Sinai. However, the foundation story featuring a burning bush draws clear parallels between the two sites. By omitting the mountain, Nea Moni's Transfiguration mosaic produces the same effect on the viewer as the Sinai metamorphosis in the apse. One must complete the image by understanding the mosaic as the mountain's peak and the physical floor of the church as the mountain's lower ridge. The lack of summit in both cases creates a meeting place, drawing the viewer into the biblical event.

Throughout the feast celebrated on August 6, the faithful could further experience the Transfiguration. Mountains and light are consistently brought to the fore by the texts employed in the liturgy for that day. These performed words invited the viewer to ascend with the apostles and Christ. For instance, the manuscript recording the Synaxarion of Theotokos Evergetis contains notes in the margins next to the dismissal hymn, known as *apolytikion*, sung at vespers on the eve of the Transfiguration. These notes state:

> Let us, the faithful, joyously celebrating the forefeast go in advance to meet Christ's transfiguration, and let us shout, "The day of divine gladness has arrived, the Master is going up to Mount Tabor to shine forth in the beauty of his divinity."[19]

The text prods the faithful to imagine Christ's ascent up the mountain. The celebration following the forefeast, which includes vespers, *orthros* (morning office), and the Divine Liturgy (Eucharist) for August 6, then asks the worshippers to see themselves continuously standing on a mountain's peak.

The Transfiguration is a major feast within the church calendar, and as such its celebration remains fairly consistent across typika and synaxaria, documents detailing what was to be read at each day's services. I use the evidence of three typika to reconstruct the liturgy for August 6 at Nea Moni: the tenth-century Typikon of the Great Church; the eleventh-century monastic Synaxarion of Theotokos Evergetis; and the twelfth-century Typikon of the Monastery of Christ the Savior from Messina, Italy.[20] While the Typikon of the Great Church is representative of the cathedral rite of Hagia Sophia and the other two typika draw on the Stoudite monastic rite introduced in the ninth century in Constantinople, all three reveal a relatively uniform shape of the liturgy for the Feast of the Transfiguration.[21] All three texts feature the same Old Testament and Gospel readings and use segments of Psalm 88(89) as a recurring motif during the feast of August 6.[22]

Nevertheless, much of my reading is drawn from the Evergetis synaxarion. As John Klentos has explained, the synaxarion from Theotokos Evergetis belongs to the last stage of typikon development: it has a full festal calendar drawn from both the Great Cathedral and Stoudite traditions and very detailed instructions on what poetic material was to be read at each point in the service. Further, the influence of this particular synaxarion was widespread, reaching even to Mount Athos, where it became a model.[23] Since the typikon of Nea Moni has not survived, I will draw on the popularity of the Evergetis synaxarion and treat it as a proxy for the rites at the monastery on Chios. Furthermore, like Athos rising through imperial sponsorship, Nea Moni, too, was an imperial foundation bound to look for a liturgical inspiration in the Evergetis ritual.[24]

Turning to the Evergetis text for August 6, Psalm 88(89) acts as a framework for the entire day's services, uniting officiates and faithful as they chanted the words together. The synaxarion lists four main services for each feast day: vespers on the eve; ***pannychis***, which took place just after vespers; *orthros*, which was begun in the early morning and ended around sunrise; and the Eucharistic liturgy.[25] Verses 12–13 of Psalm 88(89)—"The heavens are thine, and the earth is thine: thou hast founded the world, and the fullness of it. Thou hast created the north and the west: Tabor and Hermon shall rejoice in thy name"—were sung at vespers after the readings and right before the *apolytikion*, or the dismissal verse.[26] At *orthros*, Psalm 88(89):13 ("Tabor and Hermon shall rejoice in thy name") was sung once more as the *prokeimenon* verse right before the Gospel reading (Luke 9:28ff.).[27] The *prokeimenon*, like the Latin gradual, is a chanted Old Testament verse.[28] The psalm returns during the last service of the day, the Eucharistic liturgy, where it was selected as the Alleluia refrain (Ps. 88[89]:12), sung after the apostle reading (Peter 1:10ff.) and before the Gospel reading (Matt. 17:1ff.).[29] Finally, a line from the same psalm (Ps. 88[89]:16) is used a third time as the *koinōnikon*, or Communion verse.[30] Thus, Psalm 88(89) on August 6 runs as a thread throughout the feast, unifying the whole of the day's services and binding together officiates and the congregation. As chanted poetic responses, originating in the cathedral rite of the Great Church, these selected lines from Psalm 88(89) sung under the dome at Nea Moni would have activated the reverberant acoustics of the space and expressed the musical articulation of the divine word during the service.[31]

This synesthetic visual and sonic brightness emerging in the physical space under the cupola of Nea Moni gains further energy through the semantics of verse 16 of Psalm 88(89), which was used for both the Alleluia and *koinōnikon* at the climax of the Eucharist liturgy. In the Synaxarion of Evergetis this verse, used for the Alleluia, draws attention both to the act of coming near to the Lord and to seeing his luminous face: "Blessed is the people that knows the joyful sound: they shall walk, O Lord, in the light of thy countenance."[32] The snippet of the same verse, chanted as the *koinōnikon*—"in the light of your glory"[33]—further brings to consciousness light as epiphanic and unfolds it temporally through the act of chanting the words in the space under the dome. As a result, a bright, reverberant soundfield emerges in the nave, enveloping the faithful.

Psalm 88(89) helps to establish the sense of place throughout the day. The verses included in the earlier services of vespers and *orthros* (Ps. 88[89]:12–13) consistently mention mountains, in contrast to the selections for the Eucharistic liturgy (Ps. 88[89]:16), which speak of God's light. At the beginning of the feast, the verses call to mind the two peaks of epiphanies, Tabor and Hermon. They establish a sense of place and anchor the celebration in the idea of a mountainous terrain. Moreover, the refrain "on the mountain Tabor" (ἐν τῷ ὄρει τῷ Θαβώρ) sung after the recitation of Kathisma 12 (a ***kathisma*** is a division of the Psalter according to the monastic rite; in this particular case, Kathisma 12 includes Psalms 86–90) at *orthros* draws attention again to the holy mountain.[34] Like the absent mountain of the mosaic, the segments of Psalm 88(89):12–13 and other verses inserted into the service encourage the participants to think of mountains. The poetic verses thus ascend mountains and, reaching the top, look further up toward God's light (Ps. 88[89]:16).

Byzantine Psalter books confirm this close link between Psalm 88(89) and the metamorphosis of Christ. Starting with the ninth-century marginal Psalters, but repeated in the eleventh-century Theodore Psalter, these illustrations are embedded in the liturgical reality of the performed poetic texts (Figure 6.5).[35] The Theodore Psalter links Psalm 88(89):13, which constitutes the chanted responses during vespers and *orthros* for August 6, with the Transfiguration scene. Here the epiphany is topographically situated at the peak of a tall mountain. In the manuscript, image and text are right next to one another, allowing the reader to see, say, and hear all at once. The mosaic, by contrast, has no strict connection to any words other than the label "metamorphosis" above Christ. In fact, if the monks were facing the eastern altar during the service, the scene in the southwest corner would have been at their backs, and they would not have seen it. How can we relate the words spoken and heard during the service and the image that remained behind the monks, outside their field of vision? Here, three stimuli that are brought together—imagination, stirred by the story of the Transfiguration read during the day's celebration, the visual incarnation of this story in the mosaic itself, and the memory of having seen this image—could collectively produce the experience of divine epiphany in the faithful gathered for the feast on August 6.

The Transfiguration story was told twice on August 6: during *orthros* (Luke 9:28ff.) and during the Eucharistic liturgy (Matt. 17).[36] In the tale of the metamorphosis, the three apostles *see* Christ shining—transfigured—talking to Elijah and Moses, but they do not fully understand the significance of their vision. Peter tells Christ that they should build tabernacles on Tabor to all three—Elijah, Moses, and Christ—thinking they are equal. In response to Peter's mistake, God clouds the mountain, and they *hear* his voice proclaim that Christ is his son. Only then do they understand the full weight

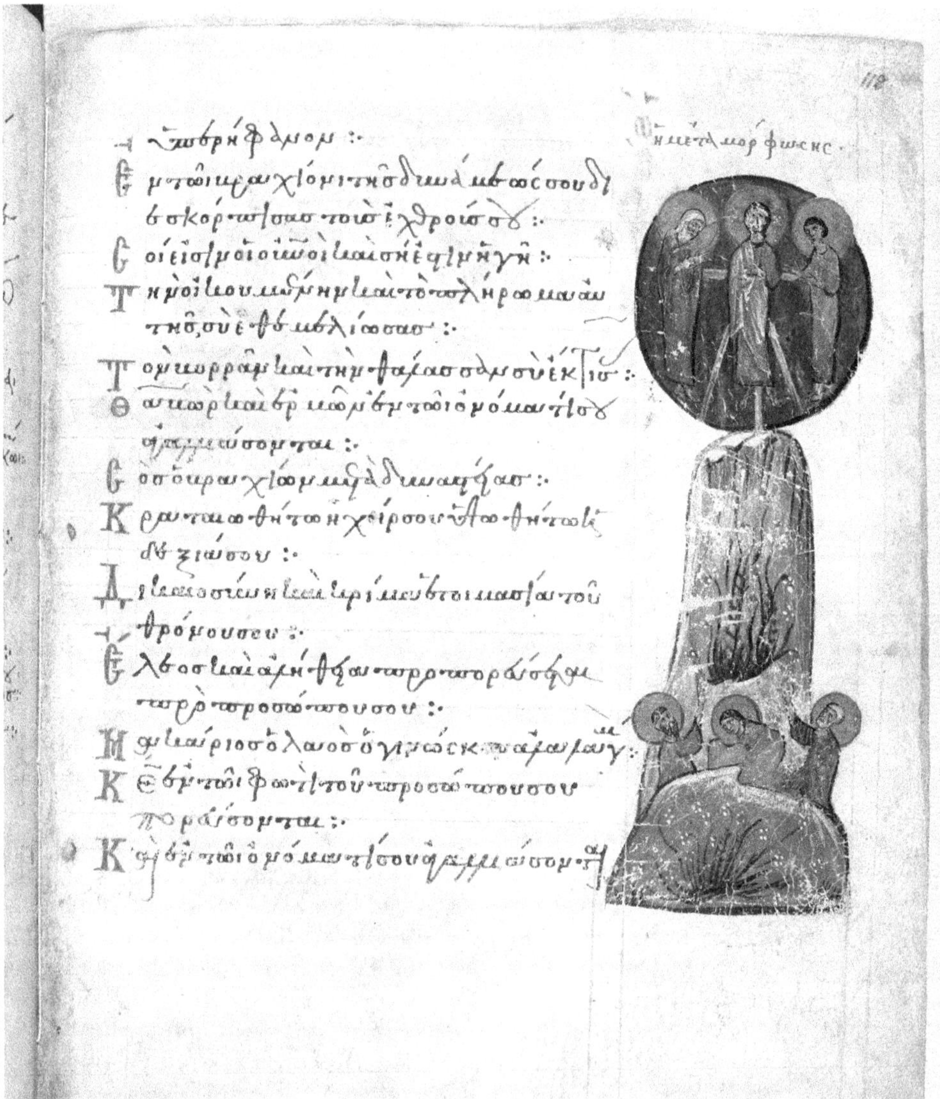

Figure 6.5 Transfiguration, Theodore Psalter, c. 1066, British Library, Add MS 19352, fol. 118r. Photo © British Library Board, Add. 19352.

of the conversation and event they have witnessed. The story alternates between vision and blindness, speaking, hearing, and seeing/non-seeing—placing the apostles on the precipice of understanding.

Similarly, the lifelong exposure of the faithful to the monumental mosaic program in the interior of the church established a familiarity with the image of the Transfiguration at Nea Moni. They knew it was behind them in the southwest when they were present at the Eucharistic liturgy, but they could not see it when they faced east toward the altar. Just as vision is denied to the apostles at the moment they understand Christ's true nature, the monks would have been denied sight of the mosaic at the moment the bread and wine transformed into Christ's present form on earth. And if the Son of God was shrouded with cloud on the mountaintop, the experience of the Eucharist in Nea Moni might have been similarly hazy in the bright August sun. Directed to the mountain and then up toward the light of God by Psalm 88(89), by the time of the late morning Eucharistic liturgy, the monks would have seen the sun streaming into the eastern windows on the Feast of the Transfiguration. Light catching on the golden mosaics, which filled the church, would have created a bright, golden haze through which tesserae-composed figures could be made out only indistinctly. The monks would have heard the service half-blinded by the intense summer sun—they would have heard the bread and wine become Christ's body without a clear view. In the Eucharist the monks transform into icons of the astonished apostles on the mountain: certain that they are experiencing something wonderful, but perhaps only through a bright cloud.

The celebration of the Transfiguration feast culminates with the Eucharist—the moment when Christ is made present on earth through bread and wine. The arc of the liturgical narrative throughout the day progresses from the Old Testament past through New Testament events, finally emerging into the present with the Eucharist. In his eighth-century commentary on the liturgy, Patriarch Germanus of Constantinople explains:

> Then the Holy Spirit, invisibly present by the good will and volition of the Father, demonstrates the divine operation and, by the hand of the priest, testifies, completes, and changes the holy gifts which are set forth into the body and blood of Jesus Christ our Lord, Who says: "For their sake I sanctify myself, that they also may be sanctified" (John 17:19), so that "He who eats my flesh and drinks my blood abides in me and I in him" (John 6:56). Thus becoming eyewitnesses of the mysteries of God, partakers of eternal life, and sharers in divine nature, let us glorify the great, immeasurable, and unsearchable mystery of the dispensation of Christ God.[37]

For Germanus it is not just important that the faithful consume the bread and wine; they must also become witnesses. Through observation they also become "partakers of eternal life" and "divine nature." For every feast, the monks may have borne witness to the Eucharist in a slightly different way as they imagined and saw the events the liturgy evokes. The transformation of the Eucharistic elements at the end of the Feast of the Transfiguration is the culmination and the fulfillment of the Transfiguration story. This *koinōnikon* (Ps. 88[89]:16) is the metamorphosis in which the monks participated. It is the event to which they became eyewitnesses.

Approaching to receive communion, the monks would have been confronted with the fragmented bread of the Eucharist. Its shape can be deduced from a sixth-century clay mold for the bread at the Royal Ontario Museum (Figure 6.6). It is round with a chrismon (the initials of Christ's name, *Chi* and *rho*) at the center. Bissera Pentcheva has argued how the cross in a circle is a magic formula identifying the process of *empsychōsis*, wherein the energy of the Holy Spirit becomes manifest in the material world.[38] This shorthand sign for the descending spirit would have been immediately recognizable and familiar to the monks. At the distribution of the Eucharist, this round bread would have been broken into pieces and mixed with wine and distributed with a spoon by the priests to the congregation. Yet the cross-in-a-circle, the sign of the Eucharist and blessing, continued to be present in the space, embedded in the narrative scene of the Transfiguration.

After taking the spoonful of the Eucharist, the monks would have turned around. Their eyes might have rested on the shining conch of the Transfiguration, understanding now more fully its import. Perhaps they would have noticed that the golden rods behind Christ bear striking resemblance to the unbroken Eucharistic bread. This last glimpse of the mosaic forges one final link between the Transfiguration and the Eucharist. Just as it is uncertain whether Christ is receding or advancing, it becomes unclear which is the prototype: Is the image on the wall a prefiguration and reminder of the Eucharist,

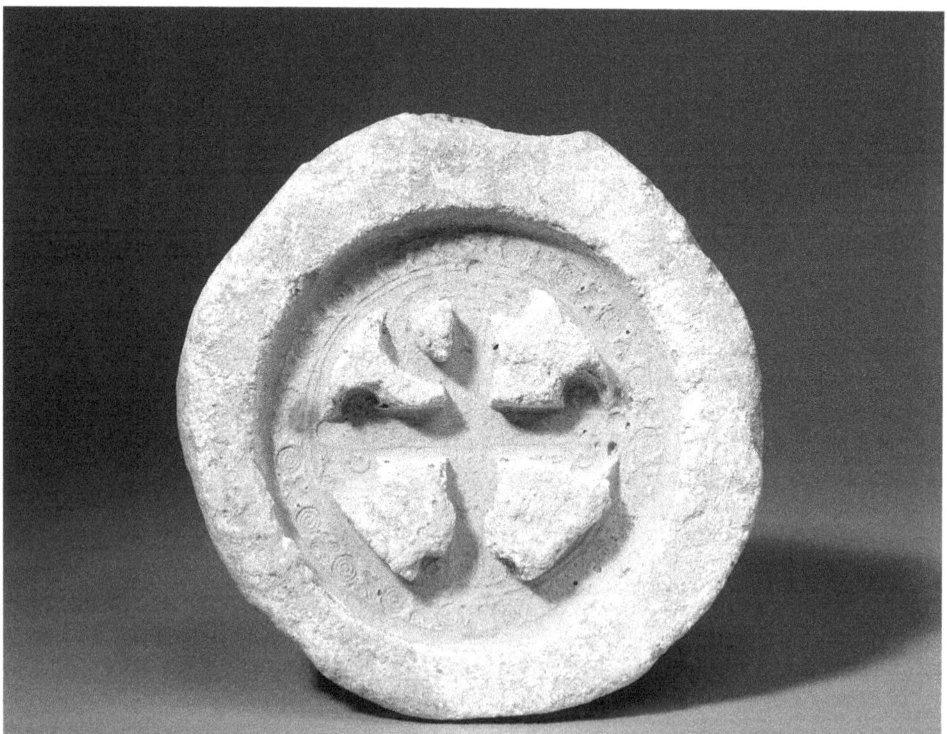

Figure 6.6 Bread mold, terracotta, fifth–sixth century. Royal Ontario Museum, ROM 986.181.122. With the permission of the Royal Ontario Museum © ROM.

or is the Eucharist an image of the event represented on the wall? To which event did the monks become witness? Intermingled like bread and wine, both of these visions are simultaneously possible. Just as Christ's Transfiguration merged three eras through the attendance of Christ, Elijah, and Moses on Tabor, the Divine Liturgy and the Eucharist bring Christ into the present and allow the faithful to partake in him through the Eucharist. The Transfiguration mosaic together with its corresponding liturgy suggest that Byzantine sacred spaces were about participation and that this process was activated by the synergies of semantics, performance, and aesthetics of the ritual and the image.

Unlike the case of the mosaic showing the Washing of the Feet and its liturgical reenactment by the hegumen, no liturgical event directly reenacts the Transfiguration. Yet having imagined the mountainous visions through the melodic fragments of Psalm 88(89) all day, the monks would have likely confronted the Transfiguration in a new and more intense way. In the end, the monks were not just encouraged to witness the Transfiguration described in the Gospels and see Christ but to touch and taste him in the Eucharist. Isolating one feast and one image narrows the field enough that we can start to grasp how different themes acquired a spatial presence and sensorial intensity on certain feast days. The celebration of the Transfiguration stresses encountering God face to face. As the monks were turning to leave the church at the dismissal, they entered the special spatial circumstance that allowed them to interlock their gaze with the Christ of the Transfiguration. The monks may have seen the same mosaics every day, and tasted the same wine and bread at every feast, but the words that whetted their palates gave new flavor to every meal.

General note

This article was originally presented at the 2015 Byzantine Studies Association Conference in New York City, and I would like to thank the audience and my fellow presenters for their many helpful questions and comments. I would like to thank Bissera Pentcheva for all of her generous help and guidance through the development and editing of this article. I am also thankful to Spiro Antonopoulous, Matt Mason, Henry Rownd, and Danny Smith for their generous help in solidifying and making clear my ideas.

Notes

1 The eight scenes in the conches, starting from the northeast, are the Nativity (now lost), the Presentation in the Temple, the Baptism, the Transfiguration, the Crucifixion, the Descent from the Cross, the Anastasis, and finally the Annunciation. Doula Mouriki, *The Mosaics of Nea Moni on Chios*, 2 vols. (Athens: Commercial Bank of Greece, 1985), 1:38–39.

2 I thank Gregory Tucker and Catharine C. Taylor for pointing out this particular aspect in discussing this paper with me.

3 William Tronzo, "Mimesis in Byzantium: Notes toward a History of the Function of the Image," *RES: Anthropology and Aesthetics* 25 (1994): 61–76; and Charles Barber, "Mimesis and Memory in the Narthex Mosaics at the Nea Moni, Chios," *Art History* 24, no. 3 (June 2001): 323–37.

4 The mosaic gives rich visual articulation of all the apostles and elaborates on Christ's preparations for the washing, which the Euchologion instructs the priest to imitate; Tronzo, "Mimesis in Byzantium," 66–67.

5 Tronzo ("Mimesis in Byzantium," 68–71) particularly explains that Psalm 50(51), which was recited in the narthex every night and addresses the theme of cleansing, was illustrated with the Washing of the Feet in five of the nine Middle and Late Byzantine marginal Psalters.

6 Barber, "Mimesis and Memory," 323–37.

7 Barber, "Mimesis and Memory," 327–30, 334–35.

8 Patriarch Germanus, *Historia ekklesiastikē kai mystikē thēoria*, trans. Paul Meyendorff (Crestwood, NY: St. Vladimir's Seminary Press, 1984) (hereafter *HE*), sec. 41, Greek and English in

St. Germanus of Constantinople: On the Divine Liturgy, trans. Paul Meyendorff (Crestwood, NY: St. Vladimir's Press, 1984), 92–95; and the essay by Walter Ray in this volume. See also Bissera Pentcheva's discussion of this phenomenon of ritual mimesis as a performative and spatial icon; Pentcheva, "Mirror, Inspiration, and the Making of Art in Byzantium," *Convivium* 1, no. 2 (2014): 10–39.

9 John Klentos, "Byzantine Liturgy in Twelfth-Century Constantinople: An Analysis of the Synaxarion of the Monastery of the Theotokos Evergetis (codex *Athens Ethnike Bibliotheke 788*)" (PhD diss., University of Notre Dame, 1995), 46, 61–63, 70–71, 78–79.

10 Alois Riegl, *The Group Portraiture of Holland*, intro. Wolfgang Kemp, trans. Evelyn M. Kain and David Britt, Texts & Documents (Los Angeles, CA: Getty Research Center for the History of Art and the Humanities, 1999), 253–54, 270–72, 282–95.

11 Otto Demus, *Byzantine Mosaic Decoration: Aspects of Monumental Art in Byzantium* (Boston, MA: Boston Book & Art Shop, 1955), 6–8.

12 Demus, *Byzantine Mosaic Decoration*, 17–35. In contrast to Demus's insistence of spatial connectivity and expansion of the mosaic into the physical space, Mouriki (*Mosaics of Nea Moni*, vol. 1) treats the mosaics more like panel paintings, detailing every part; she does not, however, describe how the viewer relates to them.

13 Robert S. Nelson, "To Say and to See," in *Visuality before and beyond the Renaissance: Seeing as Others Saw*, ed. Robert S. Nelson (Cambridge: Cambridge University Press, 2000), 143–68, esp. 156–59.

14 Demus, *Byzantine Mosaic Decoration*, 33. On "negative perspective" or "reverse perspective," see most recently Clemena Antonova, *Space, Time, and Presence in the Icon: Seeing the World with the Eyes of God* (Farnham: Ashgate, 2009).

15 Demus, *Byzantine Mosaic Decoration*, 34.

16 Jaś Elsner, "The Viewer and the Vision: The Case of the Sinai Apse," *Art History* 17, no. 1 (March 1994): 81–102, esp. 93–94.

17 For Elsner, the vision is more importantly one of a spiritual peak. For him ("The Viewer and the Vision," 94, 97), the scene represents the attainment of full vision of God and becomes a vision itself. See also Jaś Elsner and Gerhard Wolf, "The Transfigured Mountain: Icons and Transformations of Pilgrimage at the Monastery of St Catherine at Mount Sinai," in *Approaching the Holy Mountain: Art and Liturgy at St Catherine's Monastery in the Sinai*, ed. Robert S. Nelson and Sharon E. J. Gerstel (Turnhout: Brepols, 2010), 37–71.

18 Charalampas Bouras, *Nea Moni on Chios: History and Architecture*, ed. David A. Hardy (Athens: Commercial Bank of Greece, 1982), 25; Mouriki, *Mosaics of Nea Moni*, 1:21–22.

19 "Χριστοῦ τὴν μεταμόρφωσιν προϋπαντήσωμεν φαιδρῶς πανηγυρίζοντες τὰ προεόρτια πιστοὶ καὶ βοήσωμεν · Ἔφθασεν ἡ ἡμέρα τῆς ἐνθέου εὐφροσύνης, ἄνεισιν εἰς τὸ ὄρος τὸ Θαβὼρ ὁ Δεσπότης τῆς θεότητος ἀπαστράψαι τὴν ὡραιότητα." Text and translation from Robert H. Jordan, ed., *The Synaxarion of the Monastery of the Theotokos Evergetis*, Belfast Byzantine Texts and Translations, 3 vols. (Belfast: Belfast Byzantine Enterprises, the Institute of Byzantine Studies, the Queen's University of Belfast, 2000), 2:228–29.

20 Juan Mateos, Ed., *Le Typicon de la Grande Église, Ms. Sainte-Croix no. 40, Xe siècle* (Rome: Pontifical Institutum Orientalium Studiorum, 1962), 1:360–63; Jordan, *The Synaxarion of Theotokos Evergetis*, 231–35; and Miguel Arranz, Ed., *Le typicon du Monastére du Saint-Sauveur à Messine, Codex Messinensia Cr 115, A.D. 1131* (Rome: Pontificium Institutum Orientalium Studiorum, 1969), 175–77. The Evergetis MS, Athens, National Library, MS Gr. 788, is dated to the eleventh century: see Klentos, "The Byzantine Liturgy in Twelfth-Century Constantinople," 73–74, where it is dated to the twelfth century; and Robert F. Taft, "The *Synaxarion* of Evergetis in the History of Byzantine Liturgy," in *The Theotokos Evergetis and Eleventh-Century Monasticism*, ed. Margaret Mullett and Anthony Kirby (Belfast: Belfast Byzantine Enterprises, 1994), 291.

21 On the cathedral rite, S. Parenti, "The Cathedral Rite of Constantinople: Evolution of a Local Tradition," *Orientalia Christiana Periodica* 77, no. 2 (2011): 449–69; and Alexander Lingas, "Sunday Matins in the Byzantine Cathedral Rite: Music and Liturgy" (PhD diss., University of British Columbia, Vancouver, 1996). On the monastic Stoudite rite, see most recently Derek Krueger, *Liturgical Subjects: Christian Ritual, Biblical Narrative and the Formation of the Self in Byzantium* (Philadelphia, PA: University of Pennsylvania Press, 2014).

22 Both the Synaxarion of Evergetis and the Typikon of Christ the Savior specify that Luke 9:28ff. should be read for the matins Gospel at *orthros*. All three call for 2 Peter 1:10ff. for the apostle reading at the liturgy. The Gospel reading for the liturgy is Matthew 17:1ff. in both Evergetis and Christ the Savior at Messina. During *orthros* Psalm 88(89), verse 13 is used for the *prokeimenon* in both the Evergetis and Christ the Savior at Messina typika, but Evergetis then specifies verse 6, while Christ the Savior at Messina says verse 2 should be chanted. The psalm is sung as a poetic *kathisma* in the Evergetis text, but verses 2–20 are sung after the antiphon and *polyeleos* in the text from Messina. For the liturgy, all three use verses from 88 for the Alleluia: Evergetis and Christ the Savior at Messina specify verses 12 and 16, while the Great Church uses 12–13. Finally, the *koinōnikon* of Christ the Savior at Messina and the Great Church are Psalm 88(89), verses 16b–17 and 16b–17a, respectively. The *koinōnikon* of Evergetis is a slight reworking of verse 16 and reads "Ἐν τῷ φωτὶ τῆς δόξης σου." Mateos, *Le Typicon de la Grande Église*, 1:360–63; Jordan, *The Synaxarion of Theotokos Evergetis*, 231–35; and Arranz, *Le typicon du Monastére du Saint-Sauveur à Messine*, 175–77.

23 Klentos, "The Byzantine Liturgy in Twelfth-Century Constantinople," 61–63, 70–71, 78–79.

24 According to legend, the three monks who found the icon also predicted the ascendancy of Constantine IX Monomachos, and in gratitude he built them a monastery. The monastery was established sometime about 1042 as an imperial foundation. Robin Cormack, *Byzantine Art* (Oxford: Oxford University Press, 2000), 168–69; and Bouras, *Nea Moni*, 24–25. Mouriki cites thirteen chrysobulls issued by Constantine IX between January 1044 and May 1054, which attest to his interest in Nea Moni; Mouriki, *Mosaics of Nea Moni*, 1:1, 27. Henry Maguire further connects the monastery to Constantinople. He sees the mosaic program as corresponding to the most important feasts in the imperial liturgical calendar; Maguire, "The Mosaics of Nea Moni: An Imperial Reading," *Dumbarton Oaks Papers* 46 (1992): 205–14.

25 For detailed patterns of the services and liturgical terms, see Jordan, *The Synaxarion of Theotokos Evergetis*, 1:ix–x, 753–66. For the organization of the liturgical diurnal and nocturnal cycles, see Klentos, "The Byzantine Liturgy in Twelfth-Century Constantinople," 95–98.

26 "σοί εἰσιν οἱ οὐρανοί, καὶ σή ἐστιν ἡ γῆ· τὴν οἰκουμένην καὶ τὸ πλήρωμα αὐτῆς σὺ ἐθεμελίωσας. τὸν βορρᾶν καὶ τὴν θάλασσαν σὺ ἔκτισας, Θαβὼρ καὶ Ἑρμὼν ἐν τῷ ὀνόματί σου ἀγαλλιάσονται."

27 The *prokeimenon* in the Typikon of Saint-Sauveur is Psalm 88(89), verse 16b and then verse 2; Arranz, *Le typicon du Monastére du Saint-Sauveur à Messine*, 176.

28 For the definition of *prokeimenon*, see Christian Troelsgaard's entry in the *Grove Dictionary of Music*, www.oxfordmusiconline.com/subscriber/article/grove/music/40455. The liturgical terms are also explained in Mateos, *Le Typicon de la Grande Église*; and Juan Mateos, *La célébration de la parole dans la liturgie byzantine: Étude historique* (Rome: Pontificium Institutum Orientalium Studiorum, 1971). For further discussion on the *prokeimena*, see Gisa Hintze, *Das byzantinische Prokeimena-Repertoire*, Hamburger Beiträge zur Musikwissenschaft 9 (Hamburg: Verlag der Musikalienhandlung Wagner, 1973); and Christian Troelsgård, "The Prokeimena in Byzantine Rite: Performance and Tradition," *Cantus Planus VI: ēger 1993*, 2 vols. (Budapest: Hungarian Academy of Sciences Institute for Musicology, 1995), 1:65–77.

29 The Alleluia is the same in the Messina typikon, but the typikon of the Great Church specifies Psalm 88(89), verses 12–13: Arranz, *Le typicon du Monastére du Saint-Sauveur à Messine*, 177; Mateos, *Le Typicon de la Grande Église*, 1:360–63. On the Alleluias, in addition to the entry by Alexander Lingas in the *Oxford Companion to Music*, see Christian Thodberg, *Der byzantinische Alleluiarionzyklus: Studien im kurzen Psaltikonstil*, Monumenta Musicae Byzantinae Subsidia (Copenhagen: E. Munksgaard, 1966).

30 The *koinōnika* are very similar though not identical in each text. The Great Church calls for Psalm 88(89), verses 16b–17a (Mateos, *Le Typicon de la Grande Église*, 1:362–63), while Christ the Savior at Messina adds the last half of verse 17, specifying that 16b–17 should be sung (Arranz, *Le typicon du Monastére du Saint-Sauveur à Messine*, 177). Last, the Evergetis typikon provides a reworking of verse 16: "Ἐν τῷ φωτὶ τῆς δόξης σου" (Jordan, *The Synaxarion of Theotokos Evergetis*, 234–35). On the *koinōnikon*, in addition to the entry by Christian Troelsgård in the *Grove Dictionary of Music*, see Dimitri E. Conomos, *The Late Byzantine and Slavonic Communion Cycle: Liturgy and Music* (Washington, DC: Dumbarton Oaks Research Library and Collection, 1985); and Simon Harris, Ed., *The Communion Chants of the Thirteenth-Century Byzantine Asmatikon* (Amsterdam: Hardwood Academic Publishers, 1999).

31 On the chanted responses and their effect in domed interiors, see Bissera Pentcheva, "Hagia Sophia and Multisensory Aesthetics," *Gesta* 50, no. 2 (2011): 93–111; Pentcheva, "Performing the Sacred in Byzantium: Image, Breath, and Sound," *PRI Performance Research International* 19, no. 3 (2014): 120–28; and the essay by Wieslaw Woszczyk in this volume.

32 "μακάριος ὁ λαὸς ὁ γινώσκων ἀλαλαγμόν· Κύριε, ἐν τῷ φωτὶ τοῦ προσώπ ου σου πορεύσονται."

33 "ἐν τῷ φωτὶ τῆς δόξης σου." Jordan, *The Synaxarion of Theotokos Evergetis*, 234–35.

34 Jordan, *The Synaxarion of Theotokos Evergetis*, 230–35.

35 On the liturgical rooting of the miniatures in the ninth-century marginal Psalters, see Maria Evangelatou, "Liturgy and the Illustration of the Ninth-Century Marginal Psalters," *Dumbarton Oaks Papers* 63 (2009): 59–116. Tronzo ("Mimesis in Byzantium," 68) has noted that the image of the Washing of the Feet appeared alongside Psalm 50(51), which was also read in the company of the mosaic of the same subject, in five of the marginal Psalters. Pentcheva has furthered this connection between liturgical rites and the marginal psalters in her work on the consecration rite in Hagia Sophia. She has shown how verses 12–14 from Psalm 50(51) recited during the rite of consecration were accompanied by images of the descending Holy Pneuma on the altar. The words of the psalms became a way of signaling the descent of the Holy Spirit. Pentcheva, "In-Spiriting in the Byzantine Consecration (Kathierosis) Rite," *Codex Aquilarensis* 30 (2014): 83–108, esp. 49–53.

36 Verses 12 and 16 of Psalm 88(89) introduced the Gospel reading of the Eucharistic liturgy (Matthew), and a paraphrase of verse 16 concluded the reading; Jordan, *The Synaxarion of Theotokos Evergetis*, 234–35. Both Gospel selections were also read at Christ the Savior at Messina; Arranz, *Le typicon du Monastére du Saint-Sauveur à Messine*, 176–77.

37 Germanus, *HE*, bk. 41, 96–99.

38 Bissera Pentcheva, *The Sensual Icon: Space, Ritual, and the Senses in Byzantium* (University Park, PA: Pennsylvania State University Press, 2010), 17–44, esp. 28–36.

7 We who musically represent the cherubim

Laura Steenberge

Fourteenth- and fifteenth-century Byzantium was a golden era for musical composition. After centuries of development, chant notation had matured into the Middle Byzantine diastematic notation system (see Christian Troelsgård's essay in this volume), enabling composing with neumes (notational markings that designate single pitches or small clusters of notes) to become a creative musical act unto itself, rather than solely a system for documenting orally transmitted chants. The style paradoxically linked past and future, as newly composed chants were expected to borrow material from older chants, yet were unique enough that composers and arrangers began to be credited by name. In the words of Manuel Chrysaphes, the esteemed fifteenth-century composer and *lampadarios* (the patriarchal candle carrier, who became one of the cantors in the Late Byzantine period), at Hagia Sophia in Constantinople, one should "take over some melodies unchanged from tradition" and "not depart from their original melodics but follow them accurately, step by step, and retain them."[1] Understanding how such borrowing was achieved while still yielding unique compositions is a key to unlocking the medieval craft.

Of all the music in the Divine Liturgy, the Cheroubikon has the most specific compositional demands.[2] Perhaps for this reason Cheroubika were credited to composers early on, compared with simpler and more straightforward chants such as the Trisagion that remained anonymous through the fourteenth century.[3] Most likely introduced into the Divine Liturgy of the Hagia Sophia in the sixth century by Justinian and ratified by Emperor Justin II in 573–74,[4] the Cheroubikon originated as a chant to accompany the transfer of gifts, when the Eucharistic elements of bread and wine were carried to the altar.[5] By the fourteenth century, the ritual expanded and the chant along with it, engaging the congregation as the bishop and other clergy performed a lengthy sequence of prayers and actions uttered in a low voice. These events lead up to the Great Entrance, in which a procession of clergy, wearing elaborate clothing and carrying fans that symbolize wings, escort the Eucharistic gifts through the nave of the church.[6] The human action parallels the celestial, as it is the cherubim who escort the throne of God, a parallel that is further accentuated by a reference to the Thrice Holy hymn, the song sung by the cherubim. Together, the lyrics and theater merge earthly and heavenly scenes: "We who mystically represent the cherubim and sing the Thrice Holy hymn to the life-giving Trinity, let us lay aside all worldly care to receive the King of All escorted unseen by the angelic corps, alleluia, alleluia, alleluia."[7] The Great Entrance interrupts the hymn at *hōs ton Basilea hypodexomenoi* (being in a state about to receive the King of All), a lyric symbolizing the descent of God from heaven escorted by the throne-bearing cherubim, which is mirrored by the bread and wine that are ceremonially carried in procession by the clergy.[8] In its entirety, the Great Entrance prefigures the apex of the terrestrial service, the anaphoral rite that directly follows the Cheroubikon.[9] During the anaphora, the bishop prays over the unblessed elements, and at the *epiclesis* ("Send down Thy Holy Spirit upon us and upon these Gifts set forth") the Holy Spirit descends into the bread and wine, transforming them into the body and blood of Christ, which will then be received by the congregation who eats the transfigured elements.[10] The Eucharistic symbolism is made clear through the use of *hypodexomenoi*, "being in a state about to receive," which connotes both receiving and eating.[11]

The task of the composer is to construct the chant such that the music fully embodies the drama and liturgy of the ritual act.[12] Structurally, the Cheroubikon must be flexible, permitting the choir to perform the hymn in two sections of variable duration that are adaptable to the unfolding events, thus gracefully allowing for the planned interruption at the Great Entrance. Regarding semantic connections between the music and the ritual, there is much to be observed about the influence of the performed liturgy on compositional decisions. A wealth of information resides in the 1974 publication by Dimitri Conomos, *Byzantine Trisagia and Cheroubika of the Fourteenth and Fifteenth Centuries*, which contains transcriptions of seventeen medieval settings of the Cheroubikon: seven by fourteenth-century composers, eight from fifteenth-century composers, and two early settings that, at least partially, predate the fourteenth century.[13] Using these transcriptions, I have undertaken a comparative analysis of the Cheroubika, focusing on compositional techniques that musically encode the liturgy. I am striving to understand the medieval Cheroubika not as they were performed, but as they were composed, approaching the problem not as a Byzantinist or even a musicologist, but as a composer of vocal music. The interests that underlie my creative work overlap with the study of the medieval Cheroubika in a number of ways, as I create site-specific compositions for unique acoustic spaces, study how folk and ritual music preserves and develops mythology through the ages, and seek ways that music can represent concepts beyond perceptible reality. In addition to the study of music, I also have a background in linguistics, so I am particularly interested in the connections of music and language.

I am not proficient in reading the original Byzantine notation and therefore must rely on Conomos's transcriptions that use the Monumenta Musicae Byzantinae (MMB) transcription system, which in 1974 had not yet fallen out of favor. This system provides the basic melody of the chants, notated in Western staff notation with markings above the notes to provide information about the neumes, which can be useful in identifying melodic patterns across different settings. However, the rhythm of the chant is warped in the transcriptions, owing to inherent differences between the ways rhythm is expressed in neumatic versus staff notation.[14] Furthermore, these MMB transcriptions fail to communicate crucial aspects of the chant that were embedded not in the neumes themselves but in the performance practice, such as tuning systems, ornaments, and chromatic alterations, in addition to rhythm.[15] Therefore, conclusions drawn from an analysis of the transcriptions are limited to observations about the pitches in the melody, essentially disregarding both rhythm and performance practice. In the examination that follows, I use stemless noteheads so that the inaccuracies of rhythmic transcription do not provide a distraction.

When first approaching the vast amount of material offered in the seventeen medieval Cheroubika, my attention was drawn to how the melody relates to each word of the text; I looked across the settings to see if there were any similarities. At the same time, I was introduced to the mystagogical writings of Pseudo-Dionysius the Areopagite and

Germanus, in which layers of symbolism endlessly fold in on themselves.[16] The possibility that a similar layering would also be found in the musical scores was hinted at by a brief observation of Conomos, in which he proposed that a certain motif occurring at the Great Entrance could represent the notion of procession.[17] Following this lead, I found that there was much more to be studied about this motif, an investigation that resulted in the identification of additional motifs that are similarly symbolic of the performed ritual.

When Conomos wrote about this procession motif, he suggested that it could be an early example of an *idée fixe*, a term that was first coined by the nineteenth-century French composer Hector Berlioz to represent obsession through the repeated use of a musical idea. Perhaps Conomos appropriated this term to describe this musico-dramatical correspondence because there is no appropriate term specific to medieval music, although *reminiscence motif* (originally pertaining to eighteenth-century opera) or *leitmotif* (originally pertaining to Wagnerian opera) may be a better fit than idée fixe.[18] I have appropriated the term *leitmotif* for this analysis, using it to indicate a melody that has a symbolic identity. While *motif* or *melodic formula* usually suffices to discuss a particular musical event, leitmotif is occasionally employed to call attention to symbolic musical motifs that need not be linked to specific lyrics. This is in contrast to *word painting*, the practice of representing specific lyrics through the corresponding music, a related compositional device also employed in the medieval Cheroubika. Word painting is also a term appropriated from liturgical and secular classical music, but is culturally and chronologically not far from medieval Byzantium, as the practice of word painting was common in sixteenth-century secular and liturgical music and has been applied to the earlier music of Guillaume de Machaut as well, a contemporary of the medieval Byzantine composers.[19]

That medieval Byzantine composers were musically representing objects and ideas is evident in *kratemata* (long passages sung to nonsense syllables), written in the new embellished style of singing known as *kalophonic* that emerged in the fourteenth century.[20] The Cheroubika and the Trisagia were written in this new style that accentuated vocal and compositional ability, incorporating solo *kratemata* around the Great Entrance.[21] Some *kratemata* were informally named after objects that were sonically depicted in the piece, such as “trumpet,” “bell,” and “nightingale.”[22] The one that depicts a bell, for example, imitates the striking of bells by frequently repeating the same tone, leaping back and forth between two pitches that are a fifth apart, and repeating short sequences of notes.[23]

Linguistically, the *kratemata* imitate the language of the angels through the singing of nonsemantic syllables, mostly *te* and *re*, in a style thus known as *teretism*. Since the inception of Christianity, nonsemantic language has been interpreted as the expression of an angelic vocalization that extends beyond human comprehension.[24] Augustine, for example, speaks of wordless jubilation, a response of exultation that surpasses language.[25] Writing after the medieval composers, in the mid-seventeenth century the monk Gerasimos of Crete wrote an apology for the practice of singing teretism, explaining that the *terere* originates in the prophets who heard from heaven the sound of many waters, which Ezekiel describes as the voice of God and also the sound of the wings of the cherubim (Ezek. 1:24, 43:2). Gerasimos further explains that teretism is a substitute for instrumental music, and that it imitates the sound of nature, cicadas in particular.[26] The mention of cicadas reveals the influence of the early fourteenth-century Constantinopolitan music theoretician Manuel Bryennios (fl. Constantinople, ca. 1300) who defined *teretismos* as an imitation of the trilling sound of cicadas, using either instruments or the voice.[27] Bryennios, in turn, was heavily influenced by Aristides Quintilianus (second or third century CE), who wrote about the singing of the letter tau in combination with four specific vowels as being not only a vocal expression of a plucked string instrument but also a representation of the ether, and that "a cosmos of the soul is the melody of the vowels."[28] The association of the melodies and teretismatic syllables of the *kratemata* with the sounds of the soul, nature, instruments, and angelic language encourages speculation that composers were engaged in other forms of musical representation.

The stylistic boundaries of the Byzantine musical aesthetic and notational capabilities determine the materials with which the liturgy can be musically illustrated. Medieval Byzantine chant melodies can move in only a few ways: ascend, descend, dwell within a narrow range, or stagnate on a tone. They are mostly stepwise, but frequently inflected by small leaps of a third. The melodies can belong to eight different modes, and while each one has a specific musical scale associated with it, what is more important in defining mode is the application of a specific set of melodic formulas, short fragments of a few notes that form recognizable patterns.[29] Chrysaphes writes that chants are put together similar to the way words and syllables are formed of letters, and that each chant has its own particular means of doing so.[30]

The manner in which melodic formulas align with certain words, syllables, and events is the primary focus in the investigation of how the Cheroubikon represents the liturgy in musical terms. One specialized kind of melodic formula of keen interest is intonation formulas, called *ēchēmata*, that are sung at the beginning of a chant to establish the melodic shape; they can also recur throughout. The *ēchēmata* and the *apēchēmata*, melodic tails that are attached to the end of the *ēchēmata*, combine to form different melodic formulas.[31] Egon Wellesz connected these formulas to the music of heaven, calling them the "echo of the divine hymns," and described the musician's work as linking these melodies together.[32]

I propose that the liturgical content is one key feature that determines what Chrysaphes calls the "manner and practice" of setting the Cheroubikon, and that melodic formulas, like words, represent concepts in the lyrics and the ritual. As they are notated, these formulas also function much like words. Because Byzantine pitch notation is relative, a melodic formula, regardless of the starting pitch, will be notated with the same symbols in the score, thus forming a stable orthographic representation of the melody, much like a word formed from letters or syllables.[33] If a formula represents the singing of the angels, this neume-word would function as an icon of angelic song, not unlike the Chi-Rho that represents Christ and other alphabetic abbreviations painted on the icons of saints.

As performance cues, some motifs, such as intonation formulas that prepare the choir, are inherently associated with divinity because chant itself is a representation of angelic singing. Other motifs manufacture symbolic associations with the liturgy, such as Conomos's procession motif that, by cueing the beginning of the procession, becomes implicated in the earthly mirroring of the angelic escort of God. Other correspondences are achieved through metaphors that naturally belong to both music and theology, such as directional metaphors: just as heaven is up and the earth is down, so do melodies go up and down. Through such directionality, melodies can represent actions in many different ways: ascending and descending gradually or rapidly; expressing a climactic moment by expanding the range to the highest highs or the lowest lows; suddenly changing register by leaping up or down in order to change the character of the voice. The options are myriad and only limited by the imagination and skill of the composer.

Peter Jeffery points to similarities of ascending melody and modal traditions among Byzantine, Gregorian, Old Roman, Ambrosian, and Mozarabic chants that relate back to Jerusalem chant, and he connects these similarities to the text.[34] For example, a consistent rise of a fifth or a major triad on the word "Hosanna" across different traditions may be an echo of the original Jerusalem chant.[35] In another case of ascending melody, between a Byzantine and Ambrosian setting of the phrase "Christ is risen from the dead," Jeffery identifies a stability of range, mode, and a rising melodic pattern, concluding that, at the very least, by the seventh or eighth century a tradition had been established that linked text to mode and melodic style.[36]

The absence of motion also communicates action; this is the strategy that marks Conomos's procession motif. He writes that this motif, "mostly a repeated *a* but sometimes a small embellishment around this note, is used as a melodic identification for the event, which is about to take place."[37] Conomos observes that this motif appears not only in the majority of the Cheroubika at the text *hōs ton Basilea* but also in another chant that accompanies a procession: the special Asmatikon Trisagion, which is sung at the Feasts of the Cross (September 14 as well as on the third Sunday of Lent), during which the cross is carried through the church.[38] Like the Cheroubikon, the procession in the festal Trisagion is introduced by interrupting the chant; during this interruption the *domestikos* (leader of the choir) sings the intonation containing the procession motif, thereby beginning the procession and resuming the chant.[39]

Musical symbolism also appears in linguistic cues intercalated in the lyrics that invite divine participation in the vocalization of the chant; like teretism, nonsemantic intonation syllables and aspirated syllables mirroring the divine through the limited material means of the human voice. Syllables such as *ne-a-nes* or *ne-a-gi-e* accompany intonation melodies to identify the mode of the chant; these intonation syllables are also intercalated in the chants. Furthermore, interspersed in the lyrics are the nonsemantic aspirated syllables χο, χε, and χα (*ho*, *he*, and *ha*) that come out of the *asmatikon* chant tradition, florid settings from the thirteenth century that would have been sung by the elite choir at the Hagia Sophia.[40] By drawing attention to breath, these aspirated syllables present chant as exhalation, offered as the human reciprocal contribution for the gift of the Eucharist.[41] The prosody of breath, purposefully stitched into the stream of chant, is a kind of performative icon that literally breathes life into liturgical music, invoking the presence of the Holy Spirit.[42] In the figures accompanying this essay, nonsemantic syllables are italicized.

Searching the Cheroubika for musical symbolism revealed a few clear patterns, as though a similar blueprint were guiding certain compositional decisions. In the following analysis, I will first discuss the musical mirroring of the Thrice Holy in the Cheroubikon through quotation and also through a leitmotif that succinctly represents the Thrice Holy throughout the Cheroubika. This leitmotif of angelic singing combines with the processional motif marked by Conomos. I have further identified an additional melodic formula of the descending Spirit that joins with these two. These three notions of singing, escorting, and descending Spirit combine at the Great Entrance. Together, I argue, these three motifs depict musically the ceaselessly singing cherubim who escort the throne of the triune God down from heaven to earth during the performance of the Divine Liturgy. Finally, two additional cases of word painting depict the vocalizations of the low-voiced bishop and the singing choir who perform the liturgy. Altogether, these motifs musically reflect the Great Entrance ritual, representing the praying bishop and the singing choir who are invisibly joined by the angelic choirs, escorting the King of All who descends to be received by the faithful in the church.

Mirroring: the Thrice Holy in the Cheroubikon

The lyric "sing the Thrice Holy hymn to the life-giving Trinity" embeds angelic chant within mortal vocalization. In turn, the heavenly Thrice Holy hymn is itself echoed on earth in the two Byzantine Thrice Holy chants of the Divine Liturgy: the Trisagion, sung during the Little Entrance; and the Sanctus, sung during the anaphora.[43] For any composer predisposed to musical representation, such a direct reference is a gift because the object to be represented is not only a sound but already exists as music, ready to be borrowed. It is indeed the case that many Cheroubika mirror this textual reference by quoting or imitating the Thrice Holy melodies.

The Thrice Holy hymn originates with the cherubim and seraphim, who, being in the rank of angels closest to God, pass the hymns through the supercelestial ranks down to earth.[44] As for their appearance, the book of Ezekiel describes the cherubim as four in number, each having four wings and four faces: those of a man, a lion, a calf, and an eagle. At their feet are wheels covered in eyes all around, and above them is the throne on which the Lord sits (Ezek. 1:4–18). In the Revelation of John they are slightly modified, being four separate beasts that surround the throne. Each one is six-winged, full of eyes, and possesses the face of a man, eagle, lion, or calf. They do not rest either day or night, saying, "Holy, holy, holy, Lord God almighty, which was, and is, and is to come" (Rev. 4:6–8). This song is a variation of one captured in an earlier vision in Isaiah 6:3, in which the six-winged seraphim call back and forth to each other, "Holy, holy, holy, Lord of hosts: the whole earth is full of his glory."

Both the Trisagion and the Sanctus predate the Cheroubikon by at least a century; the Sanctus was incorporated into the anaphora early on, by the third or fourth century.[45] The first two lines of the Sanctus come almost verbatim from Isaiah 6:3, and the latter three lines are taken from Matthew 21:9, in which the crowds shout "Hosanna" as Jesus enters Jerusalem riding a donkey.

> Holy, Holy, Holy Lord Sabaoth.
> Heaven and Earth are filled with your Glory.
> Hosanna in the highest.
> Blessed is He who comes in the name of the Lord.
> Hosanna in the highest.[46]

Juxtaposed, the Isaiah and Matthew passages depict Christ riding a donkey in a procession as heaven and earth cry out praises, a scene not unlike that of the King of All escorted by the singing cherubim in the Cheroubikon.[47]

As for the Trisagion, according to legend, it was taught by angelic choirmasters to a child who was taken to heaven, while crowds of faithful were praying for deliverance from divine wrath in Constantinople after an earthquake.[48] This earthquake occurred between the years 434 and 446, and records indicate that the Trisagion was sung at the Council of Chalcedon in 451.[49] The lyrics echo the three holies of the angelic cries, mapping each one to an aspect of the Trinity:

Holy God,
Holy Mighty,
Holy Immortal, have mercy upon us.[50]

The reference to the Thrice Holy in the Cheroubikon lyrics could indicate either the Trisagion or the Sanctus, as both can be called *Trisagion*, literally "thrice holy."[51] For the sake of clarity, in this paper the hymn that is today known as the Trisagion is called the "Trisagion," the hymn sung during the anaphora is called the "Sanctus," and the angelic hymn from which they both derive is called "Thrice Holy."

Because the Great Entrance prefigures the anaphora, the lyric "sing the Thrice Holy hymn to the life-giving Trinity" is quite likely a reference to the singing of the Sanctus.[52] Yet when the Trisagion was sung at the Council of Chalcedon, it was implicated in the contentious debate between Monophysite and Dyophysite theologians, and at that time there were in fact two versions of the Trisagion with subtly different lyrics that either addressed the hymn to the entire Trinity (Dyophysite version) or focused attention on Jesus Christ (Monophysite version).[53] This debate continued into the sixth century, so the lyric designating that the Trisagion is sung to the Trinity could be interpreted as supporting the Dyophysite stance. A third possibility is that the reference is to neither the Trisagion nor the Sanctus but, rather, to the celestial Thrice Holy from which both hymns derive. Variations in the prayers recited when the Eucharistic gifts are uncovered and readied for the Great Entrance procession indicate that in some cases the Trisagion was recited and in other cases the Sanctus was, suggesting a degree of interchangeability between the two because they both evoke the angelic Thrice Holy.[54]

In a number of Cheroubika, music from both the Trisagion and the Sanctus is quoted to represent the Thrice Holy, and this practice is traceable to the two oldest settings, the Asmatikon Cheroubikon and that of John of Damascus. John of Damascus's setting is quite simple compared with the *asmatikon* setting, and the many transmissions of its melody suggest that it was widespread and may predate the fourteenth century, although there are no extant transmissions from before the fourteenth century. Erroneously but reverently attributed to the great eighth-century hymnographer, it is marked on manuscripts as being "old," "common," and "sung at the palace."[55]

A case of word painting is found in the John of Damascus setting, linking the singing of angels directly from text to melody through quotations of the Trisagion melody that set *eikonizontes* (representing) and *triadi ton trisagion* (to the Trinity the Thrice Holy) (Figure 7.1). The typical fourteenth-century Trisagion is set with three related melodies: an opening melody on *Amen*; an expanded version for *Hagios ho Theos* and *Hagios ischyros*; and a third expansion of the same melody for the final phrase, *Hagios athanatos eleēson hymas* (Figure 7.2). The *Amen* melody, an intonation found in many chants, is based on the melodic formula *g a b a g*. The two melodies that follow this intonation develop it by first expanding the range to include an *f*, then expand it further to include the *e* below.

Fragments from these Trisagion melodies easily fit into this Cheroubikon because they share tonal centers: the Trisagion is almost always centered on *g*, and, like many early examples of chant, the John of Damascus Cheroubikon alternates tonal centers between *e* and *g*.[56] At the onset of both *eikonizontes* and *Triadi ton Trisagion* a wide leap launches the melody upward, calling attention to the text and shifting the tonal center to *g*. In both instances, the arrival at the new tonal center is announced with the *g a b a g Amen* formula, which is then followed by melodic fragments from the

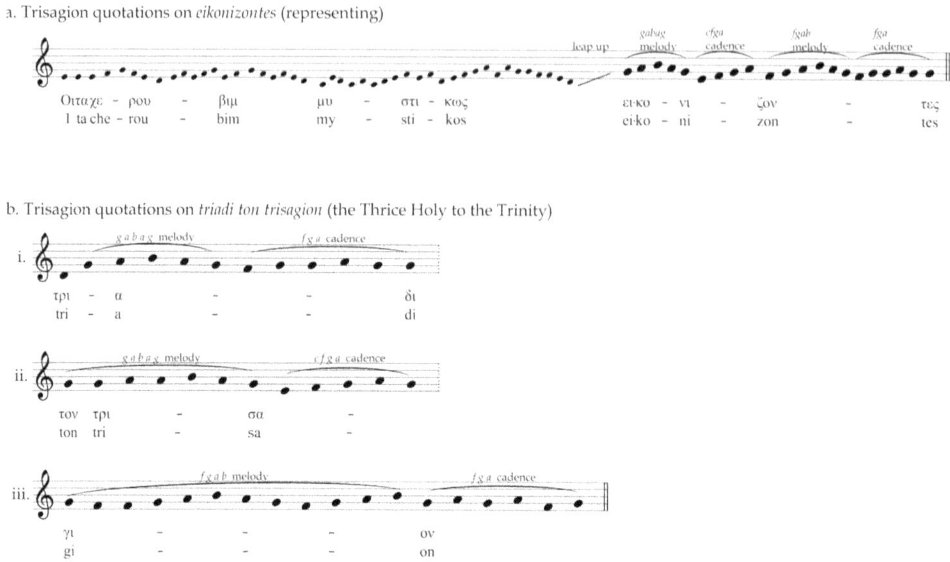

Figure 7.1 Quotations of the Trisagion in the John of Damascus Cheroubikon. Athens, National Library, MS Gr. 2458, fol. 165v; from Conomos, *Byzantine Trisagia and Cheroubika*, 143–44. (a) Cheroubikon melody leaps up to *g* to quote the Trisagion on *eikonizontes* (representing) using rearranged fragments from the Trisagion (see Figure 7.2). (b) Music on the lyric *Triadi ton Trisagion* (to the Trinity the Thrice Holy) imitates the Trisagion hymn. The melody is divisible into three parts (i, ii, iii), a form that imitates the Trisagion (see Figure 7.2). Each part contains a melody based on the *g a b a g Amen* formula cadence on *g*, and the melody is elaborated from one line to the next.

Trisagion, somewhat rearranged, but overall mirroring the expansion of range. The melody that sets *Triadi ton Trisagion* subdivides into three phrases in imitation of the tripartite melodic structure of the Trisagion.

Like this early example, many other Cheroubika musically represent the Thrice Holy in the lyric "sing to the life-giving Trinity the Thrice Holy hymn." In four of the seventeen settings, mostly from the fourteenth century, the text is set to melodies that clearly represent the notion of the Thrice Holy by imitating the general melodic contours and expansion or range of the Trisagion.[57] The fourteenth-century composer Agathon Korones used this loose imitation in one setting on *trisagion hymnon*, and a more precise quotation of the Trisagion in another. In the latter he imitates the Trisagion in full with all its three phrases, distributing the quotation over the opening text "We who mystically represent the cherubim" (Figure 7.3). As a result, the Thrice Holy quotation mirrors the act of representing the cheroubim and sets up the rest of the chant to spin out of this representative melody.

The other setting with pre-fourteenth-century origins, the Asmatikon Cheroubikon, has another kind of melodic quotation of the Thrice Holy that pervades the Cheroubika: a seven-note melodic formula that functions as a leitmotif representing the singing of the angels. Henceforth referred to as "the singing formula," it is related to the Asmatikon Sanctus and Trisagion Amen intonations (Figure 7.4). In the Asmatikon Cheroubikon, it is contained in the intonation and subsequently is intercalated into the opening chant melody, set with the intonation syllables *neanes* (identifying the *deuteros* mode within

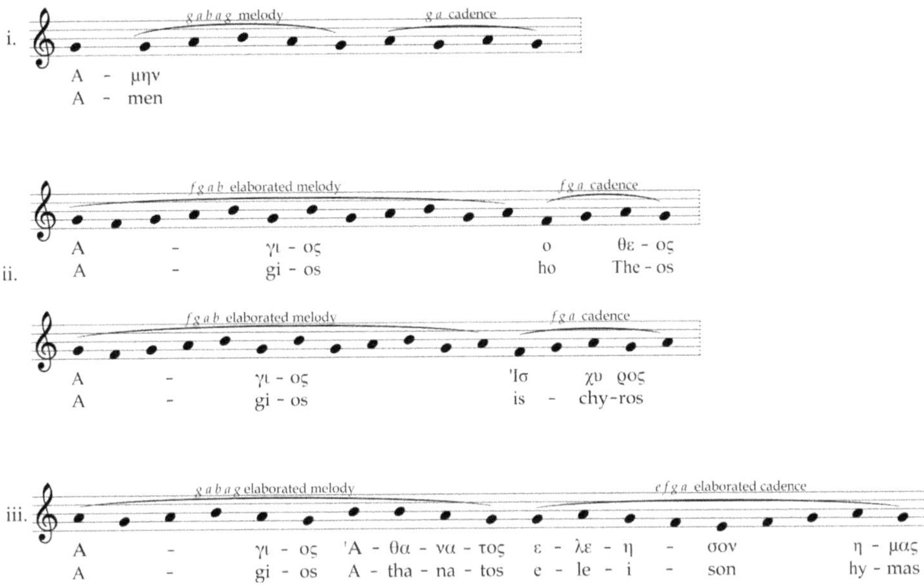

Figure 7.2 A typical fourteenth-century Byzantine trisagion with three melodies. Athens, National Library, MS Gr. 2458, fol. 144r; from Conomos, *Byzantine Trisagia and Cheroubika*, 57. (i) Opening *g a b a g* melody on *Amen*. (ii) Elaboration of *Amen* melody on *Hagios ho Theos*, repeated again on *Hagios ischyros*. Range widens to include *f g a b*. (iii) Further elaboration of melody on *Hagios athanatos eleēson hymas*. Range widens again to include *e f g a b*.

the Byzantine *oktoēchos* modal system).[58] In later settings, this singing formula persists, but after the *asmatikon*, the intercalated nonsemantic intonation syllables are less frequent.

Often the singing formula sets the lyrics that reference the Thrice Holy, but its occurrence is far more widespread, being peppered throughout the settings on many different words, fusing angelic and mortal chant. Wherever it may appear, many composers set the formula with care, demonstrating intentional links between text and melody. The clearest examples of musico-textual correspondences are found in the fourteenth-century setting by Manuel Agallianos, who attentively set specific syllables with the singing formula (Figure 7.5). In one instance of the angelic word *alleluia*, he set this formula four times in a row as an ascending sequence. Elsewhere, he used the singing formula to set both instances of the syllable *rou*, occurring in the words *cheroubim* and *doriforoumenon*.

Procession, singing formula, and descending breath

Extending the metaphor that melodic formulas function as words, multiple formulas strung together could form "phrases," combining symbols to depict celestial imagery. The following series of examples demonstrates a connection between the singing formula and Conomos's procession motif. At the Great Entrance, they are joined by a third motif that depicts the descent of the Holy Spirit. By combining three motifs (procession,

a. Loose imitation of Trisagion melodies on *trisagion imnon* (Trisagion hymn)

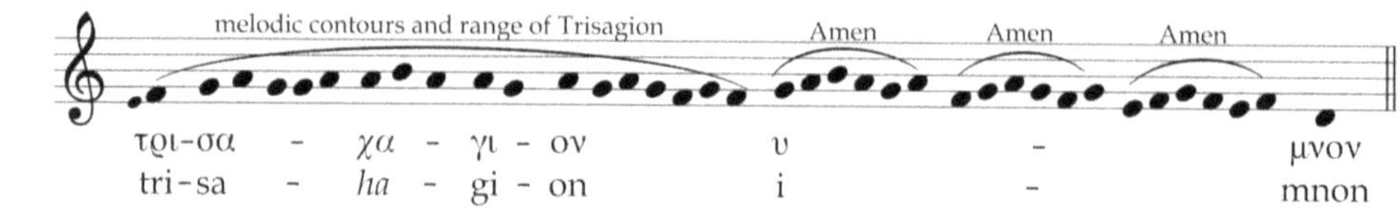

b. Tripartite Trisagion melody sets the opening of the Cheroubikon

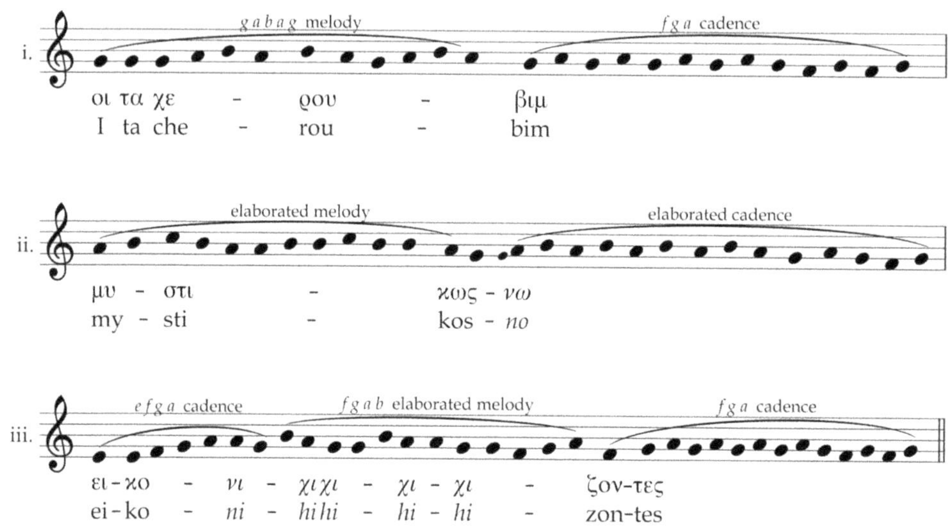

Figure 7.3 Two settings of the Cheroubikon by Agathon Korones represent the Trisagion in two different ways. (a) The melody on *Trisagion* loosely imitates the contours and range of the Trisagion hymn. *Imnon* begins with an *Amen* formula that repeats in a descending sequence. Agathon Korones, Cheroubikon 1, fourteenth century. Athens, National Library, MS Gr. 2837, fol. 160r; from Conomos, *Byzantine Trisagia and Cheroubika*, 168–71. (b) The melody on the opening lyric *I ta cheroubim mystikos eikonizontes* (we who mystically represent the Cherubim) precisely imitates the Trisagion: (i) Opening melody based on *g a b a g Amen* formula. (ii) Range expands up to include *c* and down to include *f* in an elaboration of the opening melody. (iii) Cadential melody intercalated with nonsemantic aspirations reminiscent of the descending breath motif at the Great Entrance that depicts the descent of the Holy Spirit (see Figure 7.8). Agathon Korones, Cheroubikon 2, fourteenth century. Athens, National Library, MS Gr. 2622, fol. 353v; from Conomos, *Byzantine Trisagia and Cheroubika*, 173–74.

singing formula, and descending *pneuma*), the semantic function of the melodic motifs reaches a climax of meaning and complexity at the apex of the Cheroubikon.

Conomos's procession motif begins with a repeated tone, just as in the beginning of the Great Entrance melodies. For example, in the Asmatikon Trisagion, the first four tones are *a a a b*, and at the Great Entrance in a setting by Korones, the first six tones are *a a a a b a*. Conomos points to a similar procession motif in the Asmatikon Trisagion of the Adoration of the Cross.[59]

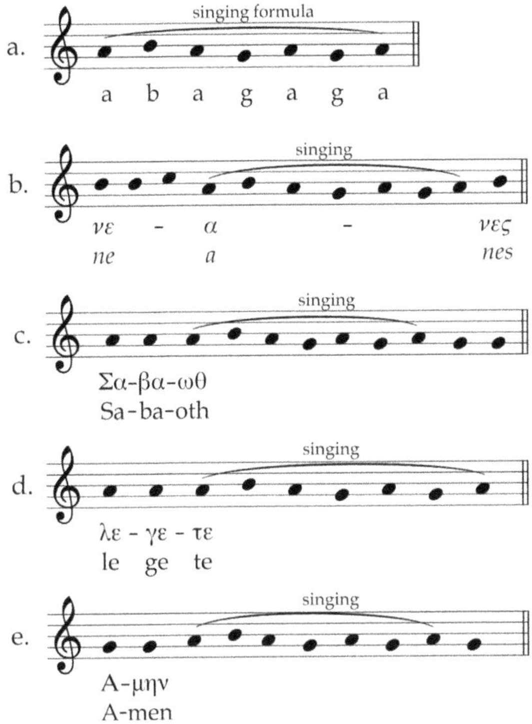

Figure 7.4 The singing formula in the Trisagion, Sanctus, and Cheroubikon. (a) Singing formula prototype with pitch names (*a b a g a g a*) indicated beneath each notehead. (b) Asmatikon Cheroubikon *neanes* intonation. Athens, National Library, MS Gr. 2622, fol. 345r; from Conomos, *Byzantine Trisagia and Cheroubika*, 124–37. (c) Asmatikon Sanctus melody on *Sabaōth* (hosts of heaven). Athens, National Library, MS Gr. 2458, fols. 167v–68r; from Levy, "The Byzantine Sanctus," 13. (d) Asmatikon Adoration of the Cross processional melody on "legete" (say!). Vienna, Österreichische Nationalbibliothek, MS Vienna theol. gr. 185, fol. 238r; from Conomos, *Byzantine Trisagia and Cheroubika*, 68. (e) Amen formula in the Trisagion. Grottaferrata, Cod. Crypt. Γ. γ. I, fol. 33v; from Conomos, *Byzantine Trisagia and Cheroubika*, 54.

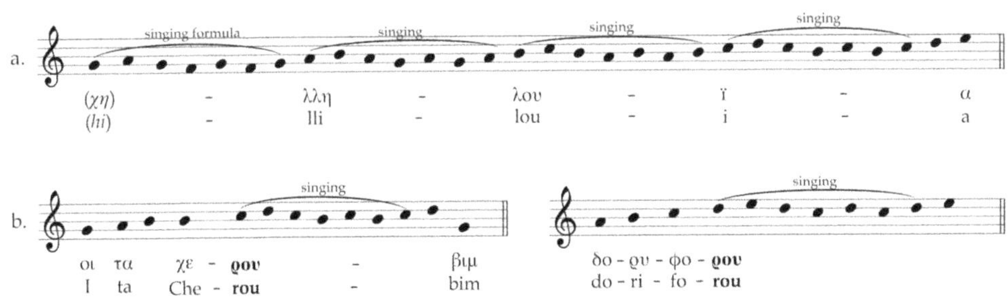

Figure 7.5 Manuel Agallianos carefully sets the singing formula set to specific syllables. (a) The singing formula repeats four times in a rising sequence on *allilouia*. (b) The singing formula occurs on both instances of the syllable *rou*, in *che<u>rou</u>bim* and *dorifo<u>rou</u>menon*. Manuel Agallianos, Cheroubikon, fourteenth century. Athens, National Library, MS Gr. 2837, fol. 156r; from Conomos, *Byzantine Trisagia and Cheroubika*, 151–55.

Although the two melodies are more distinct in the Cheroubika because the procession motif tends to be prolonged, in the Asmatikon Trisagion the singing formula overlaps with the procession motif. When Conomos discusses this instance of the procession motif, he is interested in the repeating tone but does not discuss the presence of the singing formula as a salient entity, despite his mention of the same formula in his analysis of Agallianos's Cheroubikon.[60] Instead, he refers to the entire Asmatikon Trisagion melody more generically, as material that revolves around *a*.[61]

Expanding on these observations, the processional motif is even more widespread, occurring in the Asmatikon Sanctus on *Sabaōth* (the hosts of heaven), and elsewhere in the Cheroubikon, aside from the Great Entrance. For example, there is a preponderance of the motif at the opening of the chant (Figure 7.6). This poses a slight problem because Conomos identifies the formula as an announcement of actual processions that interrupt the chant, while the motif occurring in the Sanctus (which proclaims Christ's procession into Jerusalem) or at the opening of the Cheroubikon (well in advance of the actual procession) suggests a symbolic usage of the motif, like a leitmotif that evokes the notion of procession but is not a functional announcement of a real procession.

More often than not, the procession motif is juxtaposed with the singing formula, appearing throughout the Cheroubikon as a combined motif that evokes both procession

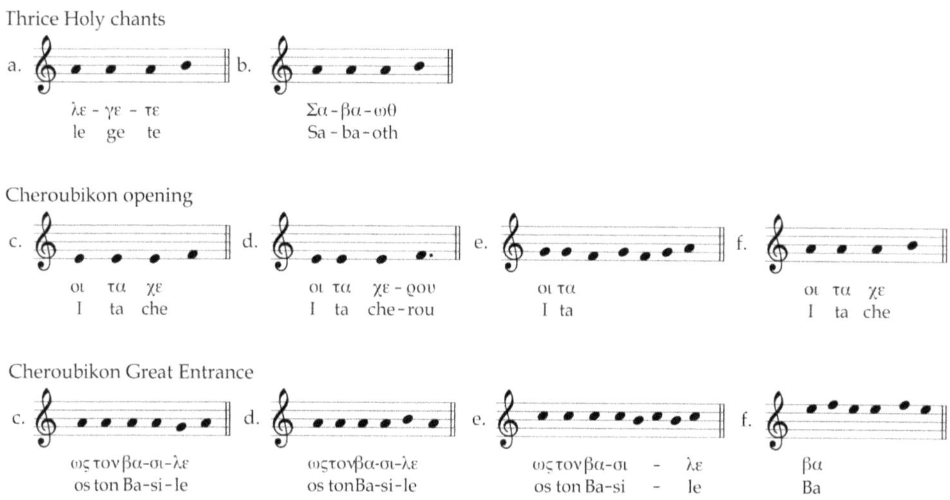

Figure 7.6 The processional motif in the Trisagion, Sanctus, and Cheroubikon. (a) Asmatikon Adoration of the Cross Trisagion solo announces the beginning of a procession. Vienna, Österreichische Nationalbibliothek, MS Vienna theol. gr. 185, fol. 238r; from Conomos, *Byzantine Trisagia and Cheroubika*, 68. (b) Asmatikon Sanctus melody on *Sabaōth* (hosts of heaven). Athens, National Library, MS Gr. 2458, fols. 167v–68r; from Levy, "The Byzantine Sanctus," 13. (c) John of Damascus, Cheroubikon, likely pre-fourteenth century. Athens, National Library, MS Gr. 2458, fol. 165v; from Conomos, *Byzantine Trisagia and Cheroubika*, 143–44. (d) Agathon Korones, Cheroubikon, fourteenth century. Athens, National Library, MS Gr. 2837, fol. 160r; from Conomos, *Byzantine Trisagia and Cheroubika*, 168–71. (e) John Koukouzeles, Cheroubikon, fourteenth century. Athens, National Library, MS Gr. 2406, fol. 465v; from Conomos, *Byzantine Trisagia and Cheroubika*, 188–89. (f) Manuel Chrysaphes, Cheroubikon, fifteenth century. Athens, National Library, MS Gr. 2406, fol. 462r; from Conomos, *Byzantine Trisagia and Cheroubika*, 193.

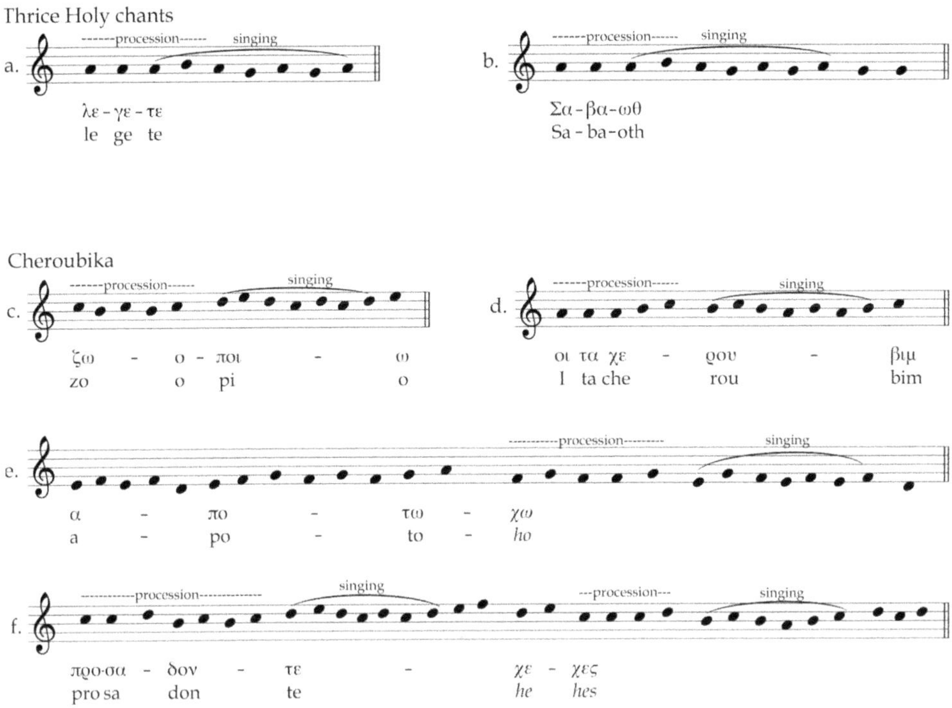

Figure 7.7 The escorting-singing motif corresponds to real and symbolic moments of processions in the Trisagion, Sanctus, and Cheroubikon that are accompanied by angelic singing. (a) Asmatikon Adoration of the Cross Trisagion. Solo melody sung to announce a procession. Vienna, Österreichische Nationalbibliothek, MS Vienna theol. gr. 185, fol. 238r; from Conomos, *Byzantine Trisagia and Cheroubika*, 68. (b) Asmatikon Sanctus. Escorting-singing motif on *Sabaōth* (hosts of heaven). Athens, National Library, MS Gr. 2458, fols. 167v–68r; from Levy, "The Byzantine Sanctus," 13. (c) John Koukouzeles, Cheroubikon, fourteenth century. The escorting-singing motif can occur throughout the Cheroubikon. Here, Koukouzeles sets it to *zoopio* (life-giving). Athens, National Library, MS Gr. 2406, fol. 465v; from Conomos, *Byzantine Trisagia and Cheroubika*, 188–89. (d) Manuel Chrysaphes, Cheroubikon, 15th century. Escorting-singing motif on *I ta cheroubim* (We the cherubim). Athens, National Library, MS Gr. 2406, fol. 462r; from Conomos, *Byzantine Trisagia and Cheroubika*, 193. (e) Xenos Korones, Cheroubikon, fourteenth century. Escorting-singing motif on the intercalated nonsemantic aspiration *ho*. Athens, National Library, MS Gr. 2837, fol. 157v; from Conomos, *Byzantine Trisagia and Cheroubika*, 158–63. (f) Markos hieromonachos sets the escorting-singing motif *prosadonte-he-hes* with two successive iterations of the escorting-motif, first set to the word "singing" and then set to nonsemantic language, depicting angelic singing that joins in with the earthly choir. Athens, National Library, MS Gr. 899, fol. 120v; from Conomos, *Byzantine Trisagia and Cheroubika*, 190–91.

and singing. The combination of procession and singing is present in the Adoration of the Cross Trisagion and the Asmatikon Sanctus melodies as well: the singing formula emerges from the repeated-tone procession motif (Figure 7.7). By combining the notion of procession with the notion of angelic singing, the cherubim are evoked, singing perpetually and escorting the throne of God. As a pair, these leitmotifs will henceforth be referred to as the escorting-singing motif.

In the Cheroubikon, the escorting-singing motif is prevalent at the Great Entrance; but the two motifs do not occur together. Instead, they are separated by some distance, which could also explain why Conomos did not mention the presence of the singing formula at the Great Entrance. The reason for this separation is a third motif inserted in the middle of the escorting-singing formula at the Great Entrance. This third motif is a melody on *Basilea* that descends stepwise by a fourth, symbolizing the descent of God from heaven. In most cases, accompanying the descent are the aspirated syllables *he-* and *ha-* that, by symbolizing breath, represent the Holy Spirit. Conomos writes about the inconsistent placement of the intercalated syllables among different transmissions of the same chant, demonstrating the randomness by listing all the variations

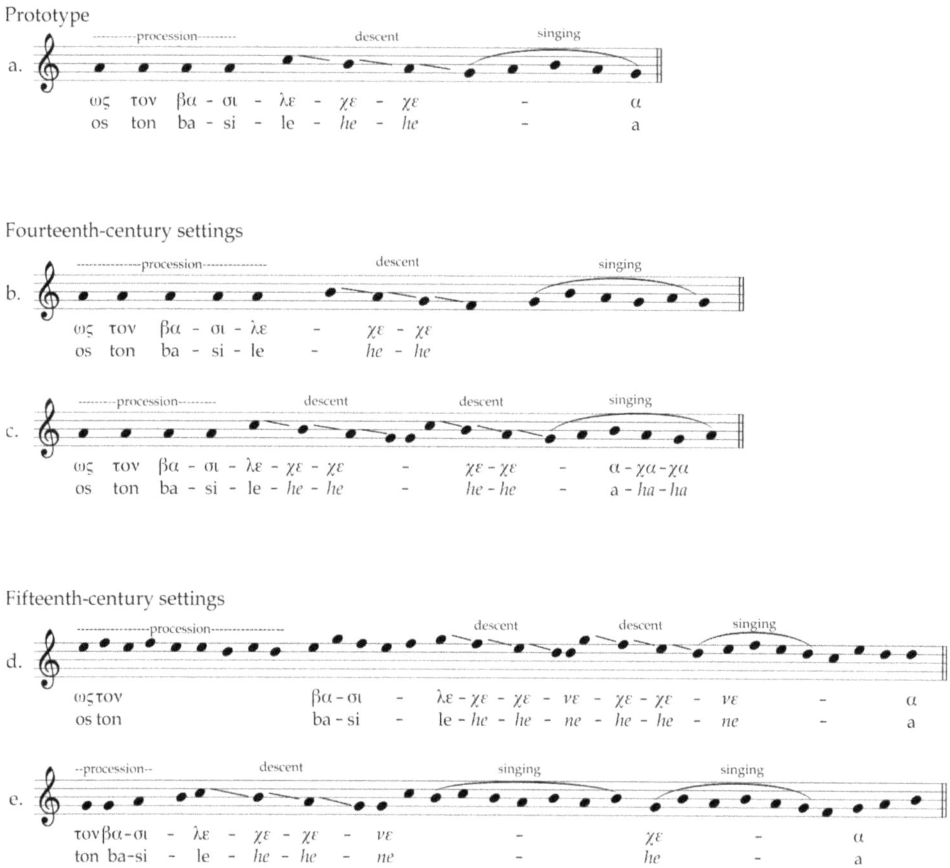

Figure 7.8 Procession + Descent + Singing motifs appear at the Great Entrance on "os ton basilea" (the King of All), depicting the descent of King of All to earth escorted by the singing cherubim. (a) Prototype of the sequence of procession, descent, and singing motifs. (b) Agathon Korones, Cheroubikon. Athens, National Library, MS Gr. 2837, fol. 160r; from Conomos, *Byzantine Trisagia and Cheroubika*, 168–71. (c) Xenos Korones, Cheroubikon. Athens, National Library, MS Gr. 2837, fol. 157v; from Conomos, *Byzantine Trisagia and Cheroubika*, 158–63. (d) Markos hieromonachos. Athens, National Library, MS Gr. 899, fol. 120v; from Conomos, *Byzantine Trisagia and Cheroubika*, 190–91. (e) Manuel Chrysaphes. Athens, National Library, MS Gr. 2406, fol. 464v; from Conomos, *Byzantine Trisagia and Cheroubika*, 196–98.

among the many transmissions of the fourteenth-century setting by Glykes. However, intriguingly, the intercalated aspirations on *Basilea* are present in every single setting, which suggests that the aspirations were a crucial detail at the Great Entrance.[62] Thus, the apex of the entire liturgy, the presence of Christ in the Eucharist, is musically represented in the Cheroubikon. When all three motifs are put together—procession, descending breath, singing—their meanings combine, mirroring the image of the trinitarian King of All who enters, escorted by the singing cherubim (Figure 7.8). This moment (*basilea tōn holōn*) frequently marks the melodic apex of the setting as well, so that the King of All descends from the highest note of the chant. One can only imagine how the impact of such a dramatic moment could have been heightened by the arrival of the procession under the dome of the Hagia Sophia with its sublime acoustics.[63]

Vocalizations of the performed rite

Two final cases demonstrate that musical symbolism is not limited to the depiction of the God and the angels, but that the human participants in the ritual are also represented through two cases of word painting that imitate the vocalizations of the bishop and the choir. The lyric "we who mystically represent the cherubim" is turned into a musical pun that mimics the voice of the bishop who recites a series of prayers in a *mystikōs* voice.[64] The word carries two meanings, being both a vocal action and a theological concept. Theologically and liturgically, the term means "mystically," connoting the Mystical Supper of the Eucharist; in the context of the performance of the rites it also means "to speak in a low or inaudible voice."[65] In twelve of the Cheroubika, the melody descends on *mystikōs*, a structural motif that could be mimicking the low voice of the priest or could also be depicting the descent of the Holy Spirit that characterizes the miracle of the Eucharist. It is even possible that both meanings are intended simultaneously: at the beginning of the Cheroubikon, the deacon, priest, or bishop recites Psalm 50(51) in a *mystikōs* voice while incensing the altar, prothesis, sanctuary, ministers, iconostasis, and congregation.[66] Verses 10–16 of this psalm are chiastic, meaning they have a mirroring structure that descends to the center. At the center is three occurrences of *pneuma*, Holy Spirit.[67] This moment brings together the descent of the choir on *mystikōs*, the *mystikōs* voice of the bishop, the evocation of descending *pneuma* through Psalm 50, and incense as a sentient presence of *pneuma*.

In the chant melody, descent occasionally occurs over the course of the word, as in the setting by John Koukouzeles, but most often the melody suddenly jumps down by either a fourth or a fifth at the onset of *mystikōs* (Figure 7.9). What marks *mystikōs* is not only its lowness but also that it does not linger in the lower range after the utterance of this word; instead, the melody rises up with *eikonizontes* (the ones who represent), depicting the clergy who are spiritually elevated as they enact the Divine Liturgy. Similarly, the choir is depicted on *prosadontes* (singing). In many settings, *prosadontes* is extended by repeating the first two syllables: *prosa-prosadontes.* The first instance of *prosa* lengthens the word, setting aside space for a melody that self-reflexively calls attention to the voice, achieved in settings by Koukouzeles, Markos hieromonachos, and Chrysaphes through a wide-ranging descending melisma. Xenos Korones exemplifies another strategy, setting *prosa* with intonation melodies and syllables that evoke singing (Figure 7.10).

Conclusion: a paradox of time and timelessness

The paradox of past and future defines the creative challenge that confronted the medieval composer, requiring simultaneous borrowing and innovation. Similarly, inherent in the Cheroubikon is a paradox of time and timelessness. The terrestrial Great Entrance is linear, moving forward in time with a dramatic apex in the second half of the piece, but in doing so it mirrors the circular, never-ending celestial ritual. For the

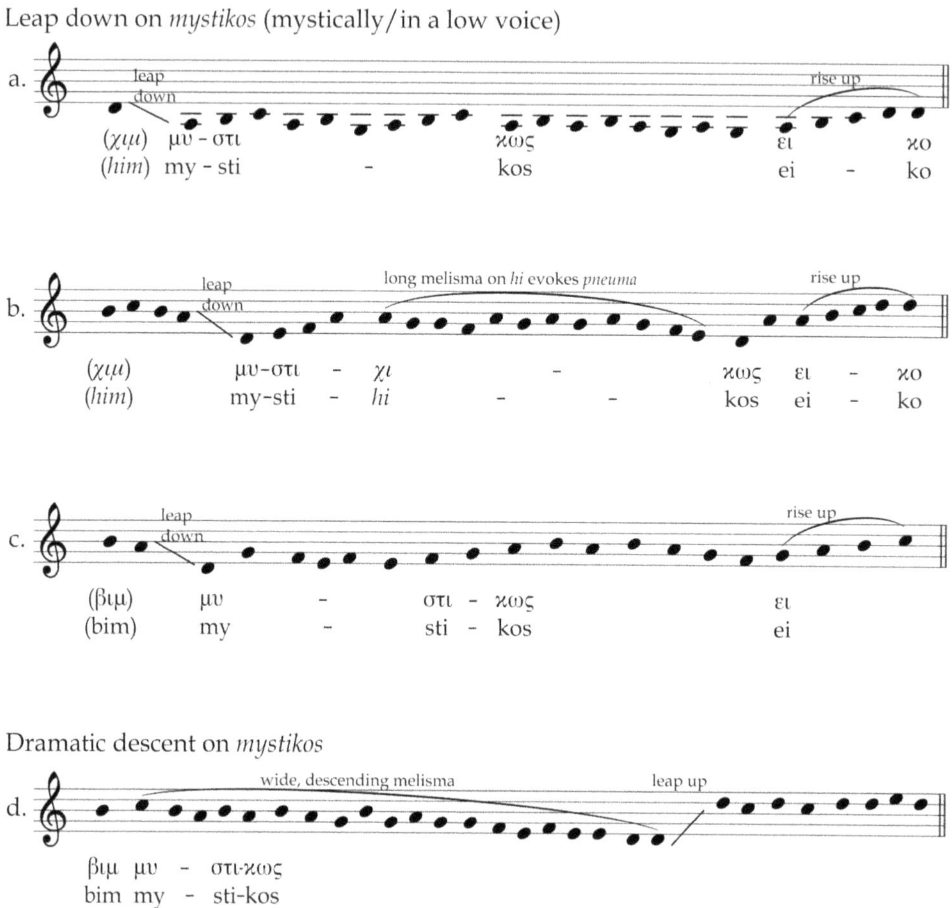

Figure 7.9 Representing a low voice and descending spirit on *mystikos*. In exx. a–c, the melody leaps down at the beginning of *mystikos* (mystically/in a low voice) and rises up on *eikonizontes* (representing). In ex. d, the melody descends over the course of *mystikos*, then leaps up at the end. (a) Xenos Korones, Cheroubikon, fourteenth century. Athens, National Library, MS Gr. 2837, fol. 157v; from Conomos, *Byzantine Trisagia and Cheroubika*, 158–63. (b) Ioannes Kladas, Cheroubikon, 14th century. Sinai, Monastery of St. Catherine's, MS Gr. 1293, fol. 256r; from Conomos, *Byzantine Trisagia and Cheroubika*, 176–82. (c) Manuel Chrysaphes, Cheroubikon, fifteenth century. Athens, National Library, MS Gr. 2406, fol. 464v; from Conomos, *Byzantine Trisagia and Cheroubika*, 196–98. (d) John Koukouzeles, Cheroubikon, fourteenth century. Athens, National Library, MS Gr. 2406, fol. 465v; from Conomos, *Byzantine Trisagia and Cheroubika*, 188–89.

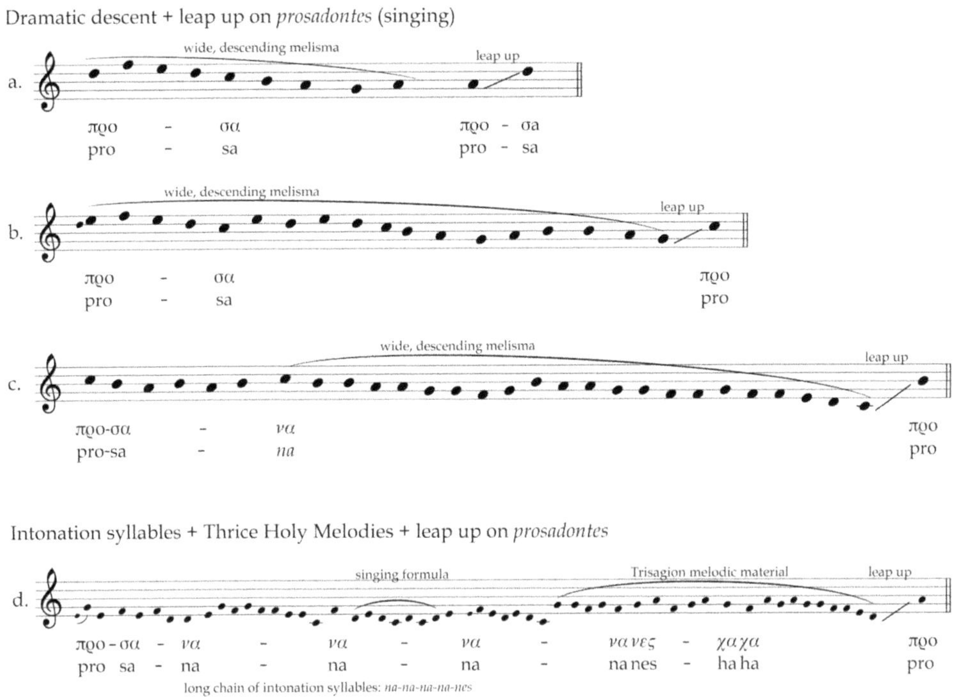

Figure 7.10 Representing a singing voice on *prosadontes* (singing). In exx. a–c, a wide, descending melody on *prosa* leaps up at the end. In ex. d, intonation syllables proliferate on *prosa* along with a singing formula and other melodic imitations of the Thrice Holy. (a) John Koukouzeles, Cheroubikon, fourteenth century. Athens, National Library, MS Gr. 2406, fol. 465v; from Conomos, *Byzantine Trisagia and Cheroubika*, 188–89. (b) Markos hieromonachos, Cheroubikon, fifteenth century. Athens, National Library, MS Gr. 899, fol. 120v; from Conomos, *Byzantine Trisagia and Cheroubika*, 190–91. (c) Manuel Chrysaphes, Cheroubikon, fifteenth century. Athens, National Library, MS Gr. 2406, fol. 464v; from Conomos, *Byzantine Trisagia and Cheroubika*, 196–98. (d) Xenos Korones Cheroubikon, fourteenth century. Athens, National Library, MS Gr. 2837, fol. 157v; from Conomos, *Byzantine Trisagia and Cheroubika*, 158–63.

twenty-first-century composer, the medieval Cheroubika offer a wealth of insight that pertains to the imitation of the eternal, as in time-based reality things must change in order to stay the same. The consistency of the chant, achieved through the repetition of melodic formulas such as the singing formula, approximates the timeless, while the lyrical and dramatic motifs that mark the energetic arc of the ritual provide the necessary scaffolding for introducing subtle changes into the flow of chant.

Notes

1 Manuel Chrysaphes, *The Treatise of Manuel Chrysaphes: On the Theory of the Art of Chanting and on Certain Erroneous Views That Some Hold about It*, trans. Dimitri Conomos (Vienna: Verlag der Österreichischen Akademie der Wissenschaften, 1985), 43. On the post of *lampadarios*, see Neil Moran, *Singers in Late Byzantine and Slavonic Painting* (Leiden: Brill, 1986), 19, 28, 90, 107, 119, 153.

2 Dimitri E. Conomos, *Byzantine Trisagia and Cheroubika of the Fourteenth and Fifteenth Centuries: A Study of Late Byzantine Liturgical Chant* (Thessaloniki: Patriarchal Institute for Patristic Studies, 1974), 35.
3 Conomos, *Byzantine Trisagia and Cheroubika*, 121.
4 Robert Taft, *A History of the Liturgy of St. John Chrysostom: The Great Entrance* (Rome: Orientalia Christiana Analecta, 1978), 68–69, 84–86, 98–112. Wolfgang Christian Schneider and Rudolf H. W. Stichel, "Der 'Cherubinische Einzug' in der Hagia Sophia Justinians," in *Performativität und Ereignis*, ed. Erika Fischer-Lichte et al. (Tübingen: A. Francke, 2003), 377–94, esp. 386–90.
5 Taft, *The Great Entrance*, 69.
6 Taft, *The Great Entrance*, 1–7. Germanus, *Historia ekklesiastikē kai mystikē thēoria*, trans. Paul Meyendorff (Crestwood, NY: St. Vladimir's Seminary Press, 1984) (hereafter *HE*), bk. 1, sects. 16, 37, 41, pp. 67, 87, 95.
7 "Οἱ τὰ χερουβὶμ μυστικῶς εἰκονίζοντες καὶ τῇ ζωοποιῷ Τριάδι τὸν τρισάγιον ὕμνον προσᾴδοντες, πᾶσαν τὴν βιωτικὴν ἀποθώμεθα μέριμναν ὡς τὸν Βασιλέα τῶν ὅλων ὑποδεξόμενοι, ταῖς ἀγγελικαῖς ἀοράτως δορυφορουμένον τάξεσιν. Ἀλληλουΐα, ἀλληλουΐα, ἀλληλουΐα." Greek and English in Taft, *The Great Entrance*, 54; and Conomos, *Byzantine Trisagia and Cheroubika*, 31.
8 Conomos, *Byzantine Trisagia and Cheroubika*, 123.
9 Taft, *The Great Entrance*, 62–68.
10 Bissera Pentcheva, "Performing the Sacred in Byzantium," *Performance Research* 19, no. 3 (2014): 120–27, esp. 120.
11 Taft, *The Great Entrance*, 64.
12 Conomos, *Byzantine Trisagia and Cheroubika*, 123.
13 The seventeen settings by century and composer, including example numbers from Conomos's *Byzantine Trisagia and Cheroubika.* Pre-fourteenth-century: ex. 2, Asmatikon; ex. 7, John of Damascus. Fourteenth century: ex. 8, Glykes; ex. 9, Agallianos; ex. 10, Xenos Korones; ex. 14, Agathon Korones; ex. 16, A. Korones; ex. 18, Kladas; ex. 21, John Koukouzeles. Fifteenth century: ex. 22, Markos hieromonachos; ex. 23, Chrysaphes; ex. 24, Chrysaphes; ex. 25 Argyropoulos/Chrysaphes; ex. 26, Theodoulos monachos; ex. 27, Athanasios monachos; ex. 28, Grigorio hieromonachos; ex. 29, Nicholas the Serb.
14 Christian Troelsgård, *Byzantine Neumes* (Copenhagen: Museum Tusculanum Press, 2011), 35.
15 Alexander Lingas, "Performance Practice and the Politics of Transcribing Byzantine Chant," *Acta Musicae Byzantinae* 6 (2003): 56–76, esp. 59.
16 See Walter Ray's essay in this volume.
17 Conomos, *Byzantine Trisagia and Cheroubika*, 69, 123.
18 Conomos, *Byzantine Trisagia and Cheroubika*, 123. *Grove Music Online*, s.v. "Idee fixe," by Hugh Macdonald, accessed March 28, 2016, www.oxfordmusiconline.com/subscriber/article/grove/music/13701. *Grove Music Online*, s.v. "Leitmotif," by Arnold Whittall, accessed March 28, 2016, www.oxfordmusiconline.com/subscriber/article/grove/music/16360. *The New Grove Dictionary of Opera*, s.v. "Reminiscence motif," accessed March 28, 2016, www.oxfordmusiconline.com/subscriber/article/grove/music/O009217.
19 *Grove Music Online*, s.v. "Word-painting," by Tim Carter, accessed March 28, 2016, www.oxfordmusiconline.com/subscriber/article/grove/music/30568.
20 Troelsgård, *Byzantine Neumes*, 88–90.
21 Conomos, *Byzantine Trisagia and Cheroubika*, 21.
22 Diane Touliatos, "Nonsense Syllables in the Music of the Ancient Greek and Byzantine Traditions," *Journal of Musicology* 7, no. 2 (1989): 231–43, esp. 241.
23 Touliatos, "Nonsense Syllables," 241–42.
24 Conomos, *Byzantine Trisagia and Cheroubika*, 273–74.
25 "Whatsoever ye cannot explain in words, you do not therefore forbear exulting: what you shall be able to explain, cry out: what ye cannot, jubilate." Augustine, "Exposition on Psalm 81," in *Nicene and Post-Nicene Fathers, First Series*, trans. J. E. Tweed, ed. Philip Schaff (Buffalo, NY: Christian Literature Publishing Co., 1888), 8:849.
26 Conomos, *Byzantine Trisagia and Cheroubika*, 275–76.
27 Touliatos, "Nonsense Syllables," 239.
28 Touliatos, "Nonsense Syllables," 235, quoted in *Aristides Quintilianus on Music: In Three Books*, trans. Thomas J. Mathiesen (New Haven, CT: Yale University Press, 1983), 201.

29 Troelsgård, *Byzantine Neumes*, 61.
30 Chrysaphes, *The Treatise of Manuel Chrysaphes*, 41–43.
31 Troelsgård, *Byzantine Neumes*, 62–65.
32 Egon Wellesz, "Words and Music in Byzantine Liturgy," *Musical Quarterly* 33, no. 3 (1947): 297–310, esp. 307.
33 Troelsgård, *Byzantine Neumes*, 23.
34 Peter Jeffery, "The Lost Chant Tradition of Early Christian Jerusalem: Some Possible Melodic Survivals in the Byzantine and Latin Chant Repertories," *Early Music History* 11 (1992): 151–90, esp.186.
35 Jeffery, "The Lost Chant Tradition," 166–70.
36 Peter Jeffery, "The Earliest Chant Repertory Recovered: The Georgian Witness to the Jerusalem Chant," *Journal of the American Musicological Society* 47, no. 1 (1994): 1–38, esp. 16–22.
37 Conomos, *Byzantine Trisagia and Cheroubika*, 123.
38 Conomos, *Byzantine Trisagia and Cheroubika*, 67. Alexander Lingas, "Sunday Matins in the Byzantine Cathedral Rite: Music and Liturgy" (PhD diss., University of British Columbia, 1996), 109.
39 Conomos, *Byzantine Trisagia and Cheroubika*, 67–69, 122–23.
40 Troelsgård, *Byzantine Neumes*, 86.
41 Bissera Pentcheva, "In-Spiriting in the Byzantine Consecration (*Kathierosis*) Rite," *Codex Aquilarensis* 30 (2014): 37–65, esp. 56–62.
42 Pentcheva, "Performing the Sacred in Byzantium," 120.
43 *The Divine Liturgy of St John Chrysostom*. English trans.: www.goarch.org/chapel/liturgical_texts/liturgy. Greek: F. E. Brightman and C. E. Hammond, *Liturgies Eastern and Western*, vol. 1, *Eastern Liturgies* (Oxford: Clarendon Press, 1896). Germanus, *HE*, 74–77 (section 25 on the Trisagion) and 88–101 (section 41 on the Sanctus).
44 Pseudo-Dionysius, *The Celestial Hierarchy*, Caput VII, sect. IV. English trans., John Parker, 30, https://en.wikisource.org/wiki/Heavenly_Hierarchy/Caput_VII.
45 Bryan D. Spinks, *The Sanctus in the Eucharistic Prayer* (Cambridge: Cambridge University Press, 1991), 60–82.
46 "Ἅγιος Κύριος Σαβαώθ πλήρις ὁ οὐρανὸς καὶ ἡ γῆ τῆς δόξης σου Ὠσαννὰ ὁ ἐν τοῖς ὑψίστοις. Εὐλογημένος ὁ ἐρχόμενος ἐν ὀνόματι Κυριου 'Ωσαννὰ ὁ ἐν τοῖς ὑψίστοις." *The Divine Liturgy of St John Chrysostom*. Greek, Brightman and Hammond, *Eastern Liturgies*, 385; and Kenneth Levy, "The Byzantine Sanctus and Its Modal Tradition in East and West," *Annales Musicologiques* 6 (1958–63): 7–67, esp. 13. English trans.: www.goarch.org/chapel/liturgical_texts/liturgy.
47 Taft, *The Great Entrance*, 107–08.
48 John of Damascus, *An Exact Exposition of the Orthodox Faith*, 3.10, in *St. John of Damascus: Writings*, trans. Frederic H. Chase Jr. (Washington, DC: Catholic University of America Press, 2014), 288–89.
49 Conomos, *Byzantine Trisagia and Cheroubika*, 25.
50 "Ἅγιος ὁ Θεός, Ἅγιος Ἰσχυρός, Ἅγιος Ἀθάνατος, ἐλέησον ἡμᾶς." Greek and English in Conomos, *Byzantine Trisagia and Cheroubika*, 25.
51 Spinks, *The Sanctus in the Eucharistic Prayer*, 11.
52 Hans-Joachim Schulz, *The Byzantine Liturgy*, trans. Matthew J. O'Connell (New York: Pueblo Publishing Company, 1986), 165. Taft, *The Great Entrance*, 65.
53 Schulz, The Byzantine Liturgy, 22. John of Damascus, An Exact Exposition of the Orthodox Faith, 3.10.
54 Taft, *The Great Entrance*, 225–26.
55 Conomos, *Byzantine Trisagia and Cheroubika*, 145.
56 Conomos, *Byzantine Trisagia and Cheroubika*, 121.
57 Conomos, *Byzantine Trisagia and Cheroubika*, 158 (ex. X. Korones, ex. 10, ll. 6–7), (A. Korones, ex. 14, ll. 5–6), 177 (Kladas, ex. 18, ll. 5–6), 202 (Theodoulos monachos, ex. 26, ll. 4–6).
58 Troelsgård, *Byzantine Neumes*, 60–63.
59 Conomos, *Byzantine Trisagia and Cheroubika*, 69.
60 Conomos, *Byzantine Trisagia and Cheroubika*, 157.
61 Conomos, *Byzantine Trisagia and Cheroubika*, 69.
62 Conomos, *Byzantine Trisagia and Cheroubika*, 264.

63 For the acoustics of the dome as well as the role of processions through the interior in activating the acoustics of the Great Church, see Wieslaw Wozsczyk's essay in this volume.

64 *The Divine Liturgy of St John Chrysostom*. English trans.: www.goarch.org/chapel/liturgical_texts/liturgy. Greek: Brightman and Hammond, *Eastern Liturgies*, 309–44, esp. 324.

65 Thanks to Bissera Pentcheva for calling my attention to the multiple definitions of this word. See *A Greek-English Lexicon*, ed. Henry George Liddell et al. (1843; Oxford: Oxford University Press, 1961), s.v. "μυστικός," 1156; *A Patristic Greek Lexicon*, ed. G. W. H. Lampe, ed. (1961; Oxford: Oxford University Press, 2001), s.v. "μυστικός" and "μυστικῶς," 893; the citation above is found in Peter Day, "Mystic Recitation," in *The Liturgical Dictionary of Eastern Christianity* (Collegeville, MN: Liturgical Press, 1993), 204.

66 Taft, *The Great Entrance*, 4–5.

67 Pentcheva, "In-Spiriting in the Byzantine Consecration (*Kathierosis*) Rite," 47–50.

8 Spatiality, embodiment, and agency in ekphraseis of church buildings

Ruth Webb

The ekphraseis of sacred buildings that are such a precious source of information about the appearance and structure of monuments and about the ways in which those monuments were seen by contemporary viewers are often, for very good reasons, associated primarily with the sense of sight. The self-proclaimed goal of ekphrasis is to bring its subject "before the eyes," and today we encounter these texts primarily through the medium of the written word laid out on the printed page. Originally, however, whether composed in poetry or in prose, they were spoken performances whose rhythms and sonorities filled the spaces in which they were recited (which were usually not the buildings described). What is more, they engaged far more senses than the sense of sight, evoking the churches as sites of multisensory experiences.[1] In this essay I will explore one particular aspect of these texts: their evocation of a perceiving body—surrounded by and moving through the space defined by the building—in order to create a sense of presence that goes far beyond pure visuality. These evocations combine appeals to different bodily senses in the context of that body's position within the space and its physical engagement with that space. That these texts do more than represent such experiences, that they may cause the listener or reader to imagine and thus to live them, is suggested by recent research on the impact of language, particularly action words, on the mind.[2]

Part of the experience of the building that the ekphraseis seek to evoke is their overall impact on the viewer-visitor, an effect that is often dramatic and that comes very close to the "captivation" analyzed by Alfred Gell in his study of agency.[3] In the second part of this essay I will follow this lead in order to consider some of the different sources of agency as presented by the authors of the ekphraseis, particularly Paul the Silentiary, ending with the agency of the authors themselves. Their contribution is important because, tempting as it is to read their texts as evidence for experience, the authors are an integral part of the networks of political and other powers that surrounded both words and monuments and, by creating virtual experiences in the reader's or listener's mind, they may contribute to the constitution of that experience.

Spatiality

A common way of organizing ekphraseis of any type of building or place is to describe a journey through it, using the form of the *periēgēsis*. That the "tour around" the place, space, or other subject is intimately and profoundly connected with ekphrasis in general is suggested by the presence of the term *periēgēmatikos* in the ancient textbook definitions of ekphrasis as a "speech that leads one around, bringing the entity shown vividly before the eyes."[4] Commenting on this definition in the mid-ninth century, John of Sardeis explained it as being "as if someone took a recent arrival in Athens and guided him around the city, showing him the gymnasia and the Peiraeus and all the rest,"[5] thus bringing out the root meaning of *periēgeomai*, "to lead" (*hēgeomai*) "around" (*peri*). The presence of the idea of *periēgēsis* in the definition of ekphrasis does not necessarily signal a generic or literary-historical connection between periegematic literature, like the work of Pausanias, and ekphrasis. Instead, I would argue that it underlines the fundamental importance to ekphrasis of movement and of a particular conception of space. An ekphrasis does not merely lead a listener or reader straight through the subject matter, as a *diēgēsis* (from *dia*, "through," and *hēgeomai*) does, but lingers, metaphorically leading him or her around to explore different aspects of the subject. The distinction is comparable to that made by Nikolaos the Sophist (fifth century CE) in his *Progymnasmata* between *diēgēsis*, which simply states that something has happened (his example is "that the Athenians and Lacedaemonians were at war") with the ekphrasis, which elaborates on the way in which the events occurred (in Nikolaos's example this takes the form of details of the weapons used and the preparations that were made), investigating various facets of the subject described.[6] As Nikolaos's analysis implies, it is not only spatial entities such as places, buildings, or objects that lend themselves to this type of exploration but also events, which can be portrayed conceptually in greater or lesser detail, depending on the degree and type of focalization (the notional standpoint from which a narrator speaks).[7]

When the subject does happen to be a building or a place, the tour can become a powerful organizing principle for the verbal representation, as in the ekphrasis of the Alexandrian Acropolis proposed as a model by Aphthonios in his *Progymnasmata* or in ekphraseis of cities contained within epideictic speeches.[8] From the perspective of the orator or poet, the *periēgēsis* format has the advantage of corresponding to the natural way in which we remember places as experienced and, as is well known, in ancient mnemotechnics the speech itself, whatever the subject matter or genre, could be recorded mentally in the form of a tour around a familiar building in which each section corresponded to a room or particular space.[9] Authors of ekphrasis vary in the details of the presentation and, in particular, in the focalization used and the role of the audience. In his description of the Church of St. Sergios at Gaza in praise of its patron, Bishop Marcian, Chorikios makes use of the *periēgēsis* format, beginning from the exterior and evoking the individual listener's experience of it through the use of the second-person singular and of verbs of motion: "as you go" (17), "as you go up" (20), "as you move through" (22).[10] We can contrast this technique, and its effect on the listener, to the very different strategy chosen by Procopius for his description of Hagia Sophia (*Buildings* 1.1.27–78), in which the periegetic element is attenuated. He starts with general comments about the overall shape and proportions of the building (an emphasis that is entirely consistent with the immediate context of the passage), moving from exterior to interior, describing from a standpoint that is never clearly defined. Although there are occasional remarks on the impression made by the whole on the visitor, defined as a generalized "someone" (*tis*), there is no apostrophe of the listener. Procopius's choices result in an overall effect of distance between speaker, audience, and building that is very different from that of Chorikios's speech. Paul the Silentiary's strategy lies between these two extremes and takes the form of an attenuated tour beginning with the eastern end of Hagia Sophia, the part directly opposite the visitor who enters via the narthex (*Descriptio S. Sophiae*, ll. 354–410). The western end is only described later (ll. 417–43), after what the manuscript notes is a pause in the original recitation, followed by an explanation of the central dome (ll. 444–550), the side aisles (ll. 550–85), and the courtyards (ll. 586–616).[11] Although Paul does not evoke the viewer's progression through the building by using verbs of motion, he does frequently use the second-person singular (for example, l. 389, νοήσεις, "you will perceive," l. 399, ἐσόψεαι, "you will see"), collapsing the distinction between the notional viewer through whom the description is focalized and the audience of the speech in order to situate the listener in his or her mind's eye within the church. The narrative of construction is present in the overall organization of the ekphrasis, as was noted by Ruth Macrides and Paul Magdalino, in that the detailed account of the marble and mosaic decoration follows that of the structure.[12]

The *periēgēsis* form is much more than a convenient organizational scheme for the orator wishing to represent entities that exist in space through the medium of language. In this context, the questions of focalization alluded to above are of far more than purely technical interest, as is suggested by the discussion of the direct address to the reader-listener in the treatise *On the Sublime*, which may well belong to the third century CE.[13] The author, to whom I will refer as Longinos, points to the strong impact that such uses of apostrophe could have on the reader (let alone the listener, in physical proximity to the describer). Commenting on a brief account given by Herodotos (*Histories*, 2.29) of the journey from Elephantine to Meroe, Longinos remarks on how the historian, by means of his words, "seizes the soul and leads it through the places, transforming hearing into sight."[14] Longinos's commentary underlines the importance of the use of the second-person singular by Herodotos, since it is through this means that the effect is achieved. The further significance of this passage lies precisely in its focus on the listener. As Longinos implies, this is a powerful rhetorical device that encourages the listener not only to visualize (as any language that emphasizes the physical attributes of an entity would do) but to visualize in a situated manner. The device is akin to that of "embedded focalization" within a narrative in which the narrator speaks from the point of view of a participant in the action, and, again, the result is far from being merely formal or constative but draws the listeners into the situation by prompting them to rehearse mentally the act of moving through the space. Ekphrasis and *enargeia* (the quality of language which appeals to the imagination) were thought to create a feeling of presence that was spatial and sonorous as well as visual, underlining a multisensory engagement with the entity perceived and to do so through the use of sound in the form of the orator's voice in interaction with the place of recitation.[15] The importance of movement in these ekphraseis suggests, however, that further modes of perception may be involved.

Chorikios's account of the visitor's path through the outer colonnades toward the Church of St. Sergios itself is a particularly good example of this periegetic organization and of the way in which this exploration in motion engages not just the sense of sight but the senses in general. He gives a technical account of the form of the construction and the interaction of its parts involving relatively precise accounts of the approximate position of the various elements of the construction (the distance between the columns and walls, the symmetry of the whole, the details of the form of the arches and ceiling). As he does so, Chorikios situates the listener in imagination within the building, explaining that "as you go up, an area measured out by four equal spaces, greets you" (*Laud. Marc.* 1.20). The emphasis is as much on the nature of the spaces delimited by the colonnades as on the building itself, the former being evoked through mentions of the "breadth" and through the reference to the encircling columns "running around" the central space. This subtle evocation of the volumes delimited by the construction and their visual and spatial interaction is followed by a startlingly sensual evocation of the wind that blows through the courtyard and its cooling effect on the body as it gently stirs the visitor's clothing:

> The long stoa to the west leads one in full of contentment as a delightful breeze blows from the western atrium, for it blows pleasantly and not excessively and penetrating inside your clothing it gently cools your body as your clothing is lifted up by its breath [*pneuma*].[16]

Chorikios's ekphrasis thus combines the sense of sight with that of touch (the experience of being touched rather than the haptic action of reaching and grasping) and with a spatial awareness that places the listener imaginatively within the building, surrounded by it and positioned within its volume along with the *pneuma*, the breath or the spirit, that also moves through it.[17]

By reproducing a personal experience of moving through space, these periegetic ekphraseis involve far more than the sense of sight. They achieve their effect by evoking an embodied viewer positioned within the building described and sensing it from different perspectives. One sense that is particularly relevant to our subject because of the emphasis on the viewer as a subject in motion is precisely the sense of position in space in relation to one's physical surroundings, an experience that is vividly and sensually evoked by Chorikios's breeze. Moreover, this detail allows him to represent the space not as empty and inert but as filled with a tangible, quasi-living entity that touches and caresses the human visitor. The space delimited by the monument is as significant as the stones and glittering surfaces that make up that monument.[18] Talking about such sensations involves a movement beyond the canonical Western aesthetic regime with its hierarchy of five senses in order to acknowledge the possibility of other modes of perception, such as kinesthesia and proprioception—the latter being the particular awareness of one's bodily position, motion, and balance explored in the domains of dance and physiology.[19] These latter studies suggest that such an awareness is a human universal, like the ability to walk upright, that is more or less recognized and expressed in different ways in different cultural contexts. Although I am not aware of any explicitly articulated ancient equivalent of proprioception or kinesthesia in Late Antiquity there was certainly an acute awareness of the importance of controlling gait, posture, and gesture, all of which demand a conscious surveillance and analysis of the subject's own bodily position and motion.[20]

Moreover, it seems that in our actual experience, we both perceive and remember objects (including monuments) as corporeally connected to us. Our relationship to the monument, as to any other physical entity, is not merely visual but corporeal; we understand and experience it with our whole bodies, in spatial relationship with it, as embodied perceivers aware not only of the visual appearance of our surroundings but of other potential interactions: the feel of the stone, mosaic, and marble to the hand or underfoot.[21] The emphasis placed in all the ekphraseis on the forms and their complex interrelationship, often evoked in terms of movement, encourages the listener to imagine his or her own bodily relation to those curves, domes, and massive pillars.

Periēgēsis disrupted

The ekphraseis thus attempt to evoke in words this bodily engagement through motion. Ironically, the impression of the listener-viewer's corporeal presence within the building emerges with greatest force at moments when the *periēgēsis* is interrupted and the viewer is said to be "struck" (*ekplettō*), rooted to the spot and unable to decide where to look. In the passage that immediately follows the evocation of the breeze quoted above, Chorikios expresses the impact of the first sight of the church interior as follows:

> As you go in, you will be made dizzy by the things seen and, striving to see everything together you will go away having seen nothing clearly, as your eyes are carried here and there as you make efforts to leave nothing unseen.[22]

This idea of the gaze being pulled in all directions is found in several church ekphraseis in both Late Antiquity and Byzantium.[23] Procopius of Caesarea makes use of it in his account of Hagia Sophia (*Buildings*, 1.1.47–48), but the most striking and most often cited formulation of the idea is to be found in Photios's tenth *Homily* on the Church of the Virgin at Pharos. Central to the effect is the physical placement of the virtual embodied viewer. The simulation of the act of looking around in all directions emphasizes the contrast between the physical limitations of the embodied viewer and the richness of the sight that surrounds him or her. At the same time, the claim that the visitor is stopped in his or her tracks, struck with awe (*ekplēxis*) or rooted to the spot at the sight of the church interior, serves, through the disruption of the ordered progression around the building, to draw attention to that movement and to the embodied nature of perception.[24] More important still is the way in which these moments of dizziness serve to present the building itself as a source of agency that works such a dramatic effect on the viewer.

Agency in church ekphraseis

The feeling of shock and awe inspired by the church corresponds remarkably closely to the captivation identified and analyzed by Gell. In Gell's account of the impact of art, any skillfully fashioned object, whether a Trobriand canoe prow or a painting by Diego Velázquez, has an immediate impact on the viewer, who responds to his or her perception of the superior abilities of the maker. The term he uses for this is "captivation," the sense of shock and awe (comparable to the Greek *thauma*) that immediately overtakes the viewer without the need for a conscious intellectual response. The captivation results from the incongruity between the viewer's understanding of his or her own capabilities and the skill that is manifested in the object. In the face of this mismatch, the viewer seeks an explanation, which, according to his or her view of such objects, can be the superior artistic skill of the maker or a divine force.[25] Photios's remarks at *Homily* 10.3 are a perfect expression of the latter type of interpretation: "Looking at it, you would not say it was the work of a human hand but that some divine power beyond us had fashioned its beauty."[26]

Photios represents a far more radical account of the collapse of the distinction between subject and object. Speaking of the atrium, he first claims that the visitor would ascribe its beauty to some suprahuman force, drawing a comparison to Orpheus, who charmed inanimate objects, making them move and act like humans or animals. He reverses the elements of the comparison to explain that whereas Orpheus made objects move, the sight of the church transforms people into trees, rooted to the spot, like columns (the last analogy is my own). Two elements are striking: the analogy with the musician, which likens the impact of sight to that of sound, and then the interchangeability between the human and the nonhuman. This latter idea is developed further when Photios recounts the impact of the sight of the interior of the church: as the viewer twists and turns to see more, he transfers this motion to the object seen until it itself seems to turn. Photios's account could be seen as a rationalization of the sense of captivation. Crucial to this explanation is that the visitor's experience is presented in terms of shared motion that breaks down any boundary between the act of perception and the thing perceived, between subject and object.[27]

The building as agent

In Chorikios's ekphrasis of the Church of St. Sergios, the describer and his listeners go from being controllers and interpreters of the space, naming the parts of the architecture that delimits it and identifying their interrelationships, to being *acted on* by the building as it draws them in (εἰσάγει), and as the breeze contained within it all but caresses the virtual visitor, "as [one's] clothing is lifted up by its breath." I would like to suggest that this identification of the building itself as not only an index of agency but also as an agent itself, capable of being the subject of both transitive and intransitive verbs, is highly significant. Such evocations of the two-way flow of interactions and exchanges between the human visitor and the building have the effect of focusing attention on the space between the two and of depicting this space as one filled with dynamic forces.

In Paul the Silentiary's text in particular, the large number of verbs whose subject is the building or its constituent parts is striking. In the first few lines of the description of the interior of Hagia Sophia three spaces (that is, conches) are said to be spread out (*anapetannumi*, l. 354), while a fourth "rises" (*anerpei*, l. 356), the effect being described through the comparison to a living being, the peacock (ll. 357–78). Then, the middle conch is said to encircle the rows of seats at its base and to draw together (*sunelkei*, l. 364) the lowest set. There may well be a rhetorical purpose to this choice, in that evocations of entities in motion are more vivid than those that have static objects as their subject matter.[28] As a device, it certainly makes the ekphrasis more lively and engaging, as a comparison between Paul, who uses it copiously, and Procopius, who tends to use passive verbs or more neutral verbs such as compounds of ἵστημι (to stand, be placed) when the subject is Hagia Sophia itself, demonstrates.[29] I would like, however, to take this interpretation further and to suggest that these verbs are part of a presentation of the building as an active entity.[30]

Paul also portrays the church building as a human body, with a head (ll. 359, 402, 503)[31] and belly (*kolpos*), which can be swollen—or pregnant—with air (ἠέρι κολπωθεῖσαν, l. 402, cf. ll. 405–6, 530), an image whose implications are brought out much later by Michael the Deacon, who speaks of the same church as being "pregnant with thousands of bodies."[32] It also has arms, which stretch out to embrace the people within (l. 374, speaking of the eastern conches) and feet (l. 590). Taken in isolation, these terms might be no more significant than our own "table legs (though even these reveal a latent desire to interpret forms in terms of the human body and human experience). Within Paul's poem, however, they are part of a systematic presentation of the building as an entity composed of parts that are in constant motion. It is surely relevant, too, that this presentation is worked out through the medium of poetic language whose tradition reaches back (mediated through Hellenistic poetry and more recent Late Antique developments) to the Homeric poems, which contain instances of marvelous objects described as if they were endowed with motion.[33] In several cases, Paul the Silentiary develops the process of personification further not just by attributing movement and action to objects but by phrasing his description in such a way that the building—or its parts—is the subject of a verb whose direct object is the human visitor. The narthex, for example, is described as "*receiving those who enter* beneath its broad openings" (*δεχνύμενος προσιόντας* ὑπ ' εὐρ[υπ]όροισι θυρέτροις, l. 426; emphasis added).

We are moving several steps further from the idea of the personification of the building, for the monuments are said to act on the human visitor. I would not like to say categorically without a good deal of further research that this usage is confined to sacred buildings, but we certainly can see how it functions to present the building as being in interaction with its human occupants, a full partner in the two-way exchange involved in seeing and experiencing. In the case of Chorikios's ekphrasis of St. Sergios, there is the further element of the vocabulary used to describe the pleasant breeze that caresses the visitor: the term *pneuma* can simply mean "breeze," but it is rich in connotations. If we take it in the sense of "breath," the building comes alive, surrounding the visitor with its vital principle, even mixing its own voice with that of the speaker. If we take it in the sense of "spirit," the implications are deeper still.[34] The space delimited by the tangible parts of the building is thus filled with active forces.

It is helpful in this light to return to Gell, whose point of departure in *Art and Agency* was precisely to point out the common human tendency to ascribe agency to objects (the car that breaks down, the computer that crashes). Again, this could be seen as evidence that presenting buildings as working on their human viewers as noted above is not restricted to the sacred. Sacredness, however, can be seen precisely as a particular interpretation placed on the spontaneous response to our perception of agency, so that the ekphraseis serve, as verbal commentaries, to channel these interpretations in a particular direction. In the case of the church ekphraseis, this interpretation involves multiple sources of agency: divine, imperial, the human intellect, and the skill of architects and (to a far lesser extent) artisans. If, with the help of the ekphraseis, we try to conceive of these forces that are active within the church (a process that one could liken to the use of ultraviolet light to reveal sights invisible in normal light), the air or space inside appears as thick with their presence, a presence made tangible in Chorikios's breeze. I would suggest that it is the multiplicity of sources of agency and the particular intensity with which they are felt that are characteristic of sacred as opposed to other spaces.

Sources of agency

The great advantage of Gell's system is its flexibility and the way in which it allows for multiple, interlocking sources of agency. Agents can be primary (the artist in Gell's own example, even when he or she is subordinate to a patron who may also be a primary agent), or secondary (the "index" or artifact),[35] moreover, the same person or entity may be simultaneously agent and patient.[36] It also allows for the fact that particular cultures or individuals attribute the power manifested in the artifact to various sources (magic, for example, or artistic skill).[37] Moreover, the primary agents can be identified differently by different viewers or at different moments. The patron may be seen as the source of agency (by the original audience, for example). In the case of images, the prototype may be seen as the primary agent (as in tales of miraculous icons), or the artist may be the primary agent (for later viewers in the context of a museum).

The ekphraseis of Hagia Sophia are notable for the way in which they represent the building as the result of divine agency working through the emperor. In the prologue to Paul's ekphrasis the personification of Rome begs Justinian to heal the disfigurement caused by the collapse of Hagia Sophia's first dome (ll. 186–285), and the emperor is, of course, omnipresent in Procopius's *Buildings*. Both authors attribute the rebuilding of the church to a combination of divine and imperial agency. For Procopius, Justinian himself found the solution to the architectural problem posed by the dome of Hagia Sophia; Paul also identifies the architects as carrying out the will of the emperor (ll. 553–55). Elsewhere, divine agency and imperial agency are presented as parallel forces. Christ the Savior, for example, is said to protect the church because of the presence of the cross in the middle of the dome (ll. 506–08). This statement is followed immediately by a renewed focus on Justinian and his intelligence (the very Homeric *mētis* or "cunning intelligence") in the construction of the building as revealed through the works of others (*aneres*, that is, the architects and/or craftsmen). This particular feat of intelligence is the roofing of the building not with wood but with stone, which Paul attributes to Justinian, the "much hymned ruler" (*anax polyhymnos*, 1 527). Consistently omitted from the account of the origins of Hagia Sophia is the agency of the craftsmen (as distinct from the architects) who cut and placed the marble and who raised the blocks of stone one on top of another to make the marvelous structure. They do appear once in Paul's poem, heroized through their epic title "skilled" men (ἄνδρες δαήμονες, l. 384), but elsewhere the agency is entirely attributed to others, even to the building itself, whose parts, in motion, are spoken of as if they spontaneously came into being untouched by human hand (arches "rise up," columns "dance," ll. 463–65, 400).

Rather than simply being a means to increase the vividness of the description, as I have suggested elsewhere,[38] these powerful metaphors also—and perhaps primarily—serve to obscure the missing agency of the craftsmen. The idea of the building as agent in itself and as an index of divine and imperial agency is certainly effective in creating an impression of it as a sacred space in which normal experience and normal modes of perception are disrupted, but this is achieved at the expense of the workmen and in contradiction to the evidence of the senses and the sensory knowledge of any visitor who had seen or heard stonemasons at work.[39] Photios similarly obscures the manual work and skill necessary to closely fit the marble floor slabs at the Church at Pharos when he remarks on the appearance of the atrium that leaves the viewer captivated. He describes how the "slabs of marble . . . by their evenness and smoothness and by their having been fitted closely conceal their placement next to one another and the juncture of their edges," thereby suggesting to the viewer's imagination (*phantasia*) that they are in fact one single piece.[40] The sense of captivation therefore depends on the omission of all traces of craftsmanship, leaving only the stones as subjects of the passive infinitive *prosērmosthai* (to have been fitted together), whose grammatical agent is not supplied, leaving unanswered the question "by whom?" The social hierarchies of the ancient and Byzantine worlds mean that these omissions are not surprising, but to ignore them is also to ignore the degree to which the writers of ekphrasis, far from reflecting a simple state of affairs, were engaged in the construction of a particular explanation of the origin of the monuments they celebrated in which some sources of agency were emphasized while others were negated.

This approach reveals to us the multiple (sometimes even competing) sources of power that could be perceived as flowing through church buildings (the power of the patron, the ability of the architects and artisans, divine power) and as fixing it within a complex network of divine and human power relations. As discourses, ekphraseis can articulate these phenomena, which were as real to the original viewers as the stones and shapes are to us. Other sources of agency emerge clearly in the language used by the authors, such as the building itself. What is more, the authors themselves, or their personified discourses, can appear, at times, as sources of agency. Far from being transparent reflections of experience, therefore, the ekphraseis were engaged in arguments and involved another source of agency: the poet or orator who acted through his words on the audience.

The agency of the poet and orator

As a poet, Paul the Silentiary is the ekphrasist who pays the most explicit attention to questions of sound and voice, referring on multiple occasions to his own voice and also to the choir within the church (ll. 429–37) as well as to the terrible sound made when the original dome collapsed (l. 176). This attention to the medium is partly explicable by the form in which he was writing, epic hexameters, which were originally designed to be sung (and which may have continued to be sung if, for example, the libretti to the popular pantomime performances that continued to draw crowds in the sixth century were composed in hexameters). The two Homeric models, of course, open with references to singing or speaking as their first verbs (*mēnin aeide*, *Iliad* 1.1; *andra moi ennepe*, *Odyssey* 1.1), but Paul was not bound to echo this emphasis on voice. This same emphasis can be seen in his mention of the songs of the church. Moreover, where the intelligence and skill of the craftsmen and of the patron are spoken of, Paul includes two virtuoso four-word lines (ll. 386, 514), thus drawing attention to his own skill and his mastery of his own material, the hexameter verse in its post-Nonnian form, and to remind us of the existence of the poet as mediator and as agent in his own right.[41] The voice of the poet, like the voice of the orator, also touches his listeners physically through the transmission of the sound (ancient warnings about education show great concern about the types of sounds, words in particular, that reach the child, penetrating the mind through the ear).[42]

I would suggest, then, that the authors of ekphrasis, in their evocations of the sentient body in motion through the church building and of the various types of interaction between the human body and the building as a quasi-human entity, are both reflecting aspects of common experience and shaping future experiences. Their ekphraseis gave rise to a multisensory impression that included, to varying degrees, the body's motion through and interaction with the building and may well have caused their audiences to enact virtually the movements and the acts of perception described, prompted by the use of apostrophe and focalization, which thus acquire an importance that goes far beyond the stylistic, as noted above.[43] In this sense, the ekphraseis are profoundly performative, bringing about what they describe, and their original audiences were both patients, acted on by the words they listened to, and, through the effect of these words, active participants in the virtual *periēgēsis*. These virtual experiences of visiting and perceiving the building could potentially make a powerful contribution to the shaping of the present and future responses of their listeners or readers.[44] The drastic selection of the details to be noted and the particular interpretation given to these details by the speech, notably the attribution of agency noted above, are all part of this highly rhetorical strategy aimed at augmenting the audience's adhesion to a thesis that they might not have held spontaneously. The fact that the ekphraseis were not recited within the buildings they describe might be said to increase their efficacy, allowing the mental impression of the building to be activated by the speech without interference from the evidence of the bodily senses.

By engaging the various senses of their listeners—sight, hearing, touch, and the all-important kinesthetic awareness—the authors of ekphraseis bring about a virtual multisensory experience of being inside the sacred space of the church. Their words are performative both in the sense that they were originally part of a live verbal performance at a specific time and place and in the more technical sense of bringing about what they describe, causing their listener to enact the movement around or through the building and to feel the various sensations involved. The fact that this enactment is a product of words means, however, that it is far from being a simple or unproblematic reproduction of a lived moment. Instead, we can see the virtual experiences offered by the texts as particular instances that privilege certain aspects over others and put forward one set of responses to the building among the many that were potentially available. These rich, vivid, and persuasive evocations of the bodily and spiritual experience that awaited the visitor represented particular readings of the buildings. This is not to relegate ekphrasis to the category of "rhetoric" in the sense of lies or empty words—they reveal some of the multilayered responses that their audiences would expect to undergo within the churches—but to attribute to them a dynamic, persuasive function that was all the more powerful for their ability to conjure a fully embodied experience and knowledge.

Notes

1 The multisensory nature of the experience depicted in ekphrasis has been the subject of recent publications. See, for example, Bissera Pentcheva, *The Sensual Icon: Space, Ritual and the Senses in Byzantium* (University Park, PA: Pennsylvania State University Park, 2010); Pentcheva, "Hagia Sophia and Multisensory Aesthetics," *Gesta* 50, no. 2 (2011): 93–111; Nicoletta Isar, "'Χορός of Light': The Vision of the Sacred in Paulus the Silentiary's Poem *Descriptio S. Sophiae*," *Byzantinische Forschungen* 28 (2004): 215–42; and Nadine Schibille, *Hagia Sophia and the Byzantine Aesthetic Experience* (Farnham: Ashgate, 2014). For a multisensory approach to archaeology, see Yannis Hamilakis, *Archaeology and the Senses: Human Experience, Sense and Affect* (Cambridge: Cambridge University Press, 2013), 75–80 on Byzantium.

2 For a survey of this research with further bibliography, see Guillemette Bolens, *The Style of Gestures: Embodiment and Cognition in Literary Narrative* (Baltimore, MD: Johns Hopkins University Press, 2012), 1–49.

3 Alfred Gell, *Art and Agency: An Anthropological Theory* (Oxford: Clarendon Press, 1998).

4 Aphthonios, *Progymnasmata*, ed. H. Rabe (Leipzig: Teubner, 1926), 36: "῞Εκφρασίς ἐστι λόγος περιηγηματικὸς ὑπ ' ὄψιν ἄγων ἐναργῶς τὸ δηλούμενον."

5 John Sardianos, *Commentarium in Aphthonii Progymnasmata*, text and translation in Ruth Webb, *Ekphrasis: Imagination and Persuasion in Ancient Rhetorical Theory and Practice* (Farnham: Ashgate, 2009), 205–06.

6 Nikolaos, *Progymnasmata*, ed. J. Felten (Leipzig: Teubner, 1913), 68–69; text and translation in Webb, *Ekphrasis, Imagination and Persuasion*, 203–04.

7 Focalization is most often discussed in the context of the novel. For our purposes, it is useful to distinguish cases where the speaker evokes a general, panoramic view of the building, implying

a distant viewpoint, from those where the speaker focuses attention on one part, describing only what could be seen from a particular point. The describer may also adopt a static viewpoint or, as in several of our texts, evoke a dynamic progression through the space. A great deal of interesting work has been done recently on space and focalization in Homeric narrative and its cognitive implications. See, for example, Jennifer Strauss Clay, *Homer's Trojan Theater: Space, Vision, and Memory in the Iliad* (Cambridge: Cambridge University Press, 2011); and Irene De Jong and René Nünlist, "From Bird's Eye View to Close-up: The Standpoint of the Narrator," in *Antike Literature in neuer Deutung*, ed. A. Bierl et al. (Munich: K. G. Saur, 2004), 63–83.

8 Aphthonios, *Progymnasmata*, ed. Rabe, 38–41. The *periēgēsis* form is not always used in ekphraseis of buildings: the model ekphrasis of the Alexandrian Tychaion attributed to Libanios describes the building from a stationary standpoint. For the text and translation, see C. A. Gibson, *Libanius's Progymnasmata: Model Exercises in Greek Prose Composition and Rhetoric* (Atlanta, GA: Society of Biblical Literature, 2008), 487–91. Procopius of Caesarea's account of Hagia Sophia in *Buildings*, ed. and trans. H. B. Dewing (Cambridge, MA: Harvard University Press, 1940), 1.1.27–78, uses a similar plan whose increased complexity reflects that of the building, combined with the ekphrasis of the process of construction (the *ekphrasis tropou*).

9 The fullest accounts are in Latin sources, *Rhetorica ad Herennium*, ed. and trans. H. Caplan (Cambridge, MA: Harvard University Press, 1954), 3.14.28–24.40; and Quintilian, *Institutes of Oratory*, ed. and trans. D. A. Russell (Cambridge, MA: Harvard University Press, 2001), 11.2. See further Jocelyn Penny Small, *Wax Tablets of the Mind: Cognitive Studies of Memory and Literacy in Classical Antiquity* (London: Routledge, 1997); and, on the connection with ekphraseis of buildings, Liz James, "Art and Lies: Text, Image and Imagination in the Medieval World," in *Icon and Word: The Power of Images in Byzantium; Studies Presented to Robin Cormack*, ed. Antony Eastmond and Liz James (Aldershot: Ashgate, 2003), 61–62.

10 Chorikios, *Laudes Marciani*, 1, in *Opera*, ed. R. Foerster and E. Richtsteig (Leipzig: Teubner, 1929), 1–26.

11 See the analysis in Paul the Silentiary, *Description de Sainte-Sophie de Constantinople*, trans. and commentary by Pierre Chuvin and Marie-Christine Fayant (Die: Éditions A Die, 1997), 34–35.

12 Ruth Macrides and Paul Magdalino, "The Architecture of Ekphrasis: Construction and Context of Paul the Silentiary's Poem on Hagia Sophia," *Byzantine and Modern Greek Studies* 12 (1988): 47–82, esp. 59.

13 As Malcolm Heath shows in "Longinus *On Sublimity*," *Proceedings of the Cambridge Philological Society* 45 (1999): 43–74, the arguments for the earlier dating are less than convincing.

14 Longinos, *On the Sublime*, ed. and comm. D. A. Russell (Oxford: Clarendon Press, 1970), 26.2: "ὁρᾷς, ὦ ἑταῖρε, ὡς παραλαβών σου τὴν ψυχὴν διὰ τῶν τόπων ἄγει τὴν ἀκοὴν ὄψιν ποιῶν." Longinos is citing Herodotos, *Histories*, 2.29, ed. N. G. Wilson (Oxford: Clarendon Press, 2015).

15 See, for example, Dionysios of Halicarnassos, *Lysias* 7, in *Critical Essays*, ed. and trans. S. Usher (Cambridge, MA: Harvard University Press, 1974); and Quintilian, *Institutes of Oratory* 6.2.32, on the more general sense of presence achieved through *enargeia* (and thus ekphrasis whose defining quality is *enargeia*).

16 Chorikios, *Laud. Marc.* 1.22: "ἡ πρὸς ἑσπ έραν εὐμήκης στοὰ μεστὸν εὐθυμίας εἰσάγει παγκάλης αὔρας ἐκ τοῦ πρὸς ζέφυρον πνεούσης προτεμενίσματος, πνεῖ γὰρ ἡδύς τε καὶ ἄλυπος καὶ τῶν ἱματίων ἐντὸς ὑποδὺς ἠρέμα διαψύχει τὰ σώματα τῶν ἐσθήτων αἰωρουμένων τῷ πνεύματι."

17 On the rich implications of *pneuma* and its interactions with sacred space, see Pentcheva, *The Sensual Icon*, 54–55, 181–82; and Pentcheva, "Performing the Sacred in Byzantium: Image, Breath, and Sound," *Performance Research International* 19, no. 3 (2014): 120–28.

18 Speaking of the empty space beneath the south and west arcades of Hagia Sophia, Paul the Silentiary, *Descriptio S. Sophiae*, l. 533, describes the air (*aer*) that spreads out: "'ὅλος δ' ἀναπέπτ[αται ἀ]ήρ."

19 Hamilakis, *Archaeology and the Senses*, 73–74, cites the importance of kinesthetic perception in some non-Western cultures. On proprioception and kinesthesia, see Susan Leigh Foster, *Choreographing Empathy: Kinesthesia in Performance* (London: Routledge, 2011), 73–125; and Barbara Tversky, "Spatial Cognition," in *The Cambridge Handbook of Situated Cognition*, ed. Philip Robbins and Murat Aydede (Cambridge: Cambridge University Press, 2012), 201–16.

20 On the second-century evidence, see Maud Gleason, *Making Men: Sophists and Self-Presentation in Ancient Rome* (Princeton, NJ: Princeton University Press, 1995). Clement of Alexandria's

Paidagogos is a rich source on bodily habitus and the control of posture and gesture. For more on Early Christianity and Late Antiquity, see Peter Brown, *The Body and Society: Men, Women and Sexual Renunciation in Early Christianity* (New York: Columbia University Press, 1988); Virginia Burrus, "'In the Theater of This Life': The Performance of Orthodoxy in Late Antiquity," in *The Limits of Ancient Christianity: Essays on Late Antique Thought and Culture in Honor of R. A. Markus*, ed. William Klingshirn and Mark Vessey (Ann Arbor, MI: University of Michigan Press, 1999), 80–96; and Jennifer Glancy, *Corporal Knowledge: Early Christian Bodies* (Oxford: Oxford University Press, 2010).

21 A. Damasio, *The Self Comes to Mind* (London: Vintage, 2012), 132–33, on perception through engagement: "The organism interacts with objects, and the brain reacts to the interaction. Rather than making a record of an entity's structure, the brain actually records the multiple consequences of the organism's interaction with the entity." The contrast drawn here by Damasio corresponds neatly with the contrast between the expectations that ekphraseis of monuments should "record structures" and the actual practice of Late Antique and Byzantine authors, who prefer to place the emphasis on such interactions.

22 Chorikios, *Laud. Marc*, 1.23: "εἰσιὼν οὖν ἰλιγγιάσεις τῷ ποικίλῳ τῶν ὁρωμένων καὶ πάντα φιλονεικῶν ἀθρόως ἰδεῖν οὐδὲν ἐναργῶς οἰχήσῃ τεθεαμένος περιφερομένων σοι τῇδε κἀκεῖσε τῶν ὀφθαλμῶν οὐδὲν ἀόρατον καρτεροῦντι παραλιπ εῖν."

23 Isar, "'Χορός of Light,'" 223–24, links these passages to the circular motion typical of the *choros* (dance).

24 The circling, darting gaze in particular has been linked to the Late Antique aesthetic of the "jeweled style," which placed emphasis on fragmentation over unity of design. Michael Roberts, *The Jeweled Style: Poetry and Poetics in Late Antiquity* (Ithaca, NY: Cornell University Press, 1989), 73–75. This is to leave aside, however, the care with which both Paul the Silentiary and Procopius evoke the overall structure of the building (the interrelationship between its colonnades, curves, solid walls, and volumes). In Ruth Webb, "The Aesthetics of Sacred Space: Narrative, Metaphor and Motion in *Ekphraseis* of Church Buildings," *Dumbarton Oaks Papers* 53 (1999): 59–74, I identified this topos as a way of conveying the three-dimensionality of the building; here I am going further in adding that the sense of the perceiving body within that space is evoked.

25 Gell, *Art and Agency*, 68–72.

26 Photios, *Homily 10* (in *Photiou Homiliai*, ed. B. Laourdas (Thessaloniki: Hetaireia Makedonikon Spoudon, 1959): "Εἴπ οις ἂν εἰς αὐτὸν ἰδών, οὐκ ἀνθρωπίνης χειρὸς ἔργον, ἀλλὰ θείαν τινὰ καὶ ὑπὲρ ἡμᾶς δύναμιν τὸ κάλλος αὐτῷ ἐπιμορφῶσασθαι." For the same thought, see Procopius, *Buildings* 1.1.61.

27 On the potential contributions of the rituals to this feeling of vertigo and of loss of self (*ekstasis*), see Pentcheva, *The Sensual Icon*, 149–54; and Isar, "'Χορός of Light.'"

28 See Aristotle, *Rhetoric* 1411b 24–25, on metaphor (Aristotle, *Rhetoric*, trans. in George A. Kennedy, *On Rhetoric: A Theory of Civic Discourse* (New York: Oxford University Press, 1991); and Lucia Calboli Montefusco, "*Enargeia* et *energeia*: L'évidence d'une demonstration qui signifie les choses en acte," in *Demonstrare: Voir et faire voir*, ed. Mireille Armisen Marchetti (Toulouse: Presses Universitaires du Mirail, 2005), 43–58.

29 He chooses a static, inanimate entity (a promontory, *skopelos*) as the vehicle of his simile at *Buildings* 1.1.38 but changes his practice somewhat when he comes to the account of the central dome, which uses more verbs of motion (particularly compounds of *bainō*, "to go, walk").

30 Irene J. Winter, "Agency Marked, Agency Ascribed: The Affective Object in Ancient Mesopotamia," in *Art's Agency and Art History*, ed. Robin Osborne and Jeremy Tanner (Oxford: Blackwell Publishing, 2007), 42–69, esp. 56–57, notes a similar phenomenon in ancient Mesopotamian descriptions of temples that make use of the distinction between agent and patient, which is an integral part of the early Sumerian language.

31 The term used here, *karē*, was used as a metaphor for mountains and buildings in Homeric poetry. In the context of a sustained personification I would argue that the root sense of "head" is activated and that the metaphor is not "dead." Paul the Silentiary, *Descriptio S. Sophiae*, l. 822, where *karē* is used for the "heads" of the human worshippers.

32 Michael the Deacon, *Ekphrasis of Hagia Sophia* 4, l. 90, in Cyril Mango and John Parker, "A Twelfth-Century Description of Hagia Sophia," *Dumbarton Oaks Papers* 14 (1960): 233–45, esp. 237.

33 See, of course, the scenes on the shield of Achilles, *Iliad* 18.483–608 but also the marvelous female statues in Hephaistos's palace in *Iliad* 6.417–20 and the moving creatures on Odysseus's brooch described at *Odyssey* 19.226–31.

34 *Pneuma* is linked to incense and to song in the hymn celebrating the inauguration of Hagia Sophia. Pentcheva, *The Sensual Icon*, 45–48; and Pentcheva, "Hagia Sophia and Multisensory Aesthetics," 104–06.

35 Gell, *Art and Agency*, 36–38.

36 Gell, *Art and Agency*, 51–65.

37 Gell, *Art and Agency*, 71.

38 Webb, "The Aesthetics of Sacred Space," esp. 68–69.

39 On the importance of taking prior bodily experience into consideration when assessing the ancient visitor's perception of a site, see Hamilakis, *Archaeology and the Senses*, 102–03.

40 Photios, *Homily* 10.4–5: "καὶ γὰρ μαρμάρων διαλεύκων πλάκες λαμπρόν τι καὶ χάριεν ἀποστίλβουσαι, ὅλην ἐναπολαβοῦσαι τὴν πρόσοψιν καὶ τὴν πρὸς ἀλλήλας θέσιν καὶ τὴν τῶν περάτων συνάφειαν τῷ ὁμαλῷ καὶ λείῳ καὶ τῷ π ροσηρμόσθαι λίαν ἀποκρύψασαι εἰς ἑνὸς λίθου συνέχειαν, γραμμαῖς εὐθείαις ὥσπερ ἐπιγεγραμμένου, θαῦμα καινὸν ὁραθῆναι καὶ ἥδιστον, τὴν τῶν ὁρώντων φαντασίαν ὑπάγουσιν." The translation is that of Mango, *The Art of the Byzantine Empire*, 185, modified.

41 On the technical aspects of the Late Antique hexameter, see Mary Whitby, "From Moschus to Nonnus: The Evolution of Nonnian Style," in *Studies in the Dionysiaca of Nonnus*, ed. N. Hopkinson (Cambridge: Cambridge Philological Society, 1994), 99–155.

42 John Chrysostom, in *On Vainglory and On the Education of Children*, ed. and trans. A.-M. Malingrey (Paris: Les Éditions du Cerf, 1972), depicts the organs of sense perception as gates through which harmful influences can come, the eyes and ears being particular points of vulnerability.

43 For the evidence that listening to or reading a verbal account of an action involves regions of the brain that are involved in performing the activity itself, see Bolens, *The Style of Gestures*, 1–49.

44 I am grateful to my colleague Yann Coello for discussion of this point in relation to the question of attention in visual perception: viewers will focus attention on the points they consider significant in a sight, opening up a space for a culturally differentiated gaze.

9 Acoustics of Hagia Sophia

A scientific approach to the humanities and sacred space

Wieslaw Woszczyk

Introduction

The Byzantine church of Hagia Sophia (Holy Wisdom) in Constantinople (now Istanbul) was commissioned by Emperor Justinian. He hired the architect Isidorus of Miletus and the mathematician Anthemius of Tralles to build a church manifesting the wisdom and perfection of God. These two highly skilled men were eager to search for solutions and open to challenging ideas.[1]

The construction, which took just five years, ten months, and four days, delivered a masterpiece of Byzantine architecture. Construction historians attribute the engineering skills of the two main architects to their training in Alexandria, where manuals for the building trade were authored and published under the name of Heron of Alexandria between the first century CE and the Byzantine period. The construction of Hagia Sophia utilized sophisticated mathematical, geometrical, and technical solutions known and practiced in Late Antiquity.[2]

The church was built using slanted bricks and mortar. Construction historians estimate that because the bricklayers used more mortar than brick, the structure was weaker than expected. Also, because of the fast pace of building, the mortar was probably not sufficiently dry before the next layers were laid, making the structure heavier than anticipated. As a result, the walls leaned outward under the weight of the dome, a situation that contributed to the collapse of the dome in 558 after two strong earthquakes in August 553 and December 557. The replacement dome was built in 558–62 by Isidorus the Younger and survived largely unchanged despite ensuing earthquakes. Isidorus the Younger made the dome taller by six meters to reduce the lateral forces acting on the supporting walls. Some bricks near the apex of the dome were hollow and made of clay, further reducing the weight.[3]

The dome is also constructed with bricks and mortar. It is 56.60 meters high and 31.87 meters at its maximum diameter (the minimum is 31.24 meters), the variance due to a slightly irregular shape caused by damage and repairs. The main cupola sits between two half-domes over the 31-meter-square nave in the center of the church supported on four pendentives that transfer its weight to four piers and the walls below.[4] Pendentives used experimentally in Roman architecture from the second to third century CE were fully developed in Hagia Sophia, which has the largest pendentive dome ever built. Forty arched windows at the base of the dome, braced with forty ribs, bring plenty of light into the dome and lighten the perceived weight of the dome. Ramps, rather than stairs, provide access to spacious upper galleries. There are sixty-seven columns in the upper gallery. The rectangular shape of the church measures 70 meters by 75 meters, not including the two narthexes and the atrium. There are five doors in each of the outer and inner narthexes, leading to nine gates that open on to the nave.

Architectural design for great acoustics?

We may be tempted to think that the architectural design of Hagia Sophia was deliberately conceived to produce the exceptional reverberant acoustics of this building. Contributing factors include the choice of large dimensions, of both the height and the overall volume; curved surfaces of the central dome and a number of semi-domes; hard, acoustically reflective stone surfaces; and the layout of various sections, including galleries, with their relative proportions and interconnections using openings between multiple pillars. However, in a likely scenario, acoustics could simply be a by-product of the engineering solution that provided a solid load-bearing structure with an enormous volume of interior space able to last for centuries without failure.

For comparison, the Süleymaniye Mosque (1550–57), built by the architect Mimar Sinan, was the largest he ever constructed. Its main dome is 53 meters high and 27.5 meters in diameter, has a volume of 115,000 cubic meters (area: 3,350 square meters), and could accommodate only 4,640 people. The interior of the mosque is almost a square, 59 meters in length and 58 meters in width. By contrast, Hagia Sophia, more than twice its volume (255,800 cubic meters) and area (7,960 square meters), could hold 15,910 people (as calculated from Odeon simulation software by CAHRISMA).[5] The enormous nave of Hagia Sophia hides its massive piers in marble revetments and opens the drum of the dome with windows, thus creating the effect of a cupola suspended from above. This ethereal appearance in the Byzantine cathedral can be contrasted with the clear structural massing of volumes in the interior of the Süleymaniye Mosque.

Hagia Sophia was an ideal setting for the religious and theatrical ritual of the Christian Orthodox liturgy, an event that usually lasted several hours and included the singing of psalmody with call-and-response dynamics involving a massive congregation, the recitation of scripture, homilies, and prayers; the use of visual tools such as candles and oil lamps that produced fragrance, smoke, and shadows; curtains gracing the intercolumniations; sumptuous crosses and liturgical vessels; and rich ceremonial attire dressing the officiating clergy. The acoustic setting was crucial in integrating these multisensory experiences because sound in a reverberant enclosure is constantly present, being an immersive medium that reaches everyone. The amplification of sound inherent in the design of Hagia Sophia was necessary to counteract the absorption of sound by the vast number of people attending the liturgy, theoretically exceeding the enclosure's capacity of nearly 16,000. The elaborate Eucharistic ritual carried out in Hagia Sophia was clearly the event not to be missed; its theatrical, dramatic, and religious impact could not be compared with any other event in scale, magnitude, and quality.

It would seem appropriate for Hagia Sophia's architecture to be designed from the start as an experiential space, with the above-mentioned factors in mind. The layout and the presence of galleries, vertical space, and hard, reflecting surfaces were likely understood to be prerequisites for the dramatic staging of high-capacity events (gatherings) with large reverberance. It was essential that everyone could clearly hear spoken and sung words. This was made possible with elevated sources on the ambo and with the audience absorbing undesirable reflections. The art historians Bissera Pentcheva and Nina Ergin point to the important notion that Hagia Sophia, both as a

Byzantine cathedral and later an Ottoman mosque, was consciously used to magnify the religious experience of the presence of the divine and in this way to increase the faith of the congregation.[6]

Acoustical support of religious ritual

In the early Byzantine church it was expected that musical performance would involve the whole congregation, not just a separate group of trained singers responsible for music.[7] Performing together with the entire assembly of people, singers commanded a powerful responsorial mass of voices, which in the highly reverberant enclosure could produce a magnificent, loud, and thrilling sonic effect and elicit a powerful religious emotion, as if everyone were joining the choir of angels. Since the floor area was absorbing the sound because of the presence of people, reverberating sound decay was perceived as lifting upward toward the dome and perhaps beyond, to heaven.

There are several important acoustic outcomes that participants in the liturgical ritual could perceive when sound was being processed by the massive enclosure of Hagia Sophia. The performance of long, sustained notes would cause a gradual buildup, an amplification effect from the accumulated sound as more sound energy was continually added than was absorbed. It is hard to estimate the total increase in sound power and perceived loudness. Suffice it to say that the elite choir, together with all the participants in musical performance, had some ability to control the perceived volume of music by using strongly sustained drones (*ison*) and the buildup of the reverberant sound energy. This was the means of controlling the volume of the sound.

The second important acoustic effect was the overlaying and dissolving of notes that if sung in unison created smoothly evolving chords, transitioning gradually between dissonant and consonant harmonies. With a reverberation time of more than 10 seconds, the monophonic melodic progressions were seemingly suspended in time and were superimposed one on another, either tonally colliding or harmonizing with the newly sung pitches. This overlapping and smearing (slurring) effect was slow enough to allow singers to gently steer the new pitch appropriately to avoid conflicting with the previously sung notes, assuring continued harmony. The melismatic singing in particular, involving the singing of many notes per single syllable of text, characteristic of Byzantine cathedral chant, swelled beautifully and was ornamented by the delayed reverberant returns from Hagia Sophia's enclosure.

An especially dramatic effect was achieved during processions within Hagia Sophia, when groups of singers moved around the church, engaging the acoustics in different sections of the church. For a stationary listener, moving sound sources appeared to change in distance and direction, illuminating the auditory presence of Hagia Sophia distinctly and differently depending on the listener's location in the church. For those in the procession, the gradually changing acoustics when passing through smaller and larger spaces produced a flowing sensation of motion in a seemingly liquid, uninterrupted space.

In our study of the acoustics of Hagia Sophia, the most dramatic effect was achieved with explosive (percussive) sources such as drums or bells (which were not used historically in the liturgy of the Great Church).[8] The direct path of sound from the source to receiver could only account for a fraction of the total acoustic energy, which, once accumulated, returned in huge quantities from all interior surfaces of the enclosure within seconds. These massive reflected returns of sound, coming from different directions and having substantial time offsets owing to the large variation of distances within the church, would overlap in a dramatic and powerful similitude of an acoustic avalanche.

Musicians and orators could be placed vertically on all four levels of the vast open space of Hagia Sophia: on the ground plane, on the lower and the upper galleries, and at the base of the dome. Research of the musicologist Neil Moran who studied the musical use of Hagia Sophia, suggests that singers, at least in the sister church of Hagia Sophia in Thessaloniki, were located at the base of the dome, standing side by side behind the metal railing.[9] Such a location could provide natural acoustical reinforcement and effective redirection of sound downward using the acoustic mirror-effect of the dome. The large boundary of the dome immediately behind and above the singers would reinforce their voices with the immediate reflection from that boundary.

Elevated sources were also possible closer to the ground level, on the ambo. Such a source, placed above the horizontal level of the floor, effectively projected sound toward people on the floor on a direct path, free of reflections. Elevated orators or singers could deliver recitations or chant with enhanced visual presence and fine sonic clarity.[10] Since most of acoustic absorption was due to people occupying the floor, the reverberant sound was perceived to decay in an upward motion toward the dome.[11] The massive volume of Hagia Sophia was a magical container of slowly decaying sounds capable of immersing large crowds of people. The acoustics of the interior could project an immense amount of sound through space, a sound with seemingly endless sustaining power and omnipresence. The more beautiful the sound generated, the greater was the perceived power and beauty of the church.

Acoustics of Hagia Sophia

The huge interior dimensions (volume exceeding 250,000 cubic meters) and high reflectivity (low acoustic absorption of marble surfaces) contribute to the prolonged duration of reverberation and its high intensity. The audible presence of reverberation is the hallmark of Hagia Sophia. Measured with a broadband sound source (producing a high sound output level between 20 Hz and 40 kHz), a reverberation time of T_{30} (in 1/3 octave bands) reaches almost 12 seconds at 500 Hz and climbs to more than 13 seconds at the lowest frequencies, with 10.35 seconds at 1 kHz (Figure 9.1). One becomes aware of the extended low-frequency reverberation time whenever one of the massive sixth-century doors (made of oak with a bronze frame and seven meters in height) is being shut when the interior is empty. The reduction of high-frequency reverberation time is consistent with the dominance of air absorption, dropping to 2.57 seconds at 8 kHz, and to 1.59 seconds at 12.5 kHz. However, because of the relative high humidity (above 50%) due to the proximity of the sea, cool temperature, and the recently discovered presence of wells and water cisterns beneath Hagia Sophia the air absorption is less than it would be in a drier environment.[12] The following graph plots the air-absorption coefficient values at different relative humidity levels (Figure 9.2).[13] The graph makes clear that acoustic absorption in air increases with frequency of sound but diminishes with relative humidity. Humid air is a better carrier of sound than dry air, allowing sound to persist longer in a humid environment, which results in a longer reverberation time, especially at high frequencies.

Hagia Sophia has many spherical- and cylindrical-shaped sections. In concert hall acoustics, large concave surfaces are considered undesirable because they focus

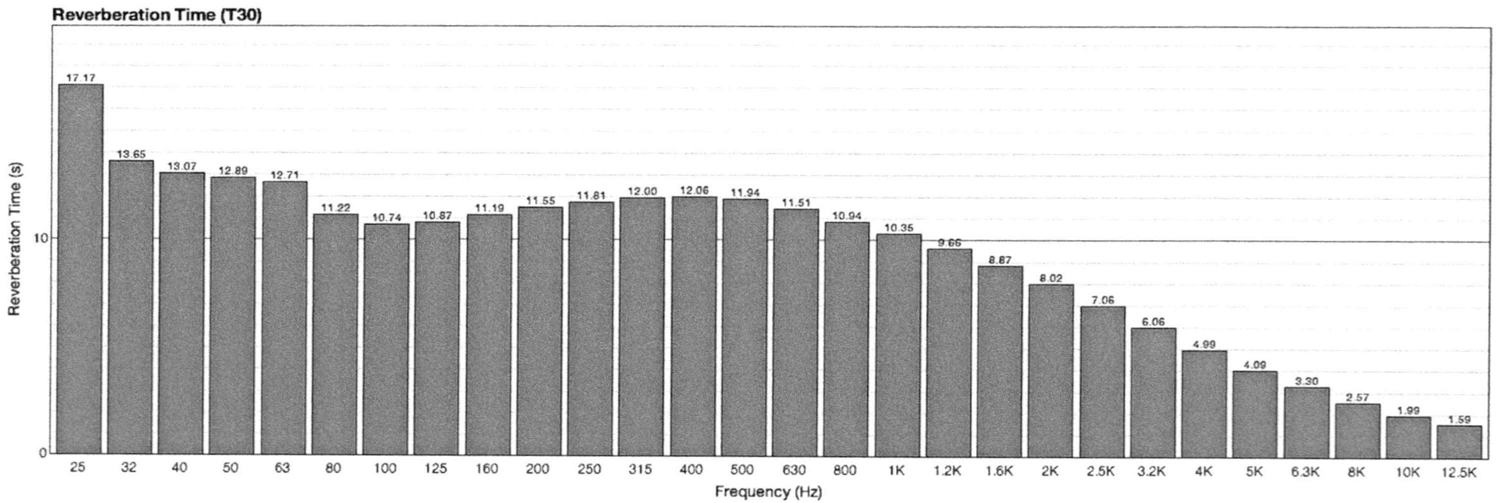

Figure 9.1 Reverberation time (T_{30}) averaged over eight positions measured with broadband sources in Hagia Sophia.
© Wieslaw Woszczyk.

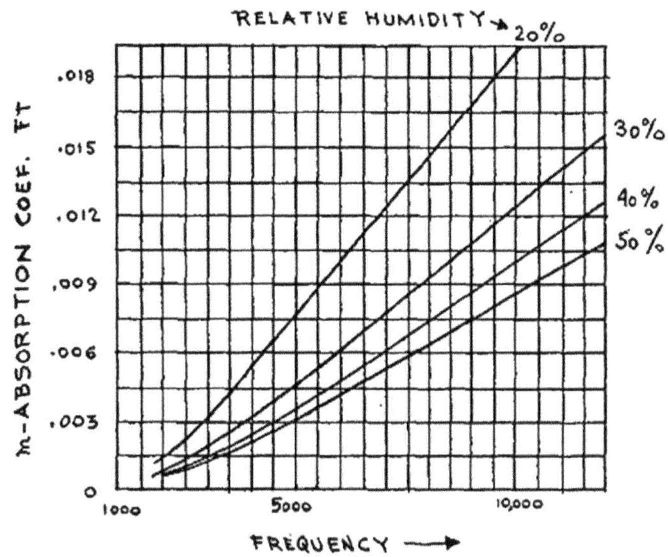

Figure 9.2 Absorption coefficient of air at different relative humidity levels and frequencies. Note the much higher absorption coefficient at high frequencies than at low frequencies, but not as much absorption at higher relative humidity.

After Michael Rettinger, "Note on Reverberation Chambers," *Journal of the Audio Engineering Society* 5 (1957): 108, fig. 1.

concentrations of energy in certain areas and prevent uniform distribution of reflected sound throughout the room, especially when the radius of such curved surfaces is comparable to acoustical wave lengths. However, spherical and cylindrical surfaces can also serve as effective scattering surfaces, producing a much better diffusing effect than a plane ceiling or wall reflector (Figures 9.3, 9.4).

In Hagia Sophia, listeners' awareness of individual acoustic reflections, even the strong and delayed reflections focused by concave surfaces, is limited owing to the presence of many other densely spaced reflections and reverberation. Therefore, it is not possible to detect echoes, that is, reflected waves appearing as distinct events standing apart from those directly received from the source. High-amplitude multiple reflections arriving along different paths can also hide flutter echoes, repetitive reflections occurring between parallel surfaces, which in isolation sound characteristically like the fluttering of a bird's wings.

Acoustics of the dome

Whether a dome concentrates the acoustic energy or scatters it, distributing it over a wide angle, depends on the location and distance of the sound source and the listener in relation to the dome. In general, if both the source and the observer are located below the circle circumscribed by the circumference of the dome, they will not experience the undesirable focusing effect that produces increased sound intensity.[14] This is the case in Hagia Sophia; the dome's height of 56.60 meters and diameter of 31.87 meters ensure that people located on the ground level are outside its focusing area.

Sound rays projected from near ground level upward toward the dome will be redirected by the curvature of the dome toward a much wider area of the floor, as they are disbursed into wider lateral angles (Figures 9.3, 9.4). The function of the dome, therefore, is to scatter the rays rather than to reflect a strong specular bundle back toward the floor. For a distant source, the focal point with the highest sound intensity is located near the dome, close to a point that is half a radial distance from the apex of the dome. A portion of the rays returns downward, directly below, since a small

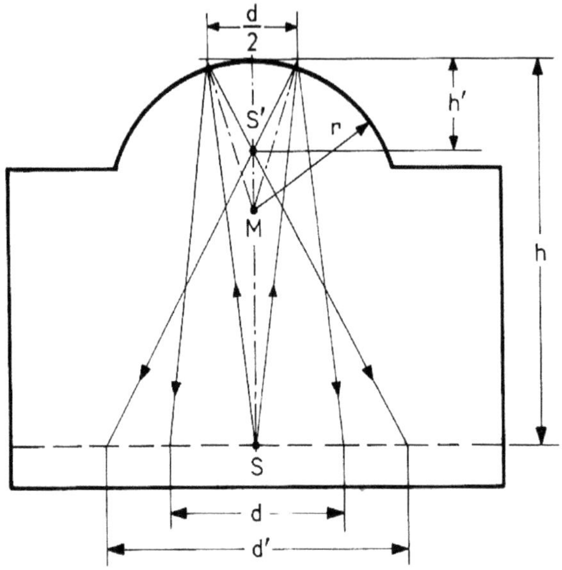

Figure 9.3 Reflections of rays from a concave ceiling in a large room.
After Lothar Cremer and Helmut A. Müller, *Principles and Applications of Room Acoustics*, vol. 1 (London: Applied Science Publishers, 1982), 59, fig. 3.5.

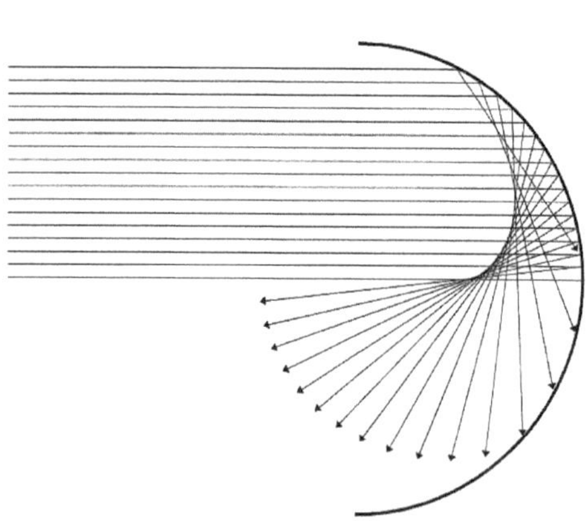

Figure 9.4 Reflections of rays from concave mirror (detail).
From Heinrich Kuttruff, *Room Acoustics* (Oxon, UK: Spoon Press, 2009), 120, fig. 4.13.

section of the dome surface near the apex is parallel to the floor. Rays reaching other areas of the dome are redirected down sideways with a horizontal component. When the source on the ground moves away from the center of the dome, some rays are reflected back to the receiver by the perpendicular section of the curvature of the dome. This tends to maintain the listener's awareness that the dome is centrally placed overhead at virtually every location. If the ceiling were a plane surface, the strongest specular reflection would follow along just above the moving source. Since the original dome was flatter at the top and lower to the ground by some 6 meters than the present dome, it reflected a stronger bundle of sound energy down toward the source standing directly below. Its sound-scattering capacity would have been weaker and less uniform than is the case now.

The additional two large half-domes and five exedrae contribute substantially to the development of ambient sound, as they redirect and disperse the sound but at different relative time intervals due to the considerable distance between them. The impression one has when listening to reflections and reverberation on-site is that multiple waterfalls are being activated at different times and places, and together converge down on the listener. In our exploration of the acoustics of Hagia Sophia, we attempted to capture this unusual effect of prolongation of the onset of the reverberation, as it is a uniquely attractive feature of the space. Whereas under the dome sound distribution and order of arrivals are more unified and uniform, some distance away from the central dome, layered contributions from the semi-domes give the acoustics a different, more intense dimension.

It is also impossible to omit the role of additional coupled volumes—side aisles and the galleries—on the overall aural impression of Hagia Sophia. They are connected acoustically to the main part of the church. The colonnades enable the nave to communicate with the side aisles; each one is composed of three serially connected volumes. Sound entering these spaces reverberates according to their acoustical characteristics, contributing unique features of depth, direction, and decay. Similarly, the galleries also lend their own decay characteristics. In all, Hagia Sophia dazzles the listener with a richly animated presence of reverberation that is distributed everywhere around and above, being different in every place. Regrettably, when visiting Hagia Sophia, one can never experience this aural architectural beauty of the church because of the constant background noise generated by the crowds of visitors.

How can Hagia Sophia immerse and overwhelm with beautiful acoustics today?

Today, Hagia Sophia is a museum and a highly important site of Turkish, European, and world cultural heritage. Millions of tourists visit the monument each year, while museum staff carries out important ongoing maintenance and conservation work, as well as research aimed at better protecting this magnificent structure from earthquakes and natural disasters. Hagia Sophia is not available as a venue for religious celebrations or staging of cultural events such as music concerts. Still, the acoustic experience within the interior space of Hagia Sophia could be overwhelming today if there were a way to hear the acoustics in conjunction with music.

Such an option becomes possible by employing digital means of active architecture, by re-creating the virtual interactive presence of Hagia Sophia in alternative spaces, for uses that are not possible within the actual structure. For this to be achievable, we opted to take precise acoustical measurements of the building with an eye to reconstructing the essential dimensions of Hagia Sophia's acoustics in a laboratory, studio, or a concert hall. With this goal in mind, researchers from McGill University made comprehensive measurements within Hagia Sophia in October 2013, with the approval and kind permission from the Turkish Ministry of Culture and Tourism, the Directorate of Museums, and the Directorate and Management of the Hagia Sophia Museum. This undertaking expanded on earlier work by Bissera Pentcheva and Jonathan Abel from Stanford University. They recorded a balloon pop in the church, from which a family of synthesized **impulse responses** was generated to auralize a vocal performance in a virtual Hagia Sophia created at Bing Concert Hall, as a component of the "Icons of Sound" project based at Stanford.[15]

Measurements versus modeling

Acoustical measurements are an alternative to computer modeling of architecture; both are able to produce auralization, that is, the ability to hear and interact with the acoustics of architectural space. Computer modeling of architectural spaces allows the locations of sources and receivers in auralization to be changed and makes it easy to derive new synthetic impulse responses when changing the acoustic properties of surfaces, for example, by introducing people or carpets on the floor or curtains on walls. This technique was used to simulate the change in the acoustics of Hagia Sophia from Byzantine church to Ottoman mosque, done with the Odeon software as a part of the CAHRISMA project.[16] The resulting quality of computer modeling depends greatly on the accuracy of the input data the software program receives, of precise dimensions, surface shapes and nonuniformities, details of surface absorption, and on the accuracy of mathematical models and numerical calculations of wave phenomena, including multiple orders of reflection, diffraction, and scattering. Simulated sound sources and receivers have idealized directional characteristics, which cannot be easily duplicated in practice. Rendering of multiple sources adds to the uncertainty and quality of results, especially when live interactive applications are considered using multiple channels. Impulse response produced by a model is free of noise and distortions, as such aspects are not included in the modeling and calculations.[17]

Acoustic measurements made on-site, by contrast, produce results that depend on a dynamic range of sources and receivers and on the background noise level within the enclosure. Care must be taken to ensure the widest possible dynamic and frequency range of the measurement, quality of signal conversion, and the off- and on-axis directional uniformity characteristics of loudspeakers and microphones interfacing with the acoustics. High-resolution acoustic measurements can produce the exact description of the acoustical environment as "illuminated" by loudspeakers and "photographed" by microphones placed in the measured space. All details of the enclosure interacting with the sound radiated by the source—the effects of propagation, reflection, absorption, scattering, and diffraction of sound in the room—are precisely recorded and converted into multiple high-resolution impulse responses.

The goal of exploration and the methodology

A preliminary visit to Hagia Sophia in June 2013 mapped out the acoustic space and identified critical zones of the architectural space that could render the most representative images of its acoustics. It resulted in choosing four source locations and a multitude of receiver locations employed in the subsequent comprehensive measurement session conducted in October 2013 (Figure 9.5).

Digital re-creation of a virtual Hagia Sophia has to be flexible enough to render multiple acoustical perspectives for listeners having different points of view and for a number of sound sources. It is impossible to assume that one measurement can capture all possible acoustical options in this enormous space, just as it is unnecessary to take measurements at hundreds of positions following an arbitrary geometric grid. We chose zones with distinct and different acoustic surroundings, and places where actual sources (singers, speakers) were likely to have been located during religious events. Two locations were near the altar and two near the center under the dome. Our source and receiver layout diagram indicates four speaker locations (SP) and twenty-nine receiver locations (MP). Altogether, eight-channel impulse-response measurements were made 180 times, producing 1,440 individual impulse responses (IRs) with 96 kHz and 32-bit resolution. This data will serve two purposes: (1) acoustic analysis of sound transformations produced by Hagia Sophia; and (2) virtual reconstruction of Hagia Sophia's acoustics in applications including live interactive performance and post-production.

In our measurements, high-resolution impulse responses were captured in three dimensions (width, depth, height) using clusters of eight-microphone receivers installed at three heights of 2 meters, 3 meters, and 4 meters in each measurement point, thus

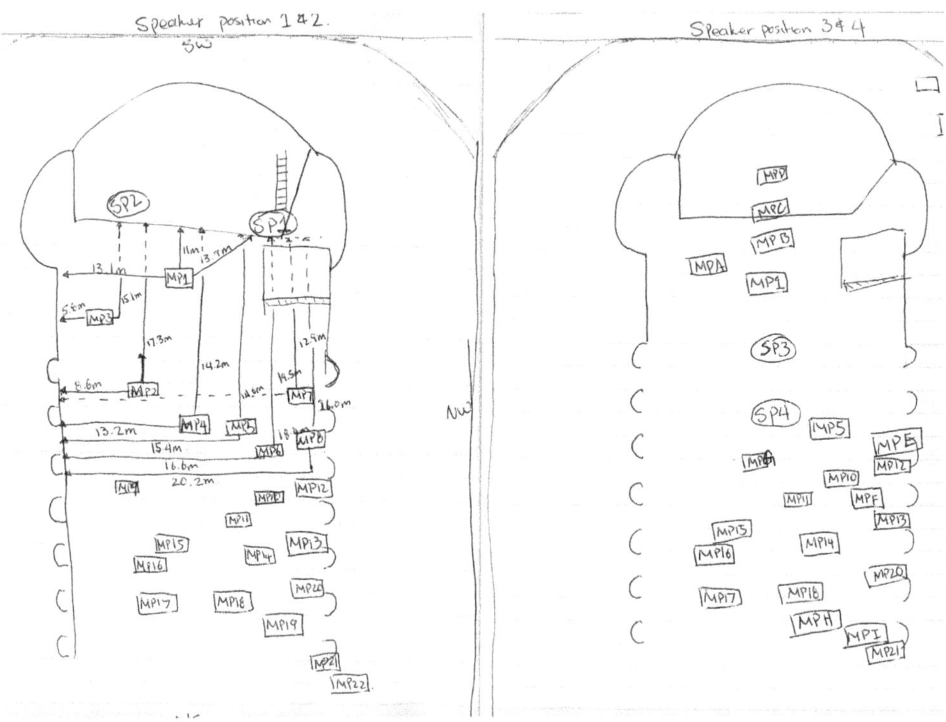

Figure 9.5 Drawing of measured positions (MP) and source locations (SP) used in measurements of the acoustics of Hagia Sophia in October 2013.

providing twenty-four independent impulse responses at each measured location. Sinusoidal sweeps 80 seconds long and a sound-pressure level of 93dB peak ensured a wide dynamic range of the measurements, suitable for the most demanding and critical applications. The use of long exponential sine sweep and high-performance loudspeakers ensures a high dynamic range of the measured impulse response, which influences the accuracy of all acoustical measurements that are derived from impulse responses. This method also removes the harmonic distortion of the loudspeaker source from the measured impulse responses, since distortion components appear separately at negative arrival times before the onset of the impulse response and are removed.[18]

Acoustic measurements

Just as a photographer moves the camera and the lights around a room to obtain the most compelling photo image of that room, the acoustician decides on the most appropriate sonic illumination of Hagia Sophia, capturing the best possible interpretation of its magnificent acoustics. This process is both artistic and technical because the choices are guided as much by the ear and aesthetic judgment as by the science and acoustical analysis of the architecture.

In our study, each cluster of measured twenty-four impulse responses represents one of the twenty-nine receiver locations on the floor of Hagia Sophia as microphone positions MP1–22 and MPA-G (Figure 9.5). Four locations of source positions are shown as SP1–SP4, each source consisting of ten transducers, together emanating a complex wave front from a virtual center of this multiple radiator. The curvature and the shape of this radiator have been adjusted to project a believable musical source; it is tested on-site by reproducing a number of monophonically recorded anechoic signals, including small chorus, trumpet, speaking voice, guitar, and drums. Once adjusted, the multispeaker source excites the interior acoustic space of Hagia Sophia with the analytical signal of a slow exponential sinusoid sweep that is recorded by the cluster of microphones at the MP locations. By moving the loudspeaker and the microphone clusters about the room, different acoustic views of the room are recorded, each giving a distinct acoustical perspective of the space.

A sample of the diversity of acoustical material collected during the measurements of Hagia Sophia is shown in the following graphs (Figures 9.6–9.12). The graphs are plotted from impulse-response data calculated from just a few locations to illustrate the complexity of the measured sound fields. A "turbulent" development of low-frequency energy (20 Hz to 300 Hz) can be observed in the room at the onset of acoustical response, within the first 300 milliseconds following the initial impulse (Figure 9.6). The visible variance of acoustic energy in level and frequency as time progresses is due to multiple low-frequency arrivals at the measurement point of waves reflected from the surrounding boundaries, some causing the level to raise, while others cause the level to drop. The acoustic waves flow in and out of the measuring point, not unlike water waves around a float in a large pool disturbed by the dropping of a large stone. The waves arrive from a multitude of directions for more than 10 seconds until the wave motion is at rest again. We can see a bigger picture of sound variance measured in the back of the nave for a source located near the center of the nave. Figure 9.7 shows the spectrum of all audio-range frequencies as their energy decays over the reverberation time, specified as the time it takes for the reverberation to become inaudible (60 decibels drop). One can clearly see that higher frequencies,

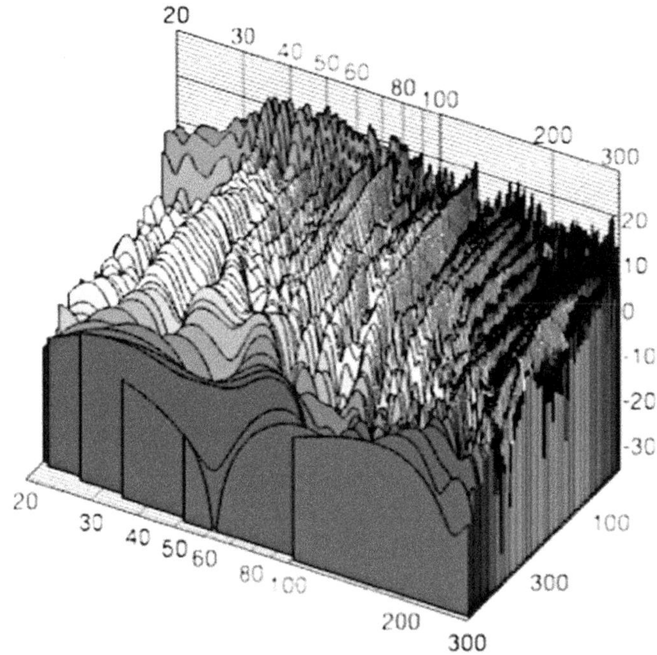

Figure 9.6 A three-dimensional graph of the low-frequency acoustic energy from 20 Hz to 300 Hz decaying in time over the initial period of 300 milliseconds from the onset of the impulse response (IR). Source and receiver positions are SP-3 (near the center of the nave) and MP-D (near the altar), the output signal of the omnidirectional microphone, Channel 1.

above 1 kHz, at the right side of the graphs, decay much faster than the low frequencies owing to their absorption by the air. Acoustic waves traveling more than three kilometers' distance in air and encountering multiple reflections lose most of their high frequencies early in the decay of reverberation. Such a decaying wave appears to move away, lose its presence, and depart into the distance. For a receiver placed closer to this source but on the altar side, the picture of the developing sound spectrum in time is somewhat different. Figure 9.8 shows more low-frequency energy at the onset (the topmost profile of the graphs) and steeper changes in level across low frequencies below 250 Hz. This is due to the close proximity of the walls, half-domes, and exedrae, because they send greater energy streams down to the measuring point. Even in a single measuring location, as shown in Figure 9.9, three microphones with different directional properties will each capture a somewhat different image of the church acoustics characterized by a different balance of energy arriving from certain directions.

There is a huge amount of important detail in the reflections and decaying reverberation of Hagia Sophia contained in the measured impulse responses. Figures 9.10, 9.11, and 9.12 only show a macro, bird's-eye view of the sonic spectacle produced by the acoustics of the church. The full analysis of all collected data is an ongoing process, as we try to better understand how best to extract, recompose, and auralize such magnificent acoustical space.

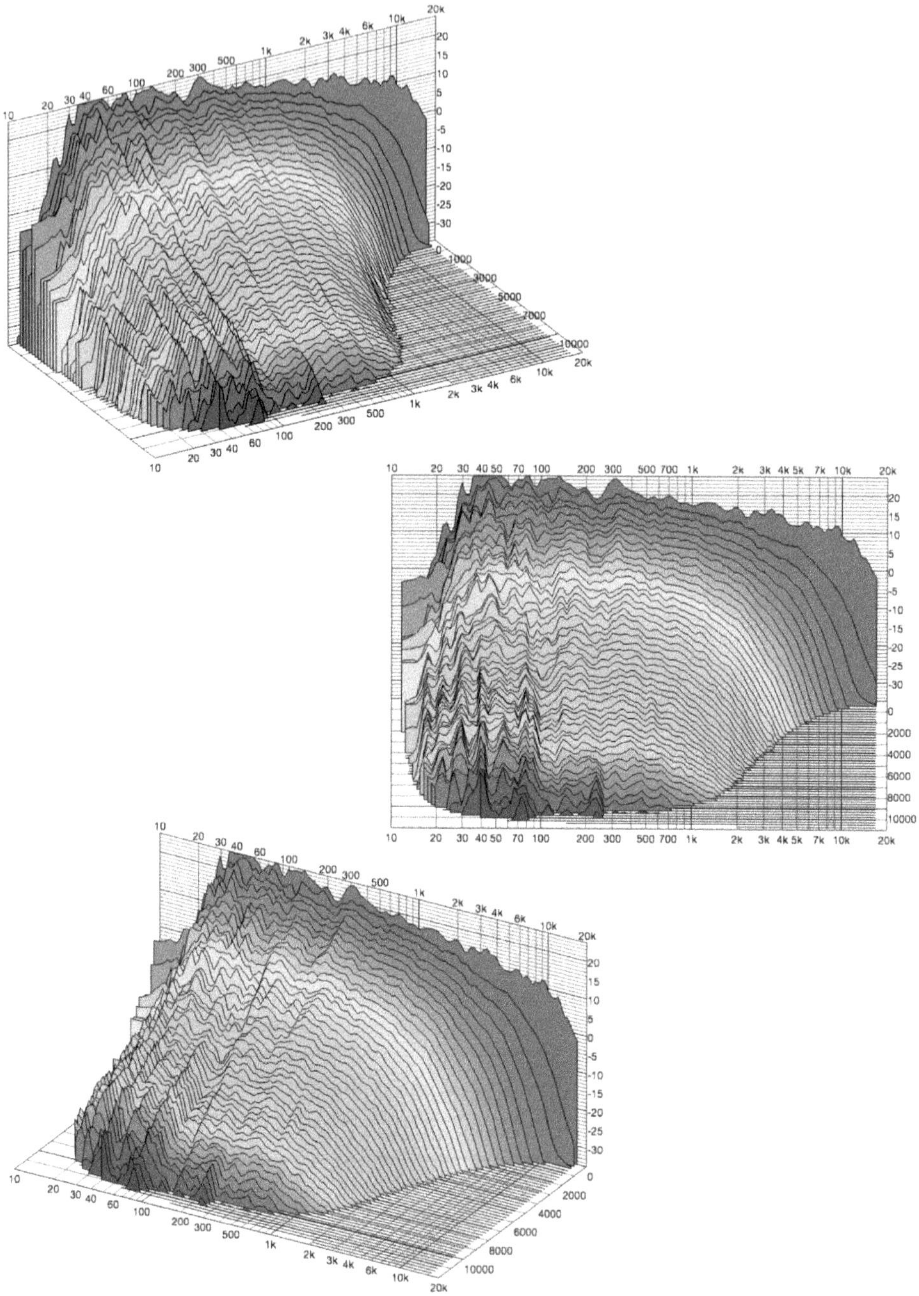

Figure 9.7 A three-dimensional cumulative spectral decay view of the full spectrum of acoustic energy (20 Hz to 20 kHz) decaying in time over ten seconds from the onset. This cumulative spectral decay is calculated with 1/12 octave smoothing and is presented from three different angles for source position SP-3 (near the center of the nave) and receiver position MP-18 (near the back of the nave), and using omnidirectional microphone signal channel 1.

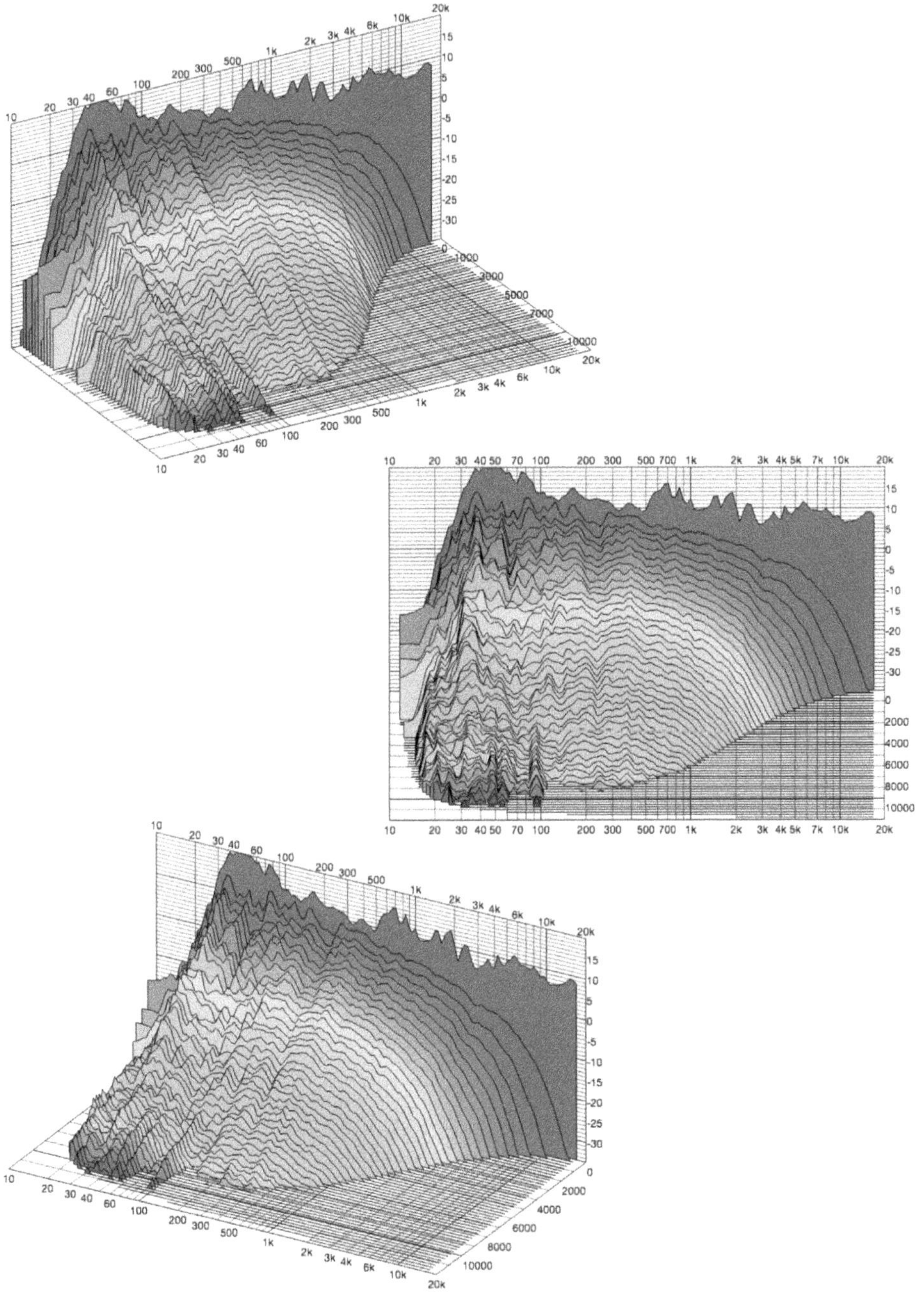

Figure 9.8 Three-dimensional cumulative spectral decay view of the full spectrum of acoustic energy (20 hz to 20 khz) decaying in time over ten seconds from the onset. This cumulative spectral decay is calculated with 1/12 octave smoothing and is presented from three different angles for source position SP-3 (near the center of the nave) and receiver position MP-D (near the altar) and using omnidirectional microphone signal channel 1.

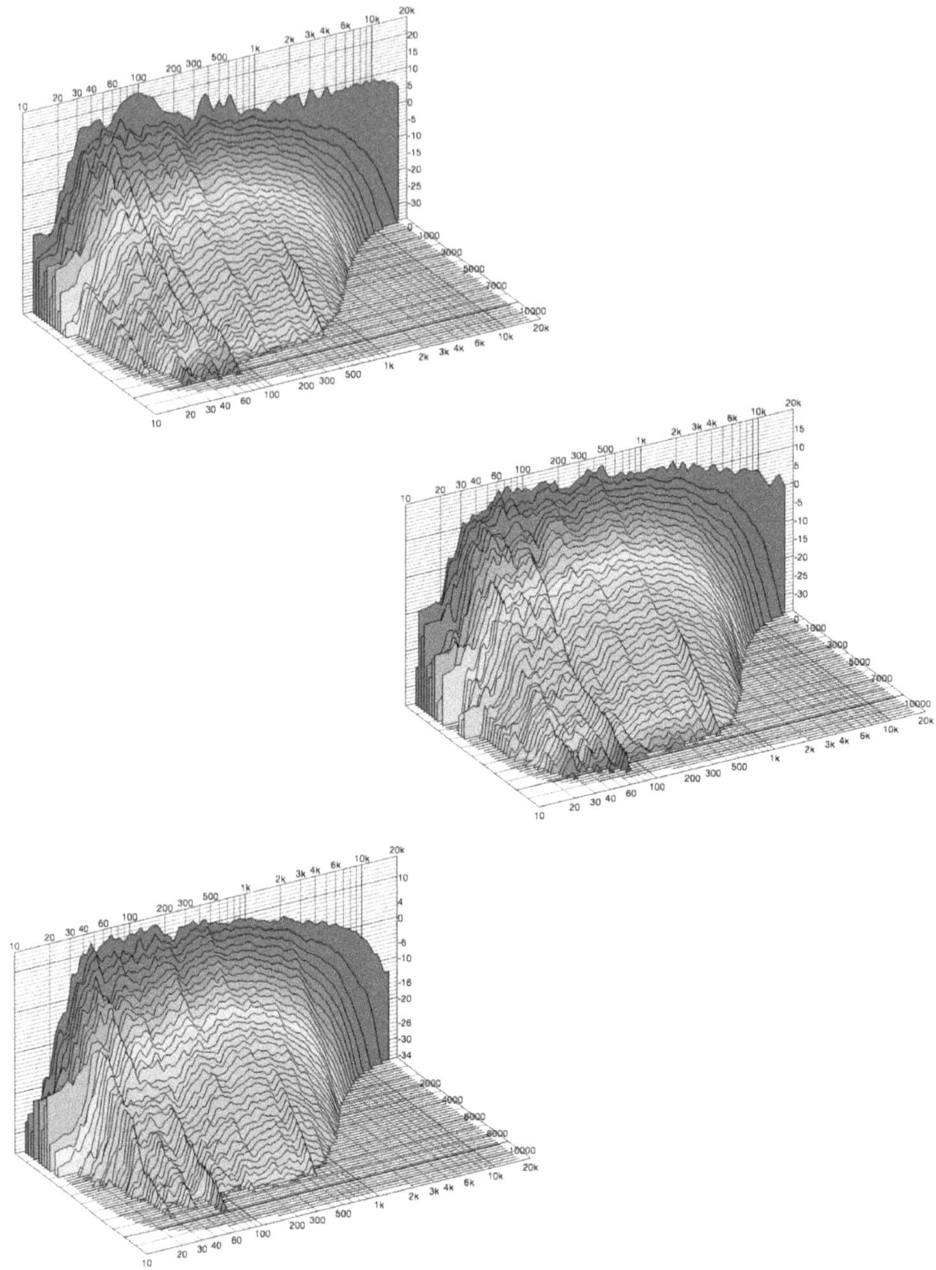

Figure 9.9 Three-dimensional cumulative spectral decay view calculated from the impulse response measured by three microphones: omnidirectional channel 1 (top), bidirectional in elevation channel 5 (center), bidirectional in azimuth channel 8 (bottom). This cumulative spectral decay is calculated with 1/12 octave smoothing and is presented for source position SP-4 (beyond the center of the nave) and receiver position MP-F (near the side of the nave). Three different microphones of eight microphones in a group (cluster) in one measurement location.

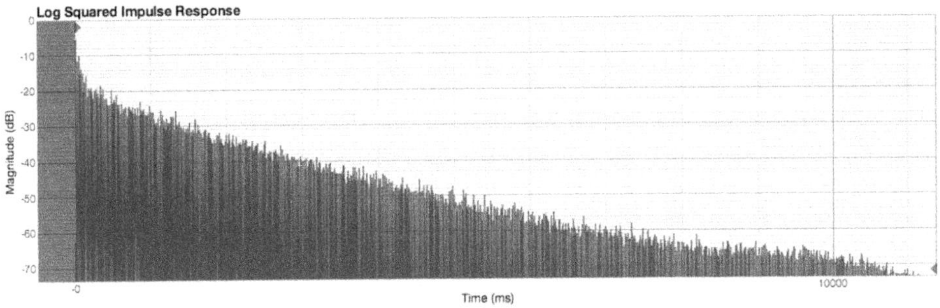

Figure 9.10 Log-squared impulse response graph from SP-2 (omnidirectional microphone signal channel 1 near Sultan's Lodge) and MP-21 (near the rear corner of the nave).

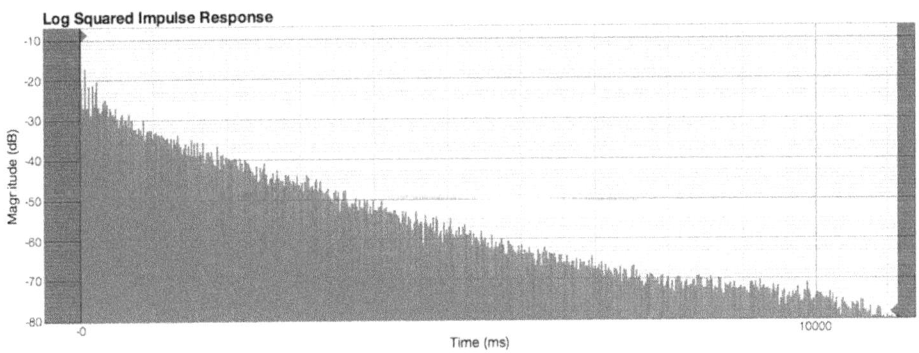

Figure 9.11 Graph of time domain magnitude (logarithm squared scale) of impulse response from SP-4 (beyond the center of the nave) and MP-1 (in the center of nave, near the altar), from omnidirectional microphone signal channel 1.

An acoustic rendering of a virtual Hagia Sophia

The resulting catalog of high-resolution impulse responses serves as the construction material for re-creating a compelling version of Hagia Sophia's acoustics in a laboratory or performance space. The unforgettable experience of hearing sound in the actual Hagia Sophia demands a complex approach to rendering a similar experience in a virtual environment. In the real space, sound evolves gradually from onset to complete decay following complex trajectories that attract the attention of listeners by way of "liquid" animations, transitions through phases, and soundscapes of varying characteristics. Hagia Sophia strongly projects its aural architecture, demanding to be heard. The interior's flowing display of sound requires that a virtual reconstruction engineer builds a constantly evolving motion of reverberation using different acoustic views captured within the clusters of impulse responses. The engineer must render dynamic

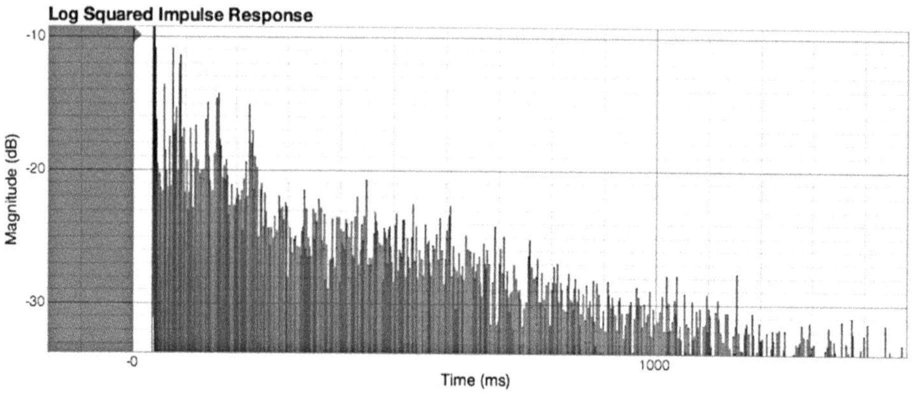

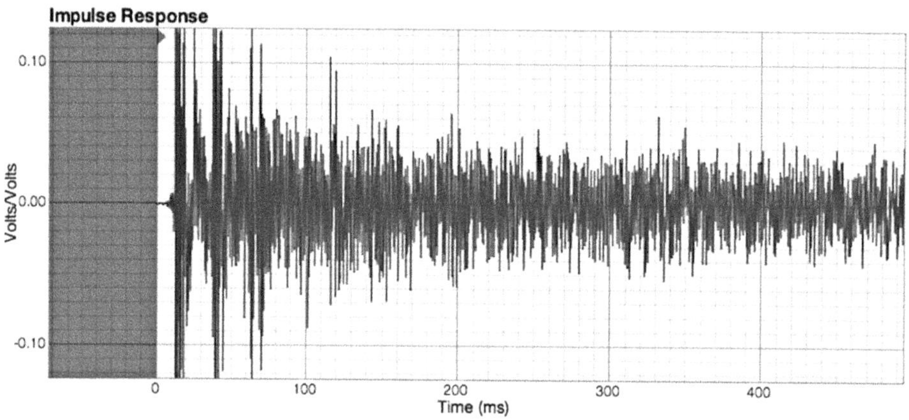

Figure 9.12 Graph of time domain magnitude (logarithm squared scale) of impulse response (visible duration 1,500 milliseconds) compared to direct impulse response (linear scale amplitude and phase) (visible duration 500 milliseconds) measured at source location SP-2 (near Sultan's Lodge) and receiver position MP-4 (near the center of the nave). The graphs show the impulse response in greater detail during the early portion of development of the sound field.

sonic animations existing in Hagia Sophia, as these are essential to forming the sensory and psychological experience of being inside the actual space.

In our experience, successful reconstruction of aural architecture is achieved when digitally rendered stimuli produce an outcome equivalent to human multisensory experience that is accumulated over several encounters with the space. The reconstruction has to be more than a simulation because it needs to unlock our memory and imagination of the space. This is why an arbitrary approach of rendering only a single-perspective view of Hagia Sophia acoustics cannot match the memory of our experience that is based on the exploration of this space. We are moving around the space, making and listening to sounds, turning about left to right and looking/listening up and down, transitioning between different sections of the enclosure. To create a compelling reconstruction of the acoustics as we remember and imagine them (this is now our reference), the engineer must incorporate active exploration of the space into the structure of the rendered acoustical response, by skillful composition of spatial layers appropriately ordered in time. It is critically important to construct and render a convincing acoustical perspective for each sound source, being especially mindful of the perceived differences in how the space reacts acoustically to each of the sources and to their spatial distribution within the enclosure. Human attention is dynamically nonlinear; it exaggerates, changes, and diminishes sensitivity to static parameters to reduce their masking power by guiding perception. The engineer must guide the listener's attention and emulate the listener's own discovery and exploration of the space.

Measuring microphones employed to capture the massive dimensions of Hagia Sophia acoustics do not have the ability to bring out the complexity of spatial transformations occurring in the expansive interior of the building. Whereas the ear-brain can sharpen and widen attention to focus on diverse aspects of spatial sound, especially shifting attention between foreground and background sounds, the microphone transfers only a fixed spatial relationship. It does not have sufficient directional selectivity to pick up specific noticeable elements of the reflected sound and reverberation. In contrast, humans using head movement and motion through space may develop a detailed multisensory image of the perceived environment. Loudspeaker and headphone reproduction further diminishes the ability of the human ear and brain to resolve sounds and focus on details contained within it, when the sound is unfavorably displayed. The dynamics of digitally rendered space, therefore, have to be augmented through spatial design and mixing by the virtual acoustics engineer to help the ear do its job.

Music performance in virtual acoustics

Digital rendering of aural architecture to serve as an interactive acoustic environment in musical performance requires fast-signal processing capable of rendering an immediate acoustic response. Fast parallel convolution of multiple-impulse responses is a hugely intensive computational task, and acoustic delays or glitches are not desirable during performance. Musicians are affected by the quality of the acoustics of the room, as they depend on it to hear their own and other players' instruments or voices and when they shape the musical balance and acoustical interactions. Research shows that good acoustics supports musical interactions, whereas poor acoustics plays a detrimental role in musical performance and collaboration.[19]

In re-creating the acoustics of virtual architecture, the acoustics engineer must have at his or her disposal a number of assets that promote musicians' aural communication. During 2005–09 the author was involved in The Virtual Haydn, a Blu-ray recording project produced at McGill University and released commercially by Naxos International containing fifteen hours of music recorded in surround sound.[20] In a laboratory, Tom Beghin performed the complete works of Haydn for solo keyboard on seven historical instruments while being immersed in the rendered acoustics of nine virtual rooms related to the music of Haydn. The principal goal was to recreate a historical and functional link between the musical piece, the instrument, and the room, which together combined to form authenticity in experiencing the works of Haydn.[21] The team experienced and measured all nine rooms on-site, then rebuilt the acoustic response in the laboratory for the solo keyboard performance and recording, and again later for the final mixing. It was critical to create, both for the performance and for mixing, the correct acoustic perspective enveloping the performer and the listener. No single capture of the original measured response was sufficient to re-create a believable immersive spatial outcome. Again, skillful balancing of room assets was critical to achieving an appropriate sensation of the acoustical surroundings in each case.

As part of the ongoing multidisciplinary research project "Icons of Sound," the January 2013 performance of Cappella Romana in Bing Concert Hall at Stanford University presented an innovative re-creation of the acoustics of Hagia Sophia in the concert hall using forty statistically independent synthetic impulse responses generated from a model derived from a recording of one balloon pop in this church.[22] Many technical challenges had to be overcome to offer a compelling experience for the singers and the audience. It is clear that for musicians, the prospect of playing or singing in a venue that is exceptionally suitable acoustically and/or historically yet is not available for concerts is hugely attractive. Many historical buildings are not open to musical uses but may exist virtually as aural architecture. Natural architecture of outdoor and enclosed spaces can also be explored virtually through active acoustics. Some buildings slated for demolition can be "saved" for posterity as virtual aural architecture.

Digital adjustability of aural architecture gives it plasticity, a potential ability to serve better than the actual architecture. For example, the total acoustic gain from the room (the level of reflected sound) can be increased or decreased to accommodate the music and the ensemble. There is no room noise in a digital rendering, unlike actual spaces, many of which are located in noisy areas. The acoustical correction of rendered space can be implemented digitally without construction costs and time constraints. For example, ceiling reflections can be attenuated, moved down to the walls and floor, or delayed (moved farther up). Acoustics can be adjusted to make playing music easier. Empowered by the increased gain from the room, musicians often play with greater ease, as they do not have to force the sound out of their instruments to reach distant parts of the audience.

Further studies of aural architecture can lead to common applications that may enrich our lives and augment our understanding of history and culture. Experiencing architecture through an aural mode using methods of digital reconstruction will bring history closer to our present sensory awareness of the world.

Conclusions

Under way at McGill University is the testing phase of a practical rendering of Hagia Sophia's acoustics in a three-dimensional (width, depth, height) laboratory containing a 22.2-channel loudspeaker system. The 22.2-channel loudspeaker configuration, developed by NHK STRL (Science and Technology Research Laboratories of Japan Broadcasting) in Japan, has been chosen as a reference system for evaluating multichannel audio technologies.[23] Hagia Sophia's measurements were used to auralize the Cappella Romana performance of *prokeimenon* recorded earlier at Stanford University as dry-only tracks, while the performers monitored their voices using a reverberation ten seconds long from the convolution of synthetic impulse responses prepared for the January 2013 concert in Bing Hall. McGill professor Richard King with PhD student Jonathan Hong mixed the dry tracks, placing voices within the acoustic space of Hagia Sophia using a multichannel tool developed at McGill University. Three different acoustic perspectives of Hagia Sophia were used. Dynamic mixing and skillful placement of sources and reverberation material rendered the performance compellingly real. Female voices were placed in elevation as if they were boys projecting from the balcony above the men. Lead and drone voices were spread around the floor at different distances to engage the vast reproduction space with overlaid perspectives of Hagia Sophia's acoustics. Two low-frequency channels helped the drones to command power and fullness. These explorations led to further perceptual studies of moving reverberation and envelopment, as these important perceptual attributes of large spaces can now be rendered experimentally.[24]

Digital technology allows us the unique advantage of studying and exploring Hagia Sophia's acoustics off-site, in a laboratory and in a performance space, using both signal analysis and rendering of acoustics of a virtual Hagia Sophia. The high-resolution data provide a thrilling option of practical explorations in different liturgical and historical uses of the church, experienced using our own present sensibility. Hagia Sophia continues to dazzle when rendered from high-resolution measurements using a 22.2-channel loudspeaker system providing immersive surround sound with height, or in other advanced multichannel reproduction systems able to render aural architecture in three dimensions, as long as the rendering includes elevation. Our further work will involve an acoustic rendering of a virtual Hagia Sophia for live instrumental and vocal performance, advancing our ability to virtually rebuild this splendid aural architecture. The scientific and artistic perspectives meet again as they did at the inception of Hagia Sophia.

Acknowledgments

The author would like to acknowledge the help of many institutions and individuals, among them the Turkish Ministry of Culture and Tourism, General Directorate for Cultural Heritage and Museums, Department of Museums, Directorate and Management of Ayasofya Museum, Prof. Nina Ergin and students in the Department of Archaeology and History of Art at Koç University, Prof. Jonathan Abel and Prof. Bissera Pentcheva of Stanford University, Prof. Alexander Lingas and Cappella Romana, Professor Ipek Tureli, Prof. Radoslav Zuk, Prof. Richard King, Doyuen Ko, Jonathan Hong, Scott Levine, Brett Leonard, Dave Benson, and students from Graduate Program in Sound Recording at McGill University. The assistance of Social Sciences and Humanities Research Council of Canada, Natural Sciences and Engineering Research Council of Canada, and of McGill University is gratefully acknowledged.

Notes

1 Note for abstract: Procopius, *De aedif.* 1.1.23ff., "Description of the Hagia Sophia in Constantinople by Prokopios, Written 544," in *The Art of the Byzantine Empire, 312–1453: Sources and Documents*, trans. Cyril A. Mango (Englewood Cliffs, NJ: Prentice-Hall, 1972, rpt. Toronto: University of Toronto Press, 1986), 72.

For the extensive literature of Isidorus of Miletus and Anthemius of Tralles, see George Downey, "Byzantine Architects and Their Methods," *Byzantion* 18 (1948): 99–118; George Huxley, *Anthemius of Tralles: A Study of Later Greek Geometry*, Greek, Roman, and Byzantine Monographs, 1 (Cambridge, MA: Eaton Press, 1959); Alan Cameron, "Isidore of Miletus and Hypathia: On the Editing of Mathematical Texts," *Greek, Roman, and Byzantine Studies* 31 (1990): 103–27; and Nadine Schibille, "The Profession of the Architect in Late Antique Byzantium," *Byzantion* 79 (2009): 27–46.

2 Helge Olaf Svenshon, "Heron of Alexandria and the Dome of Hagia Sophia in Istanbul," in *Proceedings of the Third International Congress on Construction History: Brandenburg University of Technology, Cottbus, Germany, 20th–24th May 2009*, ed. Karl-Eugen Kurrer, Werner Lorenz, and Volker Wetzk (Cottbus: Brandenburgische Technische Universität, 2009), 1387–94.

3 Rowland Mainstone, *Hagia Sophia: Architecture, Structure and Liturgy of Justinian's Great Church* (London: Thames & Hudson, 1988), 187–217; Robert Mark, Ahmet S. Carmak, K. Hill,

and R. Davidson, "Structural Analysis of Hagia Sophia: A Historical Perspective," *Transactions on the Built Environment* 3 (1993): 867–80; and Robert Ousterhout, "Holy Space: Architecture and the Liturgy," in *Heaven on Earth: Art and the Church in Byzantium*, ed. Linda Safran (University Park, PA: Pennsylvania State University Press, 1998), 81–120, esp. 89–90.

4 Umut Almac et al., "Structural Behaviour of Hagia Sophia under Dynamic Loads" (paper, Vienna Congress on Recent Advances in Earthquake Engineering and Structural Dynamics, Vienna, August 28–30, 2013).

5 Zerhan Karabiber, "A New Approach to an Ancient Subject: CAHRISMA Project," in *Proceedings of the Seventh International Congress on Sound and Vibration, ICSV-7: Garmisch-Partenkirchen, Germany, July 4–7, 2000, Volume 4* (Garmisch-Partenkirchen, Germany: International Institute of Acoustics and Vibration, 2000), 1661–68; and Patrizio Fausti, Roberto Pompoli, and Nicola Prodi, "Comparing the Acoustics of Mosques and Byzantine Churches" (paper, 19th International Committee for Documentation of Cultural Heritage (CIPA) Symposium, Antalya, Turkey, September 30–October 4, 2003).

6 Bissera V. Pentcheva, "Hagia Sophia and Multisensory Aesthetics," *Gesta* 50, no. 2 (2011): 93–111; and Nina Ergin, "The Soundscape of Sixteenth-Century Istanbul Mosques: Architecture and Qu'ran Recital," *Journal of the Society of Architectural Historians* 67 (2008): 204–21.

7 Alexander Lingas, "Sunday Matins in the Byzantine Cathedral Rite: Music and Liturgy" (PhD diss., University of British Columbia, 1996).

8 Alexander Lingas, "Sunday Matins in the Byzantine Cathedral Rite: Music and Liturgy" (PhD diss., University of British Columbia, 1996).

9 Neil Moran, "The Choir of the Hagia Sophia," *Oriens Christianus* 89 (2005): 1–7.

10 Neil Moran, *Singers in Late Byzantine and Slavonic Painting* (Leiden: Brill, 1986), 26–32.

11 This psychoacoustic effect of ascent resonates with the anagogical interpretation of the liturgy. On the latter, see Walter Ray's essay in this volume.

12 Çiğdem Özkan Aygün, "The Subterranean Tunnels, Wells and Cisterns beneath the Hagia Sophia" (paper, Twentieth Biennial New College Conference on Medieval & Renaissance Studies, Sarasota, Florida, USA, March 10–12, 2016).

13 Michael Rettinger, "Note on Reverberation Chambers," *Journal of the Audio Engineering Society* 5 (1957): 108.

14 The room acoustician Heinrich Kuttruff explains that if we consider sound propagation as rays and apply the laws of concave mirrors from optics, the relationships of the source distance **a**, the distance of the focus **b**, and the radius **r** of the mirror are related to each other by the following equation: $1/\mathbf{a} + 1/\mathbf{b} = 2/\mathbf{r}$. For distances of the source to the mirror greater than half the radius, the point of focus is located closer to the mirror than the source. The focus is at r/2 for distant sources. Kuttruff, *Room Acoustics* (Oxon, UK: Spoon Press, 2009), 117–20.

15 Jonathan S. Abel, Wieslaw Woszczyk, Doyuen Ko, Scott Levine, Jonathan Hong, Travis Skare, Michael J. Wilson, Sean Coffin, and Fernando Lopez-Lezcano, "Recreation of the Acoustics of Hagia Sophia in Stanford's Bing Concert Hall for the Concert Performance and Recording of Cappella Romana" (paper, International Symposium on Room Acoustics, June 9–11, 2013, Toronto, Canada). See also Jonathan Abel and Kurt Werner's essay in this volume.

16 Christoffer Andreas Weitze et al., "Computer Simulation of the Acoustics of Mosques and Byzantine Churches" (paper, 17th International Congress on Acoustics, Rome, Italy, September 2–7, 2001); and Christoffer Andreas Weitze et al., "The Acoustical History of Hagia Sophia Revived through Computer Simulation" (paper, Forum Acusticum, Seville, Spain, September 16–20, 2002), www.odeon.dk/pdf/ForumAcousticum2002.pdf.

17 Jens Holger Rindel, Claus Lynge Chistensen, and George Koutsouris, "Simulations, Measurements and Auralisations in Architectural Acoustics" (paper, Acoustics 2013 New Delhi, New Delhi, India, November 10–15, 2013).

18 Swen Müller and Paulo Massarani, "Transfer-Function Measurement with Sweeps," *Journal of the Audio Engineering Society* 49 (2001): 443–71.

19 Doyuen Ko, Wieslaw Woszczyk, and Song Hui Chon, "Evaluation of a New Active Acoustics System in Performances of Five String Quartets" (paper, 132nd Convention of the Audio Engineering Society, Budapest, Hungary, April 26–29, 2012); and Doyuen Ko et al., "Augmented Stage Support in Ensemble Performance Using Virtual Acoustics Technology" (paper, 21st International Congress on Acoustics, Montreal, Canada, June 2–7, 2013).

20 Tom Beghin (harpsichord) with Martha de Francisco (producer) and Wieslaw Woszczyk (virtual acoustics architect), *The Virtual Haydn: Complete Works for Solo Keyboard*, Naxos, NBD 0001–04, 2011, 4 Blu-ray discs.

21 John Irving, "Digital Approaches to Haydn's Solo Keyboard Music," *Early Music* 40 (2012): 535–38.

22 Jonathan S. Abel et al., "Estimating Room Impulse Responses from Recorded Balloon Pops" (paper, 129th Convention of the Audio Engineering Society, San Francisco, USA, November 5–7, 2010); and Jonathan S. Abel et al., "Recreation of the Acoustics of Hagia Sophia in Stanford's Bing Concert Hall for the Concert Performance and Recording of Cappella Romana" (paper, International Symposium on Room Acoustics, Toronto, Canada, June 9–11, 2013), ftp://s00279.cisti.nrc.ca/outgoing/CD_ISRA2013/Papers/P055.pdf, 2–3. On the "Icons of Sound" live auralization in Bing Hall, 2013, see the contribution of Jonathan Abel and Kurt Werner in this volume.

23 Kimio Hamasaki et al., "22.2 Multichannel Sound System for Ultra High-Definition TV," *Society of Motion Picture and Television Engineers (SMPTE) Motion Imaging Journal* 117 (2008): 40–49.

24 Jung Wook Jonathan Hong, Wieslaw Woszczyk, and Durand R. Begault, "Synthesis of Moving Reverberation Using Active Acoustics—Preliminary Report" (paper, Audio Engineering Society Convention Paper 9326, Presented at the 138th Convention, 2016 May 7–10, 2016, Warsaw, Poland).

10 Live auralization of Cappella Romana at the Bing Concert Hall, Stanford University

Jonathan S. Abel and Kurt James Werner

Introduction

The reverberant properties of a space have long been considered an acoustic signature of that space.[1] In his 2015 article "Space within Space: Artificial Reverb and the Detachable Echo," Jonathan Sterne traces out the legacy of this idea and its eventual relaxation. To describe this "separation of sounds from themselves" (their physical presence in space, that is, their reverberation), he coined the phrase "detachable echo."[2] Although the traditional association between sounds and their reverberation has been somewhat loosened in the thinking of, for example, modern architectural acoustics and artistic uses of artificial reverberation, the classical link between sound and space is an essential part of understanding the sonics of certain places, including Hagia Sophia.

Sound is intrinsically experiential; to fully understand the sound of a space, we must *hear* it. Even the most hardened signal-processing engineer would admit that studying an impulse response (the output of a dynamic system when presented with a brief input signal, or, in other words, the acoustic signature of the space) is only a supplement to, not a substitute for, listening to a piece of music in the interior characterized by that response. Space intimately relates musicians to their musical material; some music can be fully experienced only in its intended context. Consider what is lost when we try to reproduce the sound of a symphony orchestra through laptop speakers; consider how out of place a marching band might sound in a large concert hall.

This is a concern not only for listeners, but also for performers. Musicians alter aspects of their performances, including tempo, articulation, and timbre, according to the space in which they are performing.[3] A recent study showed that reverberation time significantly affected tempo and timing precision.[4] Although it also observed that effects on intonation were not very strong, the study was limited to a choir singing Western classical music (Anton Bruckner's sacred motet "Locus Iste"). Anecdotally, we have evidence that in the context of modal music, which characterizes medieval chant, both intonation and the trajectory of pitch "glide" between notes are affected by the performance space.

Singers collaborate with space itself during a performance, and properties of the space can have pronounced effects on their chant; unsurprisingly, performers decidedly favor one space over another. These subjective preferences can sometimes be linked to measurable objective features.[5] As an example, I. Nakayama studied subjective preference judgments of an alto recorder soloist in a simulated reverberation environment, one with only a single reflection. Even in a simplified environment like this, the alto recorder soloist strongly preferred certain time delays.[6]

Unfortunately, logistics, geography, and even politics can keep us from performing and experiencing music in context in its most appropriate spaces. This is the case with Hagia Sophia and Byzantine cathedral chant. The building was transformed into a mosque in 1453 following the Ottoman conquest of the city. With the establishment of the modern Republic of Turkey, the structure was secularized and made into a museum. As a result, concerts and liturgical music are no longer permitted. At the same time, there is a resurgence of interest in the elaborate Byzantine cathedral chant.[7] To overcome these limitations and place together the music within the space for which it was composed and within which it was performed, we can turn to artificial reverberation to re-create the sonic experience of the Great Church using signal processing.

The process of re-creating the aural experience of a particular space is called *auralization*.[8] Successful auralization is a complex task that involves cultural and musical sensitivity to the target space and its aural context, technical skills in signal processing, artificial reverberation research, and access to potentially complex sound-reinforcement hardware (loudspeakers).[9] Auralization has a long history, stretching back to the 1930s, although the target spaces of auralization are often much more ancient.[10]

Researchers have begun to uncover how ancient spaces are characterized by distinctive sonic properties that carry ritual significance.[11] The El Castillo pyramid in Chichén Itzá, Mexico, built sometime between the ninth and the twelfth centuries as part of a Mayan religious site, produces a chirplike sound (a "repetition pitch glide") in response to a hand clap, a result of the geometry of the pyramid's stairs.[12] This chirplike clap response approximates remarkably the sound of the quetzal bird, which held an important place in Mayan culture and religious life.[13] Similarly, the ancient (dating to at least 1200 BC) underground galleries at Chavín de Huántar, Peru have distinct acoustics. Dense and energetic early reflections, a short reverberation time, and wide soundfields create an acoustic experience of envelopment and can heighten a sense of indeterminate sound source location.[14] Along with visual and other sensory manipulations, the acoustics of the underground galleries at Chavín contributed to the ritual experience at the site.[15] After completing rudimentary acoustic measurements inside six diverse ancient structures in the British Isles, Robert Jahn, Paul Devereux, and Michael Ibison found that each sustained a strong resonance at a frequency between 95 and 120 Hz (hertz). Jahn speculates that since these prominent resonance frequencies are within the adult male voice range, the interaction between chanting and cavity resonances were invoked for ritual purposes.[16]

Researchers employ auralization to mimic the sonic properties of archaeological sites,[17] either on its own or alongside other aids, such as visual re-creations, as part of "virtual time travel" or "experiential archaeology."[18] For example, digital waveguide networks (networks of bidirectional delay lines that explicitly simulate traveling waves) have been used to imitate the acoustics of galleries at Chavín.[19] "Convolution-based approaches (see the next section, wherein convolution and its application to artificial reverberation and auralization are described) are common and have been used in particular as part of a hybrid wave field synthesis (simulation of the soundfield using multiple loudspeakers related to Hyugen's principle) and Ambisonics (full sphere surround sound based on spherical harmonics) system to create an auralization of

Stonehenge."[20] Auralization technology is mature enough that it can effectively simulate even complex concert halls.[21]

However essential the sonic signatures of spaces such as archaeological sites, concert halls, and churches may be, they are not immutable. Since ancient times, an understanding of the importance of the acoustics of spaces has been accompanied by attempts to manipulate those sonic imprints. Early techniques for affecting the auditory properties of spaces involved passive modifications to the structure. In 30 BCE, the architect Vitruvius reported on a well-established tradition of using resonating vases, which can be understood as Helmholz resonators, set up in theaters and intended to modify the acoustic properties of the space.[22] In the twentieth century elaborate signal processing-based approaches have been worked out to control the acoustics of concert halls, a process called active acoustic enhancement.[23] The goals of active acoustic enhancement may include repairing poor acoustics[24] or tailoring a concert hall's response to the divergent acoustic needs of a variety of aural situations, including spoken word, chamber music, and full-scale symphonic works.[25] These systems can even help to control the acoustics of outdoor venues; the Lares system at Chicago's Frank Gehry-designed Pritzker Pavilion is exemplary.[26] Active acoustic enhancement systems tend to be permanent installations. By contrast, the system described in this chapter is mobile, non-site-specific, and employs off-the-shelf and even open-source technology.

In the twentieth century, audio engineers in broadcast and popular music recording developed a complex set of tools for manipulating spatial impressions in audio recordings.[27] Early approaches involved merely making recordings in locations with particular sonic effects or in specially designed echo chambers; later on, researchers devised electromechanical devices, including "plate reverb" (reverberation produced using a large steel plate) and "tape echo" (a type of delay or echo processor that uses analog recording tape to achieve the effect), to add to the palette of spatial manipulations. Beginning in the 1960s, researchers began working on digital signal-processing techniques that could accomplish these manipulations entirely in the digital realm. Today, a vast literature on digital artificial reverberation techniques continues to inform new research, including approaches to auralization.[28]

Concert halls are also characterized largely by their acoustics; the sound of a concert hall is a major element in experiencing live music. Alongside investigations into and re-creations of the sonic environment of ancient archaeological sites, researchers have employed auralization to preserve the acoustics of important concert halls for posterity.[29] Among many other studies, Lamberto Tronchin and Angelo Farina explored the acoustics of the former La Fenice opera house in Venice, whose unusual elliptical shape gave it a unique sound. When the building burned down on January 29, 1996, its restoration relied in part on Tronchin and Farina's acoustic measurements.[30]

Before specific architectures for concert halls were common, churches provided such spaces. Jaime Navarro, Juan J. Sendra, and Salvador Muñoz make a case for acoustic assessment of the Western Latin church.[31] Rafael Suárez, Alicia Alonso, and Juan J. Sendra considered the Romanesque cathedral of Santiago de Compostela in Spain as a case study on the relations between architecture, music, and liturgy.[32] Deborah Howard and Laura Moretti studied the relation between sacred music and architectural design in sixteenth-century Venice.[33] Related work aimed to quantify the acoustic heritage of mosques and Byzantine churches.[34]

In this context, the acoustics of Hagia Sophia present an opportunity for fascinating case studies that range from humanistic to scientific. Building on her previous work on the phenomenology of the icon in Byzantium, Bissera Pentcheva studied the multisensory aesthetics of Hagia Sophia, focusing in part on the link between the optical shimmer of marble and gold and the acoustic reflection properties of marble.[35] Earlier studies considered the acoustics of Hagia Sophia from a computational perspective.[36] Most recently, Wieslaw Woszczyk has conducted new acoustic measurements and offered an analysis of the acoustics of the space.[37]

This essay turns to the algorithmic process, known as auralization, by means of which the acoustic signature of Hagia Sophia can be imprinted on a recorded or live performance of Byzantine chant. Using digital technology allows us to re-create in present-day performance spaces aspects of the acoustic experience of Constantinople's Great Church. After introducing some basic reverberation concepts and terminology in the next section, we focus on the idea of the room response as it relates to auralization and architectural acoustics, and then describe a study conducted to validate our method of measuring and reproducing the acoustics of a space. Preliminary experiments on auralizing Hagia Sophia are then presented and build up to a discussion of three live performances involving acoustic re-creations of Hagia Sophia.

Architectural acoustics, auralization, and the room impulse response

Particularly in reverberant spaces, listening is as much about the space as it is about the sound source. In such interiors, the vast majority of the acoustic energy arriving at a listener from a sound source has interacted with objects and surfaces on its way to the listener. In doing so, the space imprints itself on sound, modifying sound according to the geometry and materials composing the space. As a result, different interiors, having different architectures, materials, and furnishings, will "feel" different sonically: imagine the contrast between hearing singing in a large church with marble floors and walls and the same singing in a small club with hardwood floors and upholstered furniture.

In this section, we describe the mechanism by which a room manipulates sound, infusing it with the room's acoustic character. We start by considering a very simple sound, that of a popping balloon. We then introduce the "impulse response" as embodying the acoustic signature of the room, and "convolution" as the computational process of applying an impulse response to a sound. Next, we argue that an acoustic space may be simulated for any sound by the application of a room impulse response via convolution. The remainder of this section is devoted to exploring concepts of impulse response, convolution, and auralization using the example of recordings made in Stanford University's Memorial Church (Figure 10.1).

To understand how a space processes sound, let's consider a particularly simple sound source, a balloon pop. The pressure waveform generated by a balloon pop is a short-duration pulse, lasting about a millisecond (ms) for a one-foot-diameter balloon, and is roughly shaped like the letter *N*. Over time, this so-called N-wave will propagate away from the position where the balloon was popped and eventually interact with surfaces and objects in the room to create reflections, which in turn will interact with the room and other surfaces and objects, and so on, creating more and more reflections.

This process is seen in a balloon-pop recording made in Memorial Church (Figure 10.2): a listener will hear the balloon pop and its interaction with the space in three parts, first on a "direct path" from the balloon (Figure 10.2, "direct path"), then via a

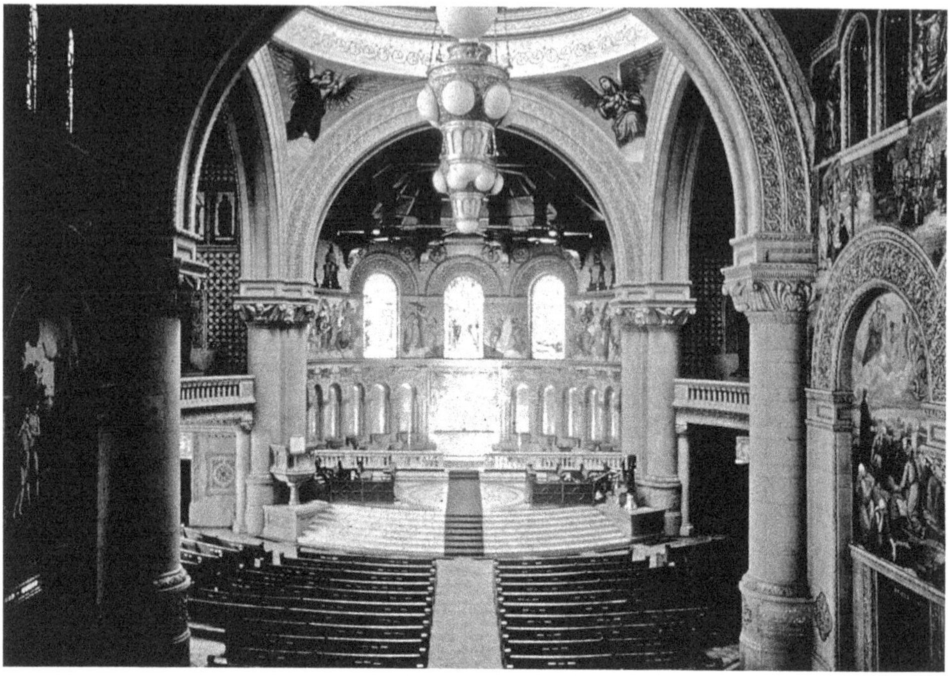

Figure 10.1 Memorial Church, Stanford University, interior.
Photograph provided by the Department of Special Collections and University Archives, Stanford University.

number of "early reflections" (Figure 10.2, "floor, chancel, dome, nave reflections"), which become increasingly dense and indistinguishable from noise, forming the "late-field reverberation" (Figure 10.2, "late-field reverberation").

The room response to the balloon pop is also illustrated by numerical simulation. Figure 10.3 presents a sequence of frames demonstrating how the balloon-pop sound (represented by a gray line) radiates from a point on the left side of a room (marked by an asterisk) and eventually fills the space with sound energy. A time stamp appears in the upper-right corner of each frame. The balloon pop is first a circle, then it becomes a folded circle after reflecting from the floor. The third frame (Frame 85) shows the wavefront passing through the listener position (marked by a circle) and has a black line between the source and listener; this is the direct path, which presents the route traversed by sound to first reach the listener from the source. The next frames (Frames 105 and 256) detail the arrival of the reflection from the floor and a reflection from the back wall and floor. Subsequent frames display the propagating wavefront being broken up into smaller and smaller segments, and tracing more and more complicated propagation paths between source and listener. In addition to the path's complexity increasing over time, the density of arrivals at the listener increases over time.

In the case of the balloon pop, as the N-wave and its reflections propagate, sound energy is absorbed through interactions with the materials and objects that compose the space (Figure 10.2). Each interaction will absorb a small fraction of the sound energy, and over time the sound energy will decay. In large spaces, absorbing surfaces

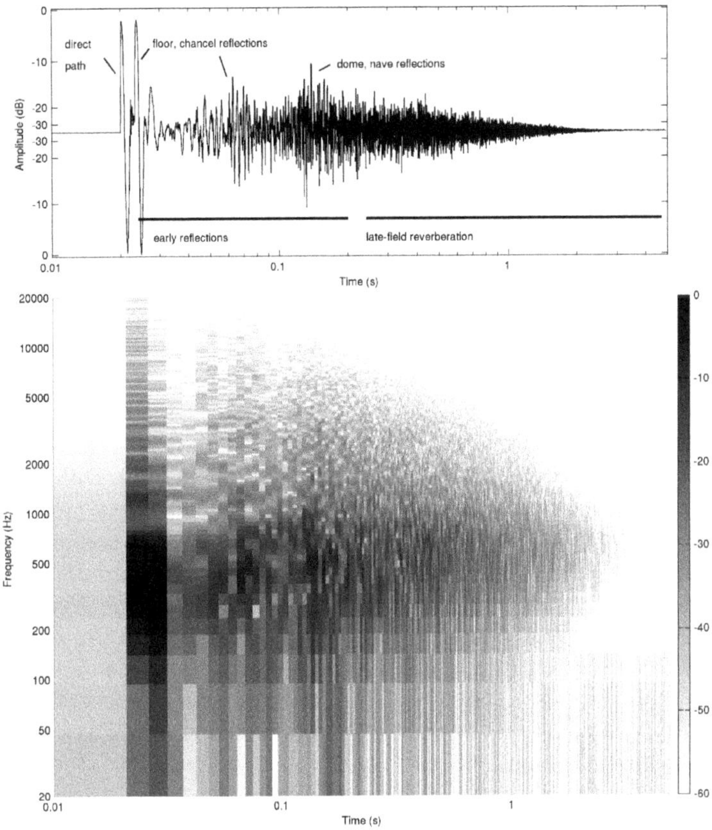

Figure 10.2 Memorial Church, Stanford University, balloon-pop response (above) and associated spectrogram (below).

Diagram by Jonathan S. Abel, 2015.

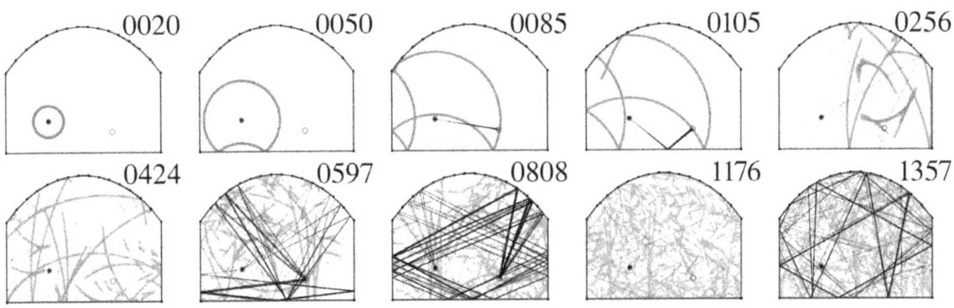

Figure 10.3 Numerical simulation of acoustic wave propagation in a domed structure, featuring wave fronts of a pressure pulse (gray) radiated from the source (asterisk) and propagation paths (black) from the source to the listener (circle), rendered at various time indices (upper right).

Diagram by Jonathan S. Abel, 2015.

are far apart and sound interacts with them relatively less frequently, leading to long decay times. By contrast, absorbing interactions occur more frequently in small spaces, and the decay times are shorter. In addition, sound energy is lost by propagating through the air, which absorbs a small fraction of the sound energy over every unit of distance traveled. Air absorption is stronger at high frequencies than low frequencies. This explains why thunder sounds more muffled the more distant the lightning strike. It also accounts for the relatively "dark" quality of large reverberant spaces, for which air takes up a greater portion of sound energy absorption, relative to reflecting surfaces.[38]

In the Memorial Church balloon-pop recording plotted in Figure10.2, the time axis is presented on a logarithmic scale to magnify details of the perceptually important balloon-pop response onset. The direct path arrival and reflection from the floor exist as clear N-waves at the beginning of the balloon-pop response. A number of early reflections, possibly from the apse and dome, appear near the 0.04 and 0.07 second marks, and another set, likely from the rear of the nave, arrive near 0.15 seconds. After these reflections, the response gives way to a noiselike late field, with frequencies in the 500 Hz range lasting nearly 4 seconds. High frequencies decay more quickly due to air absorption and a small amount of carpet, both of which preferentially absorb high rather than low frequencies. The church has a considerable number of stained-glass windows, which absorb low-frequency energy and reduce the reverberation time for frequencies below about 200 Hz. Comparing the portion of the energy arriving directly from the source (the first N-wave) with that which has been reflected (the rest of the balloon-pop response) shows that the reflected energy accounts for a sizable part of the sonic signature. This corresponds to reflected energy playing a significant role in the perception of sound in this space.

To understand how the space impacts a more complicated sound, a singer, Konstantine Buhler, was recorded in Memorial Church by two microphones simultaneously: a small headset microphone positioned next to his mouth and a room microphone on a stand in the nave (Figure 10.4, C), roughly 25 feet from the singer in the chancel (Figure 10.4, B). Because the headset microphone was so close to the singer's mouth compared to the dimensions of the church, the sound it recorded contained the singing mixed with a nearly imperceptible amount of reverberation. Such "close-miked" recordings, which capture the source essentially without room acoustics, are referred to as "dry" or anechoic; the recorded sound will feel close and clear. By contrast, most of the sound energy recorded at the room microphone was reflected many times from surfaces and objects in the church. These recordings, described as "wet," will sound distant and reverberant.

To visualize their sonic features, the dry and wet recordings were processed into spectrograms (Figure 10.5, above, dry; Figure 10.5, center, wet), which show how sound energy across the frequency range of human hearing evolves over time. Different sound frequencies will stimulate different locations along the cochlea (part of the inner ear); analogously, different sound frequencies will stimulate different "spectrogram 'bins'" along the spectrogram's frequency axis. Because of this similarity, spectrograms are widely used to analyze sound and its human perception. In fact, when the spectrogram is applied to music, aspects of the musical score are apparent. In the dry recording spectrogram (Figure 10.5, above), sung pitches generate horizontal lines at multiples of the fundamental frequency. The change in singing pitch over time is clear, as are many note onset and release times. Sibilant and other unvoiced sounds, by contrast, occupy

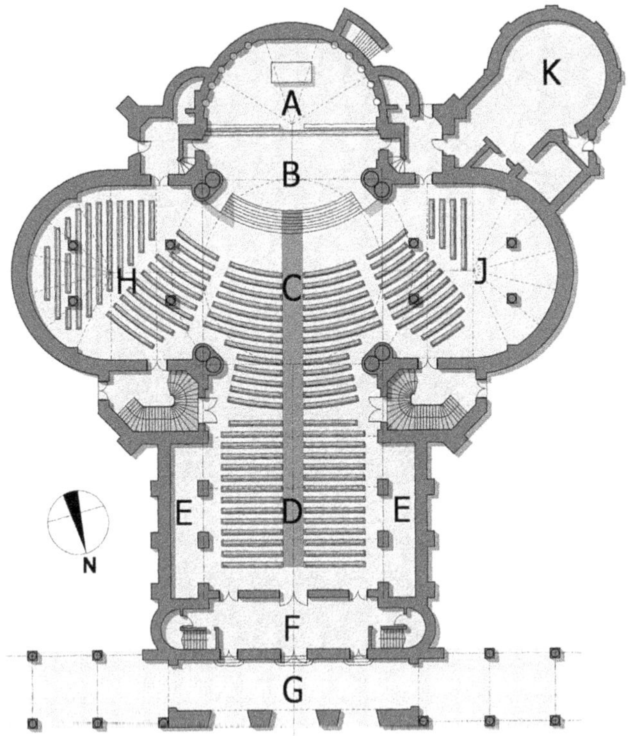

Figure 10.4 Memorial Church, Stanford University, floor plan.

Plan by Mcginnly, licensed under CC BY-SA 3.0 via Commons, https://commons.wikimedia.org/wiki/File:Stanford_Memorial_Church_Plan.jpg#/media/File:Stanford_Memorial_Church_Plan.jpg.

broad ranges of frequency and appear in the spectrogram as noiselike patches (Figure 10.5, above, just after the 1.0-second mark).

The dry (Figure 10.5, above) and wet (Figure 10.5, center) spectrograms display a number of changes that reveal the acoustic signature of the building on the singing. First, all of the wet spectrogram features have been smeared (that is, the sound is imprinted with the reverberant acoustics of the space) over time compared with those of the dry performance. This is particularly noticeable during silences and when the singer is gliding into a note, such as at 1.5 and 6.0 seconds; it is also consistent with one's experience of reverberation in the church. Second, what were solid horizontal lines tracing out a steady pitch energy in the dry recording have become scalloped and variable in intensity in the wet recording. The spectrogram illustrates that reflections in the church are combining sung notes at slightly different times in a type of chorus effect to create a complex pattern of constructive and destructive interference. This scalloping can also be observed in the dry recordings of a chorus of singers.

We now turn our attention to the process by which sound is transformed by an acoustic space. Note that irrespective of the nature of the sound—whether, for example, a balloon pop or a singing voice—the mechanism by which a space generates a set of reflections and reverberation will be the same. In other words, a space processes

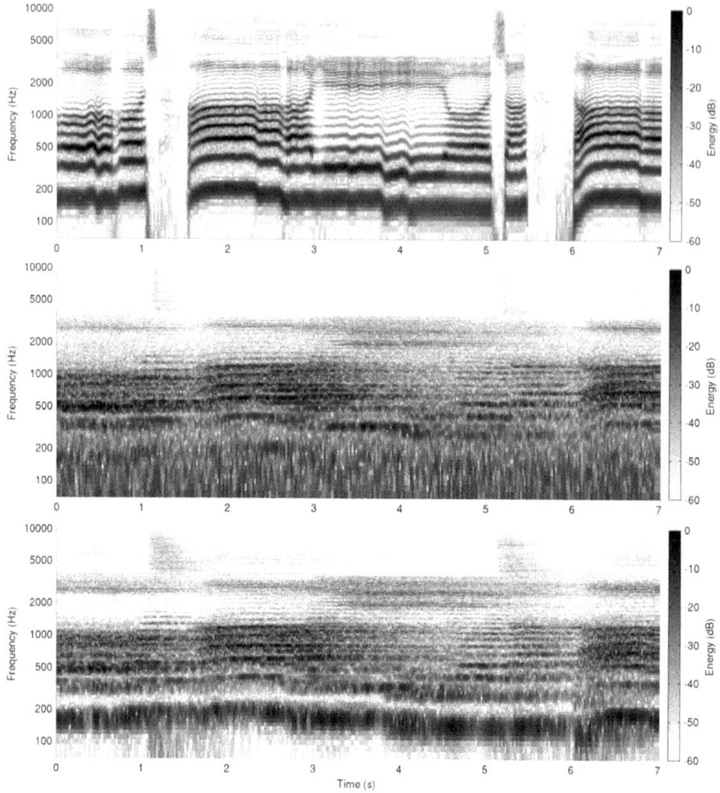

Figure 10.5 Spectrograms of chant recorded in Memorial Church, Stanford University, headset microphone (above), nave microphone (center), and simulated nave microphone (below).

Diagram by Jonathan S. Abel, 2015.

sound in a fixed way, without regard to the particulars of the sound. Consider that any sound can be thought of as being composed of a sequence of overlapping pulses, or sound “atoms,” each scaled in amplitude according to the sound pressure at that point in time. Assume that the response of the space to a single, isolated atom—say, the response to a balloon pop—is measured or otherwise known. Now, the response of the space to a given sound is simply the sum of the known responses of each of the pulses or atoms forming the sound. Put differently, each atom of sound will drag behind it an entire atom response’s worth (think balloon-pop recording’s worth) of reflections and reverberation.

This mechanism can be seen in a comparison of the dry and wet recording spectrograms in view of the balloon-pop response. What are distinct time-frequency regions in the dry spectrogram (Figure 10.5, above) have been smeared over time in the wet spectrogram (Figure 10.5, center) in exactly the same manner that the initial N-wave of the balloon-pop spectrogram (Figure 10.2, below) is smeared over time. The dry singing pitch trajectories become smeared and noiselike in the wet spectrogram. Additionally, high-frequency sibilant sounds (Figure 10.5, just after 1.0 and 5.0 seconds) reverberate for relatively shorter times, as anticipated by the balloon-pop response (Figure 10.2, below).

The sequence of pulses imprinted with the acoustic signature of the space can be expressed computationally: if the atoms are one-sample-wide unit pulses or "impulses," then the response of the space to such an atom is referred to as an impulse response. The computational process of summing such impulse responses scaled according to the signal samples is called convolution, and the sound signal is said to be "convolved with the impulse response."

The impulse response encapsulates everything relevant about the acoustics of the space for the source and listener positions used to record it. As a result, the qualities of an acoustic space are typically studied by analyzing features of impulse responses measured in the space. Moreover, convolution with the impulse response can be applied to generate the room response to any given sound. This fact means that to simulate a space, only the impulse response needs to be known. This is remarkably convenient, as the alternative of measuring the response of the space to even a small number of anticipated sounds would be prohibitively costly. For this reason, convolution with room impulse responses lies at the heart of most room acoustics simulation systems, including our live auralization system, described below.

Auralization using balloon-pop recordings

What is needed to simulate the acoustics of Hagia Sophia is a set of impulse response measurements. There are many methods available to measure room impulse responses, most commonly involving playing test signals such as swept sinusoids or Golay codes into the space from a loudspeaker and recording the room's response at a set of microphones.[39] The drawback of this approach is that it requires a significant amount of time to set up and tear down the equipment and to make the measurements. Since access to Hagia Sophia is very limited, these approaches were not available to us. In our work, which is part of the "Icons of Sound" project, we circumvented the logistical difficulties associated with loudspeaker-based measurement by relying on balloon-pop responses, recorded by Bissera Pentcheva in May and December of 2010 in Hagia Sophia. This approach calls for only a handheld recorder, a balloon, and a few minutes of time. Yet this illusory "ease" conceals the unavoidable long overseas flight and extended negotiations with the local museum authorities.

Balloon-pop responses are not impulse responses, however, and we developed a method to convert our balloon-pop recordings into impulse responses.[40] This method was unproven, and to test its effectiveness we conducted an experiment using the dry and wet recordings of a student, Konstantine Buhler, singing in Memorial Church, as described above. Leaving a microphone at position C shown in Figure 10.4, we popped several balloons from where Buhler had been singing (Figure 10.4, position B) and recorded the responses. We converted the balloon-pop recordings into room impulse responses using our method and then convolved the estimated impulse responses with the dry recording to obtain a simulated Memorial Church response at position C to Buhler's singing.

The spectrogram of the simulated position C signal is plotted (Figure 10.5, below) on the same time axis as the spectrogram of the signal actually recorded at position C (Figure 10.5, center). The spectrograms and sound of the recorded and simulated signals are very similar in the mid-frequencies and above, verifying the effectiveness of the room simulation via impulse response measurement. Where they differ is in the low frequencies, where the church heating system (which could not be turned off) added noise to the recording. It was an unexpected benefit of the simulation that, compared to the actual recording, stray environmental noises were eliminated.

Postproduction auralizations in a virtual Hagia Sophia

The convolution of the acoustic signature of Stanford University's Memorial Church, using an impulse response derived from a balloon pop, serves as a proof of concept.[41] Convolution done as a postproduction process, that is, using prerecorded sound to convolve with a room signature, does not allow musicians to interact with the space in real time.[42] To overcome this limitation, we developed a real-time method for imprinting Hagia Sophia's sonic signature onto a live performance. This system makes it possible for musicians to hear the effect of the space on their singing in real time and make adjustments accordingly. They thereby have substantially the same aural experience as performing in the real space.

As before, we started with a balloon-pop response. A balloon pop recorded by Pentcheva in Hagia Sophia, December 2010 (Figure 10.6), was converted into an impulse response suitable for live auralization (Figure 10.7).[43] This recovered impulse response has a few interesting features, not least of all its length—it takes roughly 11 seconds for the impulse response to decay from a comfortable listening level to the

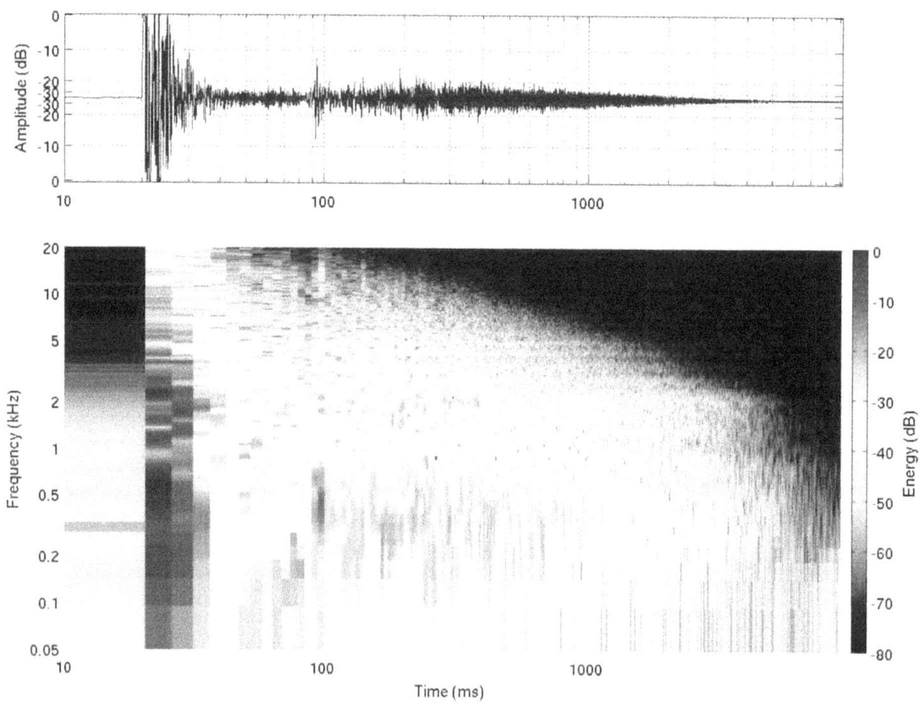

Figure 10.6 Hagia Sophia, Istanbul, balloon-pop response (above) and associated spectrogram (below), December 2010.

Diagram by Jonathan S. Abel, 2015.

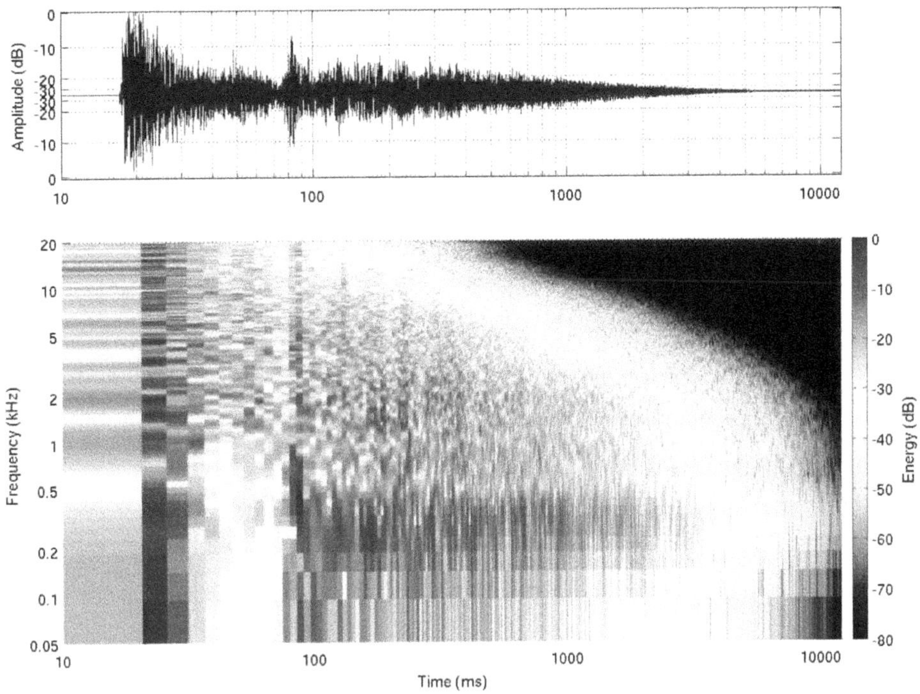

Figure 10.7 Hagia Sophia impulse response (above) and associated spectrogram (below), recovered from the December 2010 balloon-pop response recording (Figure 10.6).
Diagram by Jonathan S. Abel, 2015.

threshold of human audibility (approximately 60 dB (decibels) below conversation level). In addition, the listening position for this impulse response, about 10 meters from the balloon, is dominated by reflected energy, the "wet" portion of the impulse response. We can associate certain features of the impulse response with the geometric features of Hagia Sophia. For instance, the large spike around 100 ms (milliseconds) and the "wash" that builds to a wide peak around 300 ms issue from complex reflections produced by the dome and colonnades.

In advance of the public concert, an experimental recording session was conducted in 2011 in a small recital hall ("the Stage") at Stanford University's Center for Computer Research in Music and Acoustics (CCRMA) (Figure 10.8). It featured thirteen members of Cappella Romana, a renowned group specializing in performances of Byzantine chant. To enable these professional singers to adjust aspects of their performance, such as vocal balance and articulations, in response to the reconstructed Hagia Sophia acoustics, the auralization system needed to operate in real time and to allow the chanters to hear each other and themselves in the simulated space.

In this recording, we used Countryman B2D headset microphones (Figure 10.9), in order to capture each singer's voice on a separate track. A digital audio workstation (MOTU Digital Performer) was used to record the singers' voices, and we processed them to form a dry stereo mix. The mix was convolved in real time with stereo left

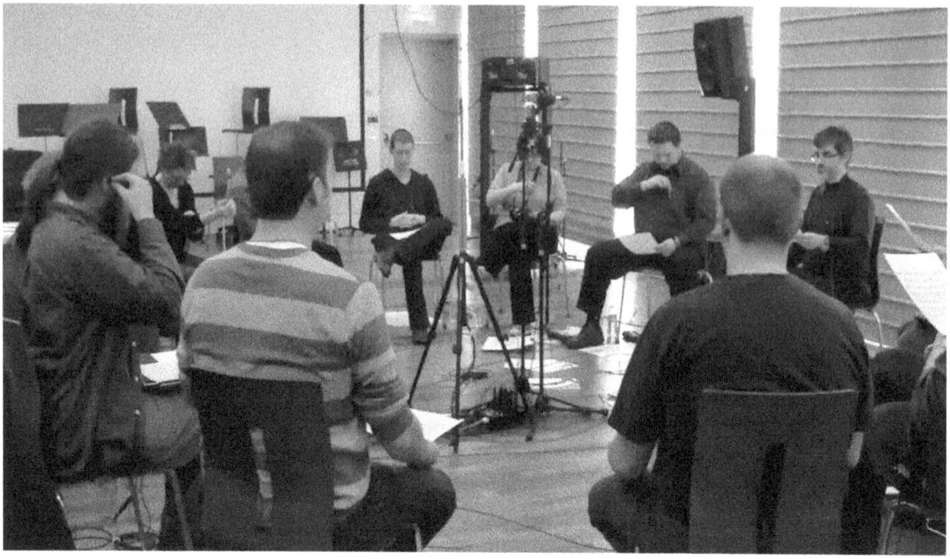

Figure 10.8 Cappella Romana at a recording session, Center for Computer Research in Music and Acoustics (CCRMA), Stage, Stanford University, March 2011.

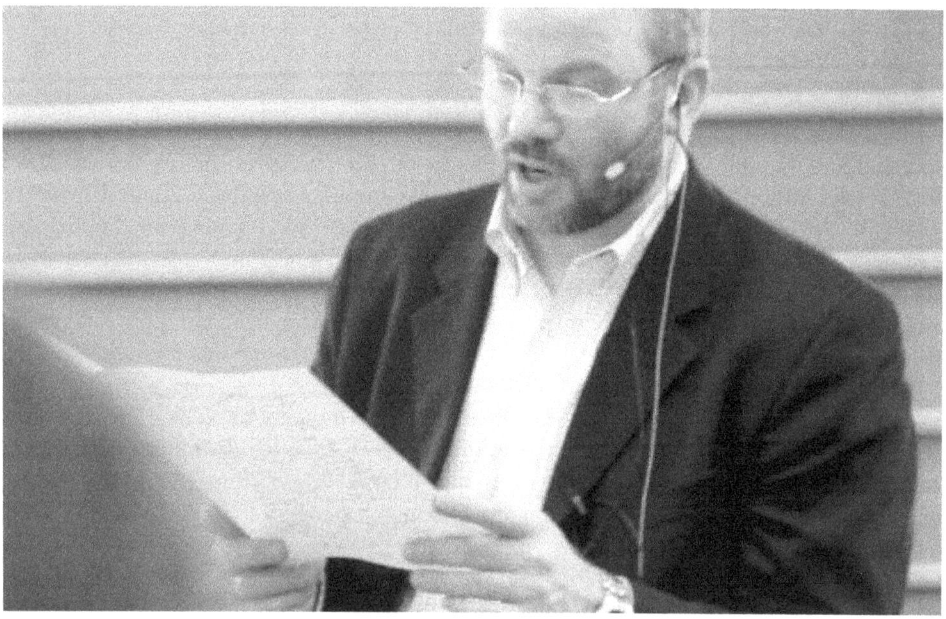

Figure 10.9 Mark Powell, a singer with Cappella Romana, outfitted with a headset microphone (Countryman B2D) and earbud headphones at the Cappella Romana March 2011 recording session (Figure 10.8).

and right statistically independent impulse responses derived from the December 2010 Hagia Sophia balloon-pop recordings, forming a wet stereo mix.[44]

The chanters were positioned in a circle so that they could see each other and the conductor (Figure 10.8). Earbud headphones made available to them the dry and wet mixes, and each performer could control the balance between wet and dry mixes and the overall level in the individual's earbud signal. In this way, the performers could hear their own singing and each other in the simulated Hagia Sophia and interact accordingly. Providing feedback over headphones made it possible to keep the microphone signals dry, absent of any of Hagia Sophia acoustics.

The group began warming up without simulated acoustics; during this time microphone levels and equalizations were set. The simulated acoustics were then enabled, placing the performers in a virtual Hagia Sophia. Next, the performers were given a chance to acclimate to the environment. Their initial reaction, expressed simultaneously by several of the chanters, was, "Oh, let's play!" The chanters reported interacting reasonably naturally with the space, slowing their tempo to accommodate Hagia Sophia's long reverberation time. The ison (the drones) particularly enjoyed singing in the virtual acoustics, because they found it easy to "ride" the resonances.

The group performed a number of chants, including: 1) a *prokeimenon* (psalmic verses prefacing the readings from the Epistle in the Divine Liturgy (Eucharistic rite), or introducing the Old Testament readings at *hesperinos* (vespers) or the New Testament at *orthros* (laudes) or morning service finishing at dawn); 2) a *kontakion* (sung sermon); and 3) a congregational setting for Psalm 140 [141] sung at vespers, all of which we processed in postproduction into stereo and surround recordings. Recording the individual tracks anechoically made a number of editing options available that would be precluded when recording such a group in the actual space using room mics, as is conventional. This is because, with room mics, the dry singing and acoustics of the space are intermixed. With the convolution applied artificially after the fact, the acoustics of the space in which the dry tracks were recorded will still appear in the final auralization. By contrast, close-miking allows us to eliminate the acoustics of the recording room as well as stray noises such as traffic sounds from outside the building. Below, we give specific examples of different musical production techniques that are afforded by this approach. These include the possibilities to reposition singers in postproduction, to sculpt and to place reverberation, and to easily overlay multiple rounds of recording.

To produce the stereo mixes, as with the live recording, melodists' and drones' (ison) dry tracks were separately panned, that is, positioned horizontally, across the stereo field. A wet stereo mix was formed by convolving the dry stereo mix with a pair of statistically independent Hagia Sophia impulse responses. The final mix was generated by balancing the wet and dry components.

Surround mixes were generated in a similar manner: each of the eight male chanters' dry signals were panned separately to a location, and clouds of reverberation formed by convolution with a number of statistically independent Hagia Sophia impulse responses were placed about those locations. In our mix of the *prokeimenon*, the eight chanters' dry signals were panned to eight equally spaced locations around the listener. The reverberation clouds were panned tightly about each melodist, in a neighborhood 30 degrees to the left and right and 15 degrees above and below. The isons' reverberation clouds were panned over a wider region, in a neighborhood of 60 degrees to their left and right. This production mix creates the effect of a ring of performers surrounding the listener.

In our mix of the *kontakion*, the dry signals of the thirteen chanters—eight men and five women—were placed on a soundstage in front of the listener, and their associated reverberation clouds were panned about them, so that they shared a wide reverberant stage. This production mix gives the impression of an ensemble playing onstage in front of the listener. The production technique for Psalm 140 was similar to that for the *kontakion*. However, many overdubs were recorded to get the effect of fifty-two singers in a much larger chorus.

Live performance in a virtual Hagia Sophia

We now turn our attention to live auralization of Byzantine chant for performance and recording in which we employ loudspeakers rather than headphones to synthesize the acoustics of Hagia Sophia. The use of loudspeakers both enables presentation to an audience and provides for a more natural interaction among the performers, allowing for freedom of movement and a truly shared performance environment. The use of loudspeakers also presents some technical issues, and in the following we discuss these challenges and outline our solutions. We begin by describing our approach to live loudspeaker-based auralization and finish by discussing live auralization over loudspeakers in three performance halls: the Bing Concert Hall, the CCRMA Stage, and the San Francisco Ritz-Carlton Ballroom.

Our technical approach was driven by the need to perform, rehearse, and record in different spaces that were not preconfigured to present virtual acoustics, and to which we have limited access. This required a virtual acoustics system that was transportable and could be quickly loaded in, installed, configured, and tuned, and also quickly torn down and loaded out. The system hardware is much like that of the headphone-based system described above, again including a set of close microphones to capture the dry voices of individual singers, as well as for recording and postproduction. A set of room microphones records the mix as heard in the space. The system also includes a set of powered full-range loudspeakers and subwoofers to present the simulated acoustics. A digital audio workstation (DAW) connects to the microphones and loudspeakers through a mixing board and audio interface. The DAW processes the close microphone signals to generate live virtual acoustics signals, rendered in the space by the loudspeakers.

The use of loudspeakers raises the issue of feedback: it is possible, particularly when simulating very reverberant spaces such as Hagia Sophia, that loudspeaker signals will find their way back into the performer microphones, forming a feedback loop. This feedback can take a mild form, in which the simulated acoustics is modified, or it can take a severe form, in which particular resonant frequencies grow in amplitude and become unpleasant "whistles."

The approach we take to minimize the possibility of feedback has two components. The first is to use many loudspeakers so that each loudspeaker signal can play a relatively quiet signal, thus minimizing feedback between any given loudspeaker and microphone. The difficulty is that there are a number of microphone–loudspeaker loops running in parallel, and their combination might create feedback if they operate coherently. This possibility is eliminated in our system by using statistically independent impulse responses—in effect, impulse responses that sound the same but do not track each other in any predictable way—to generate the simulated acoustics. Employing this approach, a significant amount of reverberation can be generated without producing perceivable feedback.

The second component that helps to eliminate feedback employs directional close microphones. Using microphones with a hypercardioid polar pattern (which is sensitive to sound coming from one particular direction) taped to the foreheads of the singers and pointed down toward their mouths (Figure 10.10) places the singers' voices well within the "main lobes" of the microphones, and therefore they will be accentuated. At the same time, the loudspeaker signals, appearing from the sides and above the singers, will arrive from outside the main lobes of the microphones, and therefore will be suppressed. We have found that this approach is sufficient to eliminate problematic feedback, even when simulating very reverberant environments such as Hagia Sophia and operating in smaller rooms such as the CCRMA Stage, where the loudspeakers are close to the singers.

Two settings were configured for virtual acoustics performance and one for rehearsal and recording. Bing Concert Hall was configured to simulate the acoustics of Hagia Sophia for the "Constantinople" portion of Cappella Romana's February 1, 2013, "From

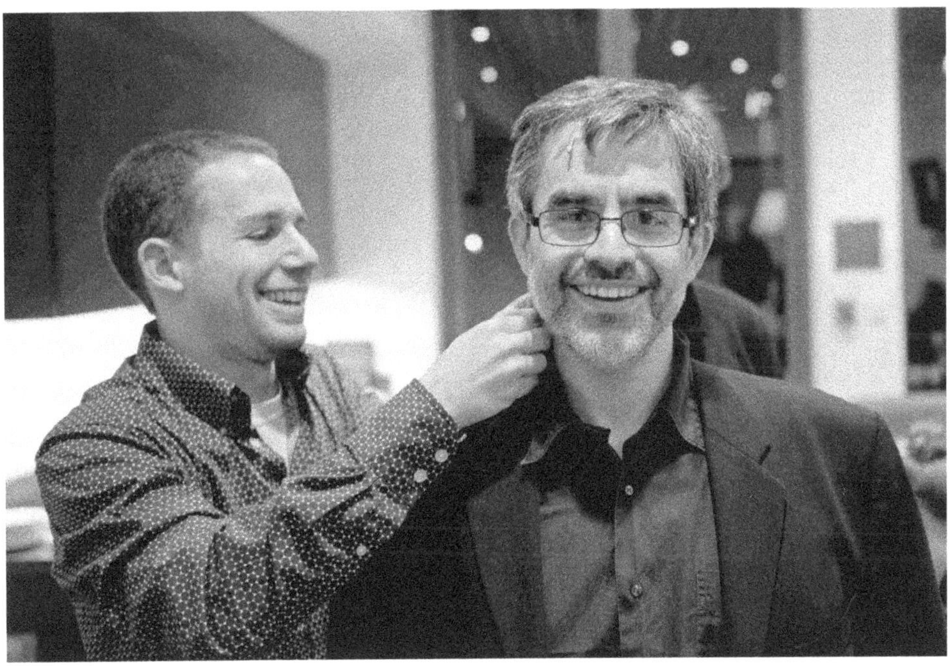

Figure 10.10 Alexander Lingas, the artistic director of Cappella Romana, being outfitted with a lavalier microphone by Scott Levine for the concert "From California to Constantinople," February 2013, at the Bing Concert Hall, Stanford. The Countryman B2D lavalier microphone has a hypercardioid polar pattern sensitive to sound coming from one particular direction, which helps in minimizing problematic feedback.

Photograph © Dave Kerr, 2013.

Constantinople to California" concert. The San Francisco Ritz-Carlton Ballroom was configured to synthesize the acoustics of Hagia Sophia and of the Church of the Holy Cross in Belmont, California (a medium-sized, reverberant church) for the September 27, 2014, Cappella Romana performance at a Patriarch Athenagoras Orthodox Institute event. In addition, the CCRMA Stage was configured to simulate the acoustics of Stanford's Memorial Church for rehearsal of Cappella Romana's February 2, 2013, "Holy Week in Jerusalem" performance in that space. Finally, the CCRMA Stage was configured to present the acoustics of Hagia Sophia for recording sessions held February 6 and 7, 2013.

We begin with our rendering of the acoustics of Hagia Sophia for live recording on the CCRMA Stage and the preparation of Memorial Church for Cappella Romana's rehearsal for its "Holy Week in Jerusalem" concert. The CCRMA Stage has sixteen full-range loudspeakers and eight subwoofers, all manufactured by ADAM Audio. Eight loudspeakers at ear level and paired subwoofers at floor level, mounted on movable stands, were placed in a ring about the Stage; the remaining eight loudspeakers are affixed to the rafters, in a ring above the singers (Figure 10.11).

The Stage's acoustics are somewhat configurable; velour and felt damping materials were deployed to reduce the Stage's inherent reverberation time to less than a third of a second, a relatively "dead" environment. In this way, the reverberant virtual acoustics generated by the loudspeakers is essentially unmodified by the room. For both the rehearsals and the recording sessions, the Cappella Romana singers arranged themselves in a circle, as before. Countryman B2D 2mm-diameter lavalier microphones with a hypercardioid were affixed to their foreheads with medical tape (Figure 10.10). The fifteen microphone signals were recorded and processed using the digital audio workstation Ableton Live, using the plug-in LAConvolver for real-time convolution, running on a laptop computer.

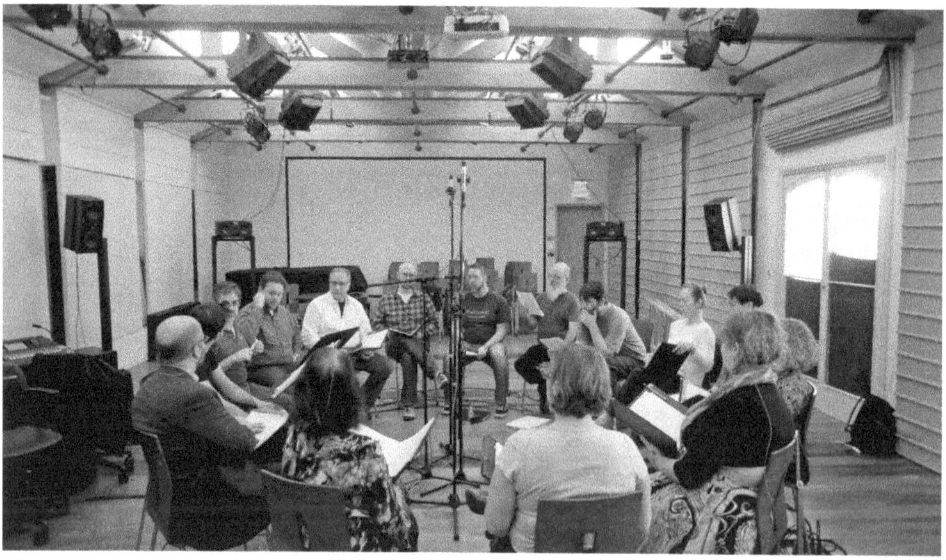

Figure 10.11 Cappella Romana at a CCRMA Stage recording session, February 2013. Photograph © Dave Kerr, 2013.

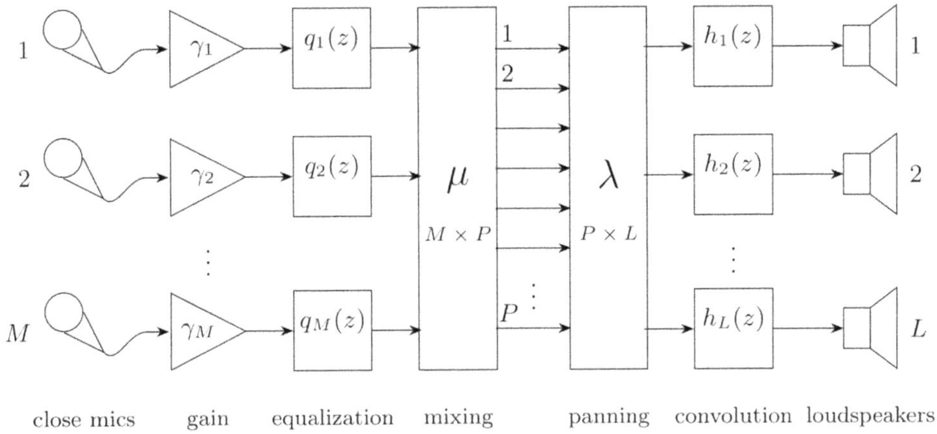

Figure 10.12 Live auralization over loudspeakers signal flow architecture, employing statistically independent loudspeaker impulse responses.

Diagram by Kurt James Werner and Jonathan S. Abel, 2016.

Each of the singer microphone channels was panned among sixteen real-time convolution channels running statistically independent Hagia Sophia impulse responses. In turn, each convolution output drove a different loudspeaker (Figure 10.12). The rafter-mounted loudspeaker signals were modestly delayed, as were the ear-level loudspeaker signals closest to the singers. The purpose was to roughly match the first reflection arrival times from the various directions about the singers. In this way, the chanters heard singing from each of the other chanters directly, as well as through the virtual acoustics of Hagia Sophia played out over the sixteen loudspeakers. The panning of the dry signals and the resulting reverberation was done in such a way as to envelop the performers in a reverberant soundfield. As with the 2011 recording session, the performers warmed up and then individually sang so that microphone levels and equalization could be adjusted. Then, with Cappella Romana chanting, the reverberation level was set, immersing them in the reverberant acoustics of Hagia Sophia.

A similar configuration was used for Cappella Romana's September 27, 2014, performance in the San Francisco Ritz-Carlton Ballroom. Our virtual acoustics system created both a virtual Hagia Sophia and a virtual Church of the Holy Cross in Belmont. The performance took place in a large ballroom venue roughly measuring 120 feet wide by 80 feet deep, with a relatively low 14-foot-high ceiling, carpeted floor, and two mirrored walls. While the mirrored walls set off a bit of flutter echo, the thick carpet produced a short reverberation time. For this concert, we arranged twelve loudspeakers on stands about the room, four across the front and back of the venue, and two on each of the sides. The performers were again outfitted with the Countryman B2D 2mm-diameter lavalier microphones, this time with radio transmitters to send the signals to the mixing board.

To generate the virtual acoustics, the singer microphone signals were first panned among the twelve loudspeakers to provide a dry signal for the venue. Another panning of the microphone signals among the twelve channels was imprinted with the needed acoustics via convolution and rendered through the loudspeakers to provide the wet signal (Figure 10.13). A challenge in configuring the system is choosing the loudspeaker wet–dry mixes to compromise among performer and audience preferences. As with the Stage and Bing performances, the choir director and some melody singers preferred a drier mix, while the ison chanters favored a wetter mix. Their differing preferences could be accommodated to a degree by reducing the wet signal level for the loudspeakers near the stage. The room was rather shallow, and we delayed the rear loudspeaker signals to give the illusion of a deeper room. In addition, we aimed a number of the loudspeakers toward the chandeliers in the ceiling in an attempt to equalize the sound level across the room, as well as to produce a modest sense of height, hoping that the dense, inverted dome-shaped chandeliers would effectively scatter sound. So as to prevent feedback, the dry signal was suppressed in the loudspeakers immediately adjacent to the stage. It turns out that this processing was rather effective, creating a sense of a much taller, enveloping space.

For Cappella Romana's "From Constantinople to California" concert, held in Bing Concert Hall just two weeks after the hall opened in January 2013, we devised a virtual Hagia Sophia using twenty-four full-range loudspeakers and six subwoofers.[45] Twelve of the loudspeakers were arranged around the perimeter of the hall and twelve were hung from the ceiling, forming a "dome" of loudspeakers (Figures 10.14, 10.15). The fifteen performers, twelve melody chanters and three ison chanters, were outfitted with B2D microphones and radio transmitters to provide dry chanter signals; for recording, there were eight room mics manufactured by DPA Microphones, two on stage and six flown in the hall.

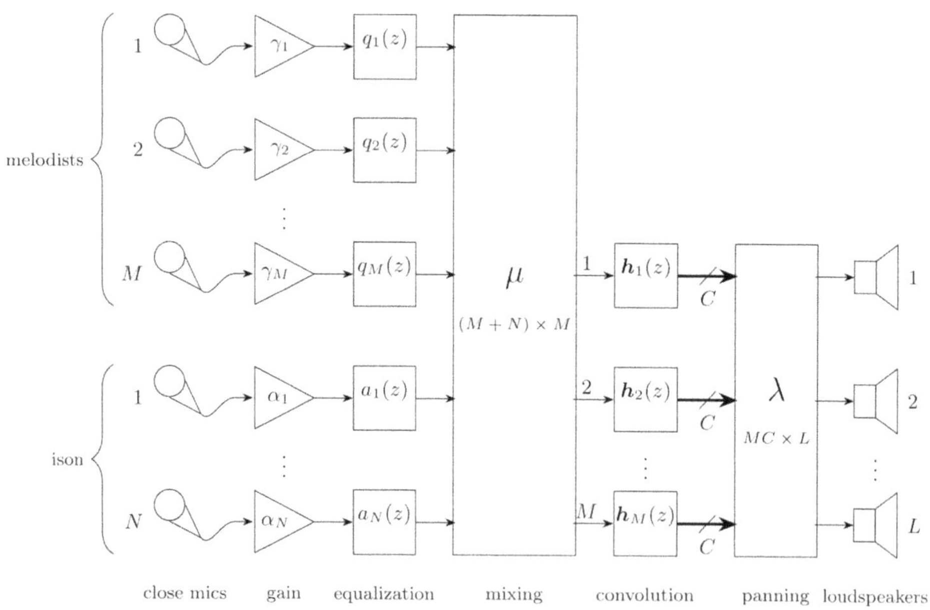

Figure 10.13 Live auralization over loudspeakers signal flow architecture, using statistically independent panned singer reverberation clouds.

Diagram by Kurt James Werner and Jonathan S. Abel, 2016.

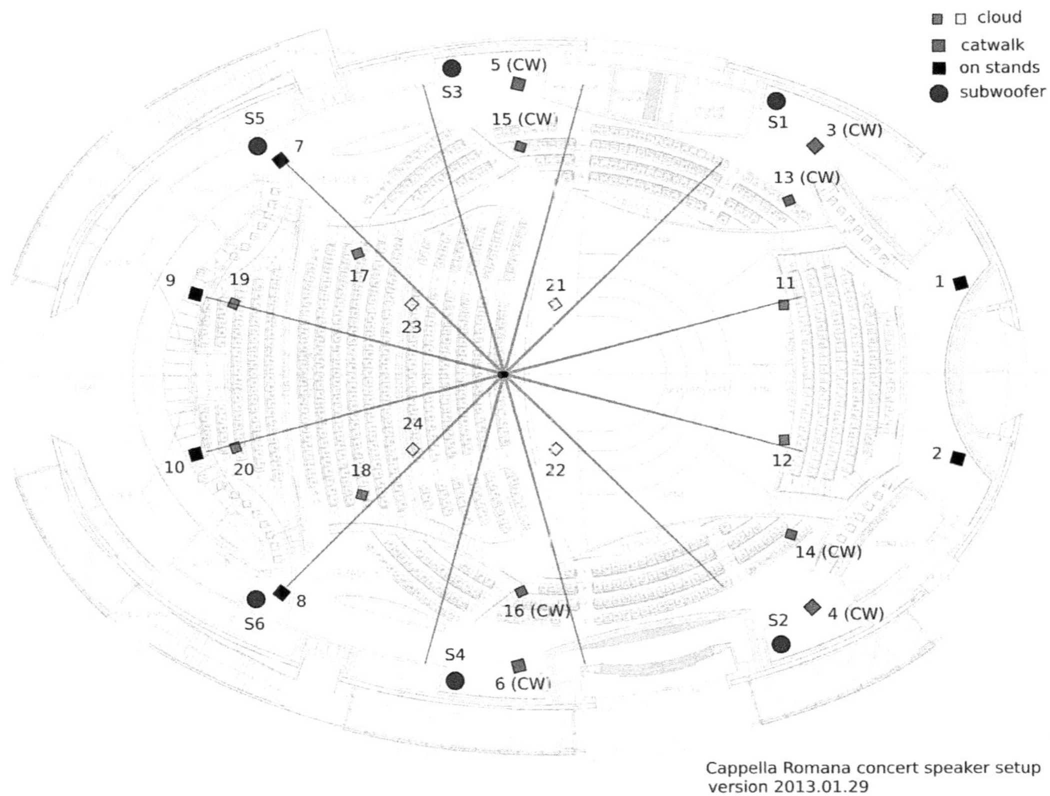

Figure 10.14 Bing Concert Hall, Stanford University, live auralization loudspeaker layout for the performance "From Constantinople to California," February 2013.

Diagram by Fernando Lopez-Lezcano, 2016.

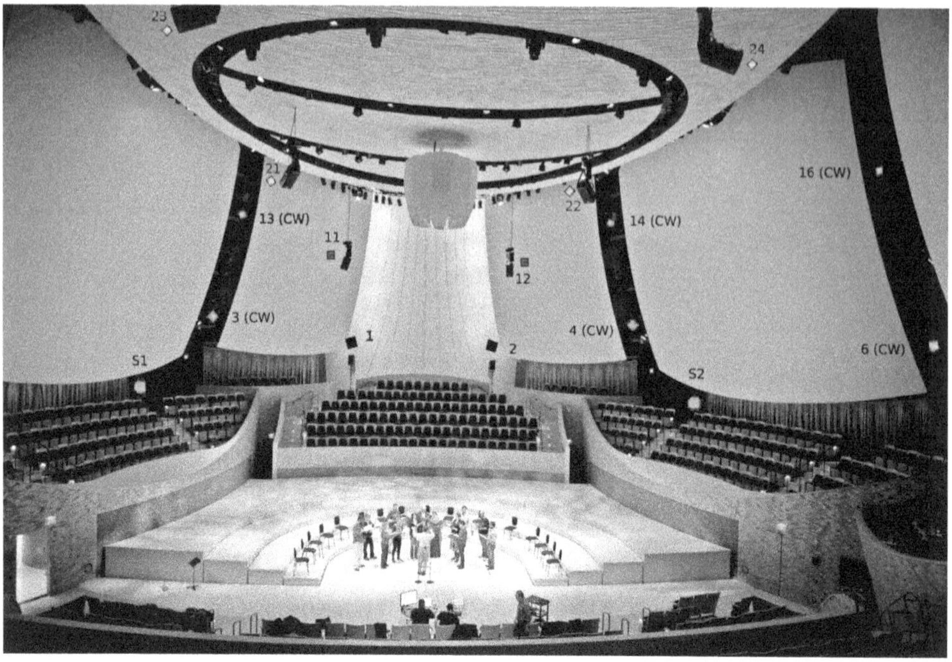

Figure 10.15 Cappella Romana at a dress rehearsal for the concert "From Constantinople to California" at Bing Concert Hall, showing selected loudspeakers (Figure 10.14).

A slightly different virtual acoustics processing approach was employed for the Bing performance. To begin with, Bing Concert Hall is a rather reverberant space, having a 2.5-second-long reverberation time with its dampening curtains deployed. Accordingly, the auralizing impulse responses were adjusted so that the resulting acoustics—the actual acoustics of Bing Concert Hall combined with the virtual acoustics presented over the loudspeakers in the concert hall—produced a faithful rendering of Hagia Sophia.

Second, rather than panning the dry microphone signals among a set of loudspeaker Hagia Sophia convolutions as above, a set of Hagia Sophia impulse responses was imprinted on each dry signal, forming a "cluster" of reverberated signals for each performer. These clusters were then placed about the venue to create a set of overlapping regions in space for the performers' reverberated singing. Specifically, the virtual acoustics were generated by separately processing each of the twelve melody chanters' dry microphone signals. Four statistically independent Bing-corrected Hagia Sophia impulse responses were employed to form a wet (that is, reverberation) signal cluster for each melody chanter. These clusters were then sized and positioned about the hall according to the singers' onstage arrangement using Ambisonics.[46] Each of the three ison chanters was associated with a group of four melody chanters. Their singing was auralized by simply mixing their dry microphone signals with that of the four associated melody chanters prior to convolution and Ambisonics processing. In this way, their bass contribution would occupy a large region in space, as tends to happen with low-frequency sounds. The result was a reverberant soundfield placed about and above the audience, with the men and women of the group layered about the dome of loudspeakers.

We now give an account of the experience of performing and listening in the virtual environments we created, as reported by Cappella Romana and concert reviewers. In the CCRMA Stage rehearsal and recording sessions, the performers found the live auralization over loudspeakers to be significantly better and more natural than the auralization over headphones. The ison chanters particularly enjoyed singing in the virtual Hagia Sophia, saying that it was very easy to find the building's resonances and grow a "wave" of reverberated ison. We should note that the ison singers preferred a wet-dry mix that was wetter than that preferred by the director, and likely somewhat wetter than that in the actual building. They also were enthusiastic about the rich, clear set of high-frequency harmonics they could generate and "play" with, reporting that the virtual building responded in a "very natural" manner. Perhaps in part owing to the significance of the simulated venue, the ison chanters described singing in the virtual Hagia Sophia as a deep, emotional, and transporting experience. A number of Cappella Romana members volunteered that the virtual Hagia Sophia sounded and responded like a real space, in contrast with their prior experiences with artificial reverberation. Many group members also commented that the virtual Hagia Sophia "held its pitch."

For the performance in the Ritz-Carlton Ballroom, Cappella Romana had about an hour to acclimate to the two virtual environments, Hagia Sophia and Belmont's Church of the Holy Cross. Despite the lack of rehearsal time, their performance was spectacular, particularly in the virtual Holy Cross. The performance and virtual acoustics were very well received; a number of audience members had the sense of being completely immersed and swept up in the performance. It is interesting that though we did not disclose the identity of the "medium-sized church" (Holy Cross) to the audience, Fr. Peter Salmas, Holy Cross's priest, recognized its acoustic signature.

For the "Constantinople" portion of the Bing concert, Cappella Romana performed selections from the Divine Liturgy at Hagia Sophia and chanted in the acoustic simulation of Hagia Sophia described above. Cappella Romana performers and several of their board members, as well as composers attending the concert, reported that the space sounded and reacted in a very natural way; they found it easy to forget that the acoustics were simulated. The ison chanters again thoroughly enjoyed the responsiveness of the virtual Hagia Sophia to high-frequency harmonics and the ease with which the drone could be built and sustained.

During the performance of the liturgy selections, the audience was asked to hold its applause so as to create and sustain an immersive experience. As Jason Serinus related in his concert review for *San Francisco Classical Voice*:[47]

> It is impossible to describe the experience objectively; to even attempt to do so would miss the point of a sensual experience meant to induce a transcendent state. Throwing all caution to the winds, as it were, the "performance" was the closest to lift-off I have experienced short of chemically enhanced listening sessions or the final hours of a seven-days meditation intensive.
>
> Closing my eyes, it was not hard to imagine that the singers' voices were actually reverberating back and forth through Hagia Sophia's enormous sanctuary and remarkable 50-meters high dome. The constant bass drone heard in much of the music helped ground an experience that sent the senses soaring, as the entire space seemed to fill with voices proclaiming the glories of the Christian God.

To the skeptical, or those allergic to computer-simulated environments, this may sound like so much caca or the hallucinations of an aging '60s tripper. But I know no other way to describe an experience that, in Pentcheva's words, was intended "to create the sensation that you were standing in a space that was neither in this world nor in heaven, that you were hovering in between." Even without the additional visual and olfactory elements of the church—the incense, glittering walls, stunning mosaics, now absent tapestries, and floors of marble whose bookmatched pieces suggest the uninterrupted waves of the sea—the sonic environment was unique. Working in consort, the team of Abel, Pentcheva, and Lingas created an experience that is sure to make history, and spearhead further projects on multiple continents in the years ahead.

Conclusion

In this chapter, we have detailed several approaches to auralizing Byzantine chant in a virtual Hagia Sophia, bringing its acoustic response to recorded audio and real-time performance applications. Responses to these systems have been positive. The performers of Cappella Romana found the virtual acoustics presented over loudspeakers to be natural and to "feel like a real space." The interactive nature of the real-time auralization was also well received; performers reported that it "responds like a real room." Ison singers asserted, "It's easy to find resonances," also noting that it "holds its pitch."

Notes

1 Emily Ann Thompson, *The Soundscape of Modernity: Architectural Acoustics and the Culture of Listening in America, 1900–1933* (Cambridge, MA: MIT Press, 2002).
2 Jonathan Sterne, "Space within Space: Artificial Reverb and the Detachable Echo," *Grey Room* 60 (2015): 110–31, esp. 110.
3 A. Harold Marshall, D. Gottlob, and Herbert Alrutz, "Acoustical Conditions Preferred for Ensemble," *Journal of the Acoustical Society of America* (*JASA*) 64, no. 5 (November 1978): 1437–42; Kanako Ueno, Kosuke Kato, and Keiji Kawai, "Effect of Room Acoustics on Musicians' Performance, Part I: Experimental Investigation with a Conceptual Model," *ACTA Acustica United with Acustica* 96 (2010): 505–15; and Jude Brereton, Damian T. Murphy, and David M. Howard, "Evaluating the Auralization of Performance Spaces and Its Effect on Singing Performances,", in *Proceedings of the 130th Convention of the Audio Engineering Society (AES)*, London, 2011 (New York: Curran, 2011). www.aes.org/e-lib/browse.cfm?elib=15847.
4 Timo Fischinger, Klaus Frieler, and Jukka Louhivuori, "Influence of Virtual Room Acoustics on Choir Singing," *Psychomusicology: Music, Mind, and Brain* 25, no. 3 (2015): 208–18.
5 Wieslaw Woszczyk and William L. Martens, "Evaluation of Virtual Acoustic Stage Support for Musical Performance," in *Proceedings of Acoustics '08*, Paris, 2008, www.conforg.fr/acoustics 2008/cdrom/data/articles/000968.pdf.
6 I. Nakayama, "Preferred Time Delay of a Single Reflection for Performers," *Acustica* 54 (1984): 217–21.
7 Alexander Lingas, "Sunday Matins in the Byzantine Cathedral Rite: Music and Liturgy" (PhD diss., University of British Columbia, 1996); Lingas, "Performance Practice and the Politics of Transcribing Byzantine Chant," *Acta Musicae Byzantinae* 6 (2003): 56–76; Lingas, "From Earth to Heaven: The Changing Musical Soundscape of Byzantine Liturgy," in *Experiencing Byzantium*, ed. C. Nesbitt and M. Jackson (Farnham, UK: Ashgate, 2013), 311–58; and Bissera V. Pentcheva, "Hagia Sophia and Multisensory Aesthetics," *Gesta* 50, no. 2 (2011): 93–11.

8 Michael Vörlander, *Auralization: Fundamentals of Acoustics, Modelling, Simulation, Algorithms and Acoustic Virtual Reality* (Berlin: Springer-Verlag, 2008).

9 Vesa Välimäki, Julian D. Parker, Lauri Savioja, Julius O. Smith, and Jonathan S. Abel, "Fifty Years of Artificial Reverberation," *IEEE Transactions on Audio, Speech, and Language Processing* 20, no. 50 (July 2012): 1421–48.

10 Mendel Kleiner, Bengt-Inge Dalenbäck, and Peter Svensson, "Auralization—an Overview," *Journal of the Audio Engineering Society* (*JAES*) 41, no. 11 (November 1993): 861–75.

11 Chris Scarre and Graeme Lawson, eds., *Archaeoacoustics* (Cambridge: McDonald Institute for Archaeological Research, 2006).

12 Nico F. Declercq, Joris Degrieck, Rudy Briers, and Oswald Leroy, "A Theoretical Study of Special Acoustic Effects Caused by the Staircase of the El Castillo Pyramid at the Maya Ruins of Chichen-Itza in Mexico," *Journal of the Acoustical Society of America (JASA)* 116, no. 6 (December 2004): 3328–35; and Frans A. Bilsen, "Repetition Pitch Glide from the Step Pyramid at Chichen Itza (L)," *Journal of the Acoustical Society of America (JASA)* 120, no. 2 (August 2006): 594–96.

13 Declercq et al., "Staircase of the El Castillo Pyramid."

14 Jonathan S. Abel, John W. Rick, Patty P. Huang, Miriam A. Kolar, Julius O. Smith, and John M. Chowning, "On the Acoustics of the Underground Galleries of Ancient Chavín de Huántar, Peru," in *Proceedings of Acoustics '08*, Paris, 2008, https://ccrma.stanford.edu/groups/chavin/publications/Acoustics08.pdf

15 Abel, "On the Acoustics of . . . Chavín de Huántar"; Miriam Anna Kolar, "Archaeological Psychoacoustics at Chavin de Huantar, Peru" (PhD diss., Stanford University, 2013); and Kolar, "Tuned to the Senses: An Archaeoacoustic Perspective on Ancient Chavín," *Appendix* 1, no. 3 (July 2013), http://theappendix.net/issues/2013/7/tuned-to-the-senses-an-archaeoacoustic-perspective-on-ancient-chavin.

16 Robert G. Jahn, Paul Devereux, and Michael Ibison, "Acoustical Resonances of Assorted Ancient Structures," *Journal of the Acoustical Society of America* (*JASA*) 99, no. 2 (February 1996): 649–58.

17 Damian T. Murphy, "Archaeological Acoustic Space Measurement for Convolution Reverberation and Auralization Applications," in *Proceedings of the 9th International Conference on Digital Audio Effects (DAFx-06)*, Montreal, 2006 (Montreal: McGill University, Schulich School of Music, Music Technology Area, 2006), 221–26, www.dafx.ca/proceedings/papers/p_221.pdf or http://ant-s4.unibw-hamburg.de/paper-archive/2006/papers/p_221.pdf.

18 Eugene Ch'ng, "Experiential Archaeology: Is Virtual Time Travel Possible?," *Journal of Cultural Heritage* 10 (2009): 458–70.

19 Regina E. Collecchia, Miriam A. Kolar, and Jonathan S. Abel, "A Computational Acoustic Model of the Coupled Interior Architecture of Ancient Chavín," *Proceedings of the 133rd Convention of the Audio Engineering Society (AES)*, San Francisco, 2012 (New York: Curran, 2013), www.aes.org/e-lib/browse.cfm?elib=16439. On waveguard networks, see Julius O. Smith III, "Physical Modeling Using Digital Waveguides," *Computer Music Journal* 16, no. 4 (1992): 74–91.

20 Bruno Fazenda and Ian Drumm, "Recreating the Sound of Stonehenge," *ACTA Acustica United with Acustica* 99 (2013): 110–17. On convolution-based approaches, see Murphy, "Archaeological Acoustic Space Measurement for Convolution Reverberation and Auralization Applications," 221–26. On ambisonics, see Michael A. Gerzon, "Ambisonics in Multichannel Broadcast and Video," *Journal of the Audio Engineering Society (JAES)* 33, no. 11 (November 1985): 859–71.

21 Jukka Pätynen and Tapio Lokki, "Evaluation of Concert Hall Auralization with Virtual Symphony Orchestra," in *Proceedings of the International Symposium on Room Acoustics (ISRA)*, Melbourne, 2010 (Melbourne: Acoustical Society, 2010), www.acoustics.asn.au/conference_proceedings/ICA2010/cdrom-ISRA2010/International%20Symposium%20on%20Room%20Acoustics.pdf.

22 Vitruvius, *The Ten Books on Architecture*, trans. Morris Morgan (Cambridge, MA: Harvard University Press, 1914; reprint, New York: Dover Publications, 1960), bk. 5, sec. 8. On resonating vases in Greek theaters, see Jens Rindel, "Echo Problems in Ancient Theatres and a Comment on 'Sounding Vessels' Described by Vitruvius," In *The Acoustics of Ancient Theatres Conference*, Patras, September, 18–21, 2011, University of Patras, www.odeon.dk/pdf/AA2011_9_Rindel.pdf

23 Ron Bakker and S. Gillian, "The History of Active Acoustic Enhancement Systems," in *Proceedings of the Institute of Acoustics: Reproduced Sound Conference* (Birmingham, UK) 36, no. 2 (2014): 56–65.

24 Alan Hardiman, "Electronic Acoustic Enhancement Systems: Part One," *Lighting & Sound America*, March 2009, 88–96.
25 Alan Hardiman, "Electronic Acoustic Enhancement Systems: Part Two," *Lighting & Sound America*, April 2009, 74–79.
26 Hardiman, "Electronic Acoustic Enhancement Systems: Part One."
27 Peter Doyle, "From 'My Blue Heaven' to 'Race with the Devil': Echo, Reverb and (Dis)ordered Space in Early Popular Music Recording," *Popular Music* 23, no. 1 (January 2004): 31–49; and Doyle, *Echo and Reverb: Fabricating Space in Popular Music Recording, 1900–1960* (Middletown, CT: Wesleyan University Press, 2005).
28 Välimäki et al., "Fifty Years of Artificial Reverberation."
29 Michael A. Gerzon, "Recording Concert Hall Acoustics for Posterity," *Journal of the Audio Engineering Society (JAES)* 23, no. 7 (1975): 569–71; Richard C. Heyser, "More on 'Recording Concert Hall Acoustics for Posterity,'" *Journal of the Audio Engineering Society (JAES)* 23, no. 7 (1975): 571–72; and Angelo Farina and Regev Ayalon, "Recording Concert Hall Acoustics for Posterity," in *Proceedings of the Audio Engineering Society (AES) 24th International Conference on Multichannel Audio*, Banff, Canada, 2003 (New York: Audio Engineering Society, 2003).
30 Lamberto Tronchin and Angelo Farina, "Acoustics of the Former Teatro 'La Fenice' in Venice," *Journal of the Audio Engineering Society (JAES)* 45, no. 12 (December 1997): 1051–62.
31 Jaime Navarro, Juan J. Sendra, and Salvador Muñoz, "The Western Latin Church as a Place for Music and Preaching: An Acoustic Assessment," *Applied Acoustics* 70 (2009): 781–89.
32 Rafael Suárez, Alicia Alonso, and Juan J. Sendra, "Intangible Cultural Heritage: The Sound of the Romanesque Cathedral of Santiago de Compostela," *Journal of Cultural Heritage* 16 (2015): 239–43.
33 Deborah Howard and Laura Moretti, *Sound and Space in Renaissance Venice: Architecture, Music, Acoustics* (New Haven, CT: Yale University Press, 2010).
34 Patrizio Fausti, Roberto Pompoli, and Nicola Prodi, "Comparing the Acoustics of Mosques and Byzantine Churches," in *Proceedings of the 19th International Symposium of the International Committee for Documentation of Cultural Heritage (CIPA)*, Antalya, Turkey, 2003 (Istanbul: CIPA 2003 Organizing Committee, 2003).
35 Bissera V. Pentcheva, *The Sensual Icon: Space, Ritual, and the Senses in Byzantium* (University Park, PA: Pennsylvania State University Press, 2010); and Pentcheva, "Hagia Sophia and Multisensory Aesthetics."
36 Christoffer Andres Weitze, Jens Holger Rindel, Claus Lynge Christensen, and Anders Christian Gade, "The Acoustical History of Hagia Sophia Revived through Computer Simulation," in *Proceedings of Forum Acusticum*, Seville, Spain, 2002 (Stuttgart: Hirzel, 2002).
37 See the essay by Wieslaw Woszczyk in this volume.
38 Audio engineers use "dark" and "bright" to refer to sounds dominated by low and high frequency energy, respectively.
39 For the use of sinusoidal sweep in the acoustics measuring of Hagia Sophia in 2013, see Wieslaw Woszczyk's essay in this volume.
40 Jonathan S. Abel, Nicholas J. Bryan, Patty P. Huang, Miriam A. Kolar, and Bissera V. Pentcheva, "Estimating Room Impulse Responses from Recorded Balloon Pops," in *Proceedings of the 129th Convention of the Audio Engineering Society (AES)*, San Francisco, 2010 (New York: Curran, 2011), www.aes.org/e-lib/browse.cfm?elib=15594.
41 Abel et al., "Estimating Room Impulse Responses."
42 Anders C. Gade, "Investigations of Musicians' Room Acoustic Conditions in Concert Halls, Part I: Methods and Laboratory Experiments," *Acustica* 69 (1989): 193–203; Angelo Farina, Paolo Martingnon, Andrea Azzali, and Andrea Capra, "Listening Tests Performed inside a Virtual Room Acoustic Simulator," in *Proceedings of I Seminário Música Ciênca e Tecnologia "Acústica Musical,"* São Paolo, 2004, http://gsd.ime.usp.br/acmus/publi/seminario2004.htm; and Brereton et al., "Evaluating the Auralization of Performance Spaces."
43 Abel et al., "Estimating Room Impulse Responses"; and Nicholas J. Bryan and Jonathan S. Abel, "Methods for Extending Room Impulse Responses beyond Their Noise Floor," in *Proceedings of the 129th Convention of the Audio Engineering Society (AES)*, San Francisco, 2010 (New York: Curran, 2011), www.aes.org/e-lib/browse.cfm?elib=15590.

44 Abel et al., "Estimating Room Impulse Responses."

45 Jonathan S. Abel, Wieslaw Woszczyk, Doyuen Ko, Scott Levine, Jonathan Hong, Travis Skare, Michael J. Wilson, Sean Coffin, and Fernando Lopez-Lezcano, "Recreation of the Acoustics of Hagia Sophia in Stanford's Bing Concert Hall for the Concert Performance and Recording of Cappella Romana," in *Proceedings of the International Symposium on Room Acoustics (ISRA)*, Toronto, 2013 (Essex: Multi-Science Publishing, 2014), www.caa-aca.ca/conferences/isra2013/en/index_en.html; and Fernando Lopez-Lezcano, Travis Skare, Michael J. Wilson, and Jonathan S. Abel, "Byzantium in Bing: Live Virtual Acoustics Employing Free Software," in *Proceedings of the Linux Audio Conference (LAC)*, Graz, Austria, 2013, http://lac.linuxaudio.org/2013/papers/52.pdf.

46 Gerzon, "Ambisonics in Multichannel Broadcast and Video."

47 Jason Victor Serinus, "Cappella Romana: Time-Travel to Constantinople," *San Francisco Classical Voice*, February 1, 2013, www.sfcv.org/reviews/stanford-live/cappella-romana-time-travel-to-constantinople, accessed February 21, 2013.

Glossary

akolouthiai. Musical manuscripts recording the order and performance of primarily the proper chants of vespers, *orthros*, and the three Divine Liturgies, emerging in Late Byzantium. They are written in the Middle Byzantine intervallic notation.

anechoic. Sound with a nearly imperceptible amount of reverberation; it bears minimal acoustic imprint because the space in which it was propagated was extremely dry or the sound was recorded through close-miking, that is, the microphones were set very close to the sound source.

apēchēmata. A simple or elaborate intonation formula in a given mode that precedes a chant in the same mode.

apolytikon. Dismissal hymn sung at vespers, *orthros*, or after the Eucharist in the Divine Liturgy.

asmatic syllables. Nonsemantic vocalisms inserted to extend and beautify the chanted texts of the cathedral rite.

asmatikon. Choir book for the *ekklēsiastēs* (cathedral rite of Constantinople) that contained the chants for the elite choir of Hagia Sophia.

auralization. The process of re-creating the aural experience of a particular space. It includes the rendering of audio data by digital means in order to achieve a virtual three-dimensional sound space.

automela. See *heirmologion*.

Cheroubikon. A hymn accompanying the procession with the Eucharist gifts in the Byzantine rite, introduced at the time of Justinian and ratified in 573–74.

contrafacta. See *heirmologion*.

convolution reverberation. A process for producing artificial reverberation, which uses a software (algorithm) in order to convolve the impulse response of the targeted space being modeled with the incoming audio signal, which can be either an anechoic recording or live sound. *See also* impulse response.

diataxis. A type of liturgical book (*see* typikon), which gives the rubrics regulating the ordinary structure of the services.

ēchēmata. Intonation formulas written in the middle of a chant or in between major sections of longer chants. In the musical manuscripts these are sometimes written out with intervallic neumes or, alternatively, are abbreviated as modal signatures. See also *apēchēmata*.

euchologion. A manuscript containing the priest's prayers for the Divine Services, first surviving in MS Barberini 336 (seventh century; Vatican City, Biblioteca Apostolica Vaticana).

heirmologion. A type of musical manuscript dating as early as the tenth century that functioned as a reference book for heirmoi. It contains automela, or "model melodies," that serve as a model to produce derivatives known as proshomoia or contrafacta texts within the strophic hymn genre of the Kanon.

heirmos. A hymn with a fixed rhythm and melody that is used as the standard rhythmic and melodic pattern for other hymns (*troparia*) in the *Kanon* of the morning office (*orthros*) in the Eastern Church.

hypakoai. Identifying short monostrophic hymns similar to Western responsories. In the cathedral rite (*ecclēsiastēs*), the *hypakoē* (sing.) could be melismatic and lengthy.

Iadgari. The Georgian *tropologion* (i.e., book of *troparia* hymns), which survives in different forms in at least seven codices dating as early as the tenth century, representing fourth-to-ninth-century Greek practices of the Cathedral of the Anastasis in Jerusalem and the Great Lavra of St. Sabas in Palestine.

impulse response. The acoustic signature of a space or the output of a dynamic system (that is, an architectural interior) when presented with a brief input signal such as a pistol shot, a balloon pop, or sinusoidal sweep. Impulse response can be likened to an indexical "snapshot," which records the acoustic signature of a space measured as the decay of –60 dB in seconds of an impulsive signal fired in that space. A broadband sound source is usually used because it produces a wide range of sound waves from low to high frequencies (20 Hz to 40 kHz), thus giving an exhaustive "picture" of the way the particular space imprints itself on sound. *See also* reverberation time *and* convolution reverb.

kalophonic chant. A distinct musical style of Byzantine chant that flourished from the late thirteenth through fifteenth centuries, featuring virtuosic vocal phrases, text troping, and sections of nonsemantic text. See *teretismata* and *kratemata*.

kathisma. One of the meanings of this term identifies the divisions of the Psalter (twenty in all) according to the Rite of the Cathedral of the Anastasis in Jerusalem and, later, the Neo-Sabaitic Rite of Late Byzantium. Each *kathisma* consists of one to five psalms.

kondakar. In Slavonic Church chant tradition, this musical codex of melismatic chants contains the repertory of the Greek models of the *psaltikon* and the *asmatikon*. It is written in Old Church Slavonic notation, which resembles the prediastematic notations found in some *psaltika*.

kontakarion. See *psaltikon* and *kondakar*.

kontakion. Liturgical poems that developed from the fifth through the seventh centuries in Greek-speaking Syria, Palestine, and Constantinople that were sung as "chanted sermons" probably after the morning service and before the Divine Liturgy on great feasts in urban centers. It consists of an introduction (*prooimion*), followed by a varying number of stanzas (*oikoi*), connected to the *prooimion* by a refrain. The *oikoi* are linked by acrostic. A model stanza (*heirmos*), one each for the *prooimion* and the *oikoi*, signals the appropriate melodies to be used in their respective signing.

kratemata. Independent melodic units sung to nonsense syllables known as *teretismata* and used to prolong a hymn. A collection of *kratemata* is called a *kratematarion*. A *kratematarion* held such melodic units arranged according to the eight modes. See also *teretismata* and kalophonic.

melisma. The singing of many notes on a single syllable.

Missal (Roman). The liturgical book that contains the texts and rubrics for the celebration of the Mass in the Roman rite of the Catholic Church.

mode. See *octoechos*.

neumes. Early system of musical notation designating single pitches or small clusters of notes.

octoechos. The eight modes in Byzantium start with the four authentic (Gk. *kyrios*, "dominant") followed by the four plagal (Gk. *plagios*, "oblique") modes; the third plagal is also known as *barys* (Gk., "heavy," "deep"). Ascribing a melody to a specific mode is based not on a specific scale structure, but on the application of a specific set of melodic formulas, considered to belong to one mode in one of the chant genres.

The eight musical modes (*octoechos*) are derived from the recitation of the Resurrection account of the four Gospels, each split into two parts for the resulting succession of eight weeks. Tenth-century Georgian manuscripts, purporting to preserve eight-century evidence from Greek liturgical manuscripts recording the Jerusalem rite, offer model psalms sung according to each musical mode. In the Byzantine rite, the Resurrection Gospel readings increase from eight to eleven, and a further eleven hymns, or *troparia heothina*, were added to the celebration of *orthros*. This new hymnody has been attributed to Emperor Leo VI (866–912).

orthros. Morning office or matins, celebrated at daybreak to consecrate the day to God. Along with the evening office (*hesperinos* or vespers), *orthros* was one of the two principal hours of the cathedral and monastic offices.

pannychis. Vigil service.

periēgēsis. The structure of ekphrasis of a building organized as a tour around and a journey though it.

prokeimena. These chants (melismatic in the cathedral rite of Constantinople) precede the readings from the Epistle in the Divine Liturgy (Eucharist rite) or introduce the Old Testament readings at *hesperinos* (vespers) or the New Testament at *orthros* (morning service). They correspond to the place of the gradual in the Roman rite.

proshomoia. Contrafacta melodies based on a musical-poetic model (*automelon*). See also *heirmologion*.

psaltikon. Also known as the *kontakarion*, the book for the precentor or soloist, which contains the most virtuosic chants of the cathedral rite of Constantinople.

reverberation time (RT). The time elapsed after an impulsive signal has been fired and its energy has decayed by –60 dB from its initial level. It is only measured in the interval –5 dB to –35 dB and doubled to match measurements over a decay of –60 dB in seconds. It could thus be described as RT T_{30}.

stational liturgy. Peripatetic liturgy exemplified with stages of the rite celebrated at different churches along a preserved urban itinerary, with Mass celebrated at a selected church, known as the *statio* of the liturgy for that day. The richest evidence for the stational liturgy comes from Rome and Constantinople.

sticheraria. Musical manuscripts dating to the eleventh century containing (mostly) proper chants called *stichera idiomela* for the liturgical feasts of the calendar year. *Stichera idiomela* are chants that have their "own melody," that is, they are not modeled after other melodies.

sticheron. A hymn genre performed at the morning (*orthros*) or evening (*hesperinos*) services as inserts in the recitation of the psalms. *Stichera* are further divided into *sticheron idiomelon*, hymns that have their own melody, as opposed to *stichera proshomoia*, which are new texts composed to a melody from a limited repertoire of well-known models.

synaxarion. A church calendar for the fixed (proper) feasts, with the appropriate readings indicated for each one, but no further texts.

teretismata. Musical vocalizations set to the nonsemantic syllables *te te te*, *to to to*, and *ri ri ri*, which appeared in certain thirteenth-century manuscripts and flourished in the fourteenth century as fundamental components of the *kalophonic* chant idiom. They recall the earlier practice of intercalating aspiratory nonsemantic syllables.

tituli. The specific churches in Rome, which claim to have their origins as house-churches of the pre-Constantinian era.

Trisagion. One of the central chants of the Divine Liturgy, praising the Trinity and chanted before the Apostle and Gospel readings. It is testified to as early as the fourth century.

troparion. The word *troparion* typically identifies a hymn, but it can also refer to a refrain.

typikon (liturgical). A typikon constitutes one of the two books regulating the services (see *diataxis*). It functions as a liturgical calendar with instructions for the propers for each day. In Byzantium there were three types of typika: that of Hagia Sophia; that of the monastic Sabaitic rite; and that of the monastic Neo-Sabaitic rite.

Bibliography

Primary Sources

Ananias of Širak. *The Geography*. English translation and commentary by Robert H. Hewsen. *The Geography of Ananias of Širak: Ašxarhacoyc, the Long and the Short Recension*. Wiesbaden: Dr. Ludwig Reichert, 1992.

Aphthonios. *Progymnasmata*. Edited by Hugo Rabe. *Aphthonii Progymnasmata*. Bibliotheca scriptorum Graecorum et Romanorum Teubneriana, Rhetores Graeci, vol. 10. Leipzig: Teubner, 1926.

Aristides Quintilianus. *On Music: In Three Books*. English translation by Thomas J. Mathiesen. New Haven, CT: Yale University Press, 1983.

Aristotle. *Rhetoric*. English translation by George A. Kennedy. New York: Oxford University Press,1991.

Augustine. *Exposition on the Psalms*. English translation and edited by Philip Schaff. Peabody, MA: Hendrickson Publishers, 1994.

Cabasilas, Nicholas. *A Commentary on the Divine Liturgy*. Greek edition and French translation by René Bornert, Jean Gouillard, and Pierre Périchon. Edited by Sévérien Salaville, *Nicholas Cabasilas: Explication de la Divine Liturgie*. Sources Chrétiennes 4. Paris: Éditions du Cerf, 1967. English translation by J. M. Hussey and P. A. McNulty. *Nicholas Cabasilas: A Commentary on the Divine Liturgy*. London: SPCK, 1960.

Cabasilas, Nicholas. *The Life in Christ*. Greek edition and French translation by Marie-Hélène Congourdeau. *Nicolas Cabasilas: La vie en Christ*. Paris: Éditions du Cerf, 1989. English translation by Carmino de Catanzaro. *Nicholas Cabasilas: The Life in Christ*. Crestwood, NY: St. Vladimir's Seminary Press, 1974.

Choricius Gazeus. *Laudes Marciani*. Edited by R. Foerster and E. Richtsteig. *Opera*. Leipzig: Teubner, 1929.

Cicero. *Rhetorica ad Herennium*. English translation by Harry Caplan. Cambridge, MA: Harvard University Press, 1954.

Clement of Alexandria, *The Instructor*. French translation and edited by Henri Marrou and Marguerite Harl. *Clément d'Alexandrie: Le Pédagogue*. Paris: Éditions du Cerf, 1960. English translation by Alexander Roberts and James Donaldson. *Ante-Nicene Fathers*, vol. 2. 1885. Peabody, MA: Hendrickson, 1994.

Constantinus Rhodius. *Description of the Church of the Holy Apostles*. English translation in Cyril A. Mango. *The Art of the Byzantine Empire, 312–1453: Sources and Documents*. Englewood Cliffs, NJ: Prentice-Hall, 1972, rpt. Toronto: University of Toronto Press, 1986.

Councils, *Oecumenical*. Edited by von Hefele, Karl Joseph, and Charles Leclerq. *Histoire des conciles d'après les documents originaux*. Paris: Letouzey et Ané, 1907.

Diēgēsis of Hagia Sophia. *Diēgēsis peri tēs oikodomēs tou naou tēs megalēs tou Theou ekklēsias tēs eponoazomenēs Hagias Sophias*. Greek edition in Theodor Preger. *Scriptores Originum Constantinopolitanarum*. Leipzig: Teubner, 1901, 74–108.

Dionysius of Halicarnassos. *Lysias*. English translation by S. Usher. *Dionysius of Halicarnassus: Critical Essays*, vol. 1. Cambridge, MA: Harvard University Press, 1974.

Dionysius Aeropagita. *The Divine Names; The Celestial Hierarchy; and The Ecclesiastical Hierarchy*. Greek edition in Beate Regina Suchla, *Corpus Dionysiacum I: De divinis nomibus*. Patristische Texte und Studien 33. Berlin: De Gruyter, 1990. Edited by Günther Heil, Adolf M. Ritter. *Corpus Dionysiacum II: De coelesti hierarchia, De ecclesiastica hierarchia, De mystica theologia, Epistulae*. Berlin: De Gruyter, 2012. English translation in Colm Luibheid. *Pseudo-Dionysius: The Complete Works*. London: SPCK, 1987.

Egeria. *Itinerum*. English translation by John Wilkinson. *Egeria's Travels*. 3rd ed. Oxford: Aris & Phillips, 2006. French translation and edited by Pierre Maraval. *Journal de Voyage (Itinéraire)*. Sources Chrétiennes 296. Paris: Cerf, 1982.

Euchologion. Barberini gr. 336. Greek edition in Stefano Parenti and Elena Velkovska. *L'eucologio Barberini gr. 336*. Rome: Edizioni Liturgiche, 2000.

Eusebius of Caesarea. *Life of Constantine*. French translation by Marie-Josèphe Rondeau. Edited by Friedhelm Winkelmann and Luce Pietri. *Vie de Constantin*. Sources Chrétiennes 559. Paris: Éditions du Cerf, 2013.

Germanus of Constantinople. *On the Divine Liturgy and Historia mystica ecclesiae catholicae*. English translation and edited and by Paul Meyendorff. Crestwood, IL: St. Vladimir's Seminary Press, 1984. Greek and Latin editions in Nilo Borgia. *Il commentario liturgico di S. Germano Patriarca Costantinopolitano e la versione latina di Anastasio Bibliotecario*. Grottaferrata: Italo-orient, 1912.

Gregory the Great. *Homilies on the Gospels*. French translation and edited by Raymond Étaix, Charles Morel, and Bruno Judic. *Homélies sur l'évangile*. 2 vols. Sources Chrétiennes 522. Paris: Éditions du Cerf, 2008.

Homer. *The Odyssey*. English translation and edited by Augustus T. Murray and George E. Dimock. Cambridge, MA: Harvard University Press, 1995.

John Chrysostom. *On Vainglory and On the Education of Children*. French translation by A.-M. Malingrey. *Jean Chrysostom: Sur le vaine gloire et l'éducation des enfants*. Sources Chrétiennes 188. Paris: Éditions du Cerf, 1972.

John Chrysostom. *The Divine Liturgy of St. John Chrysostom*. English translation and edited by Frank E. Brightman and Charles E. Hammond. *Liturgies, Eastern and Western*. Oxford: Clarendon Press, 1896.

John of Damascus. *An Exact Exposition of the Orthodox Faith*. English translation by Frederic H. Chase. *Saint John of Damascus, Writings*. New York: Fathers of the Church, 1958.

John Sardianus. *Commentarium in Aphthonii Progymnasmata*. English translation by Ruth Webb. *Ekphrasis: Imagination and Persuasion in Ancient Rhetorical Theory and Practice*. Farnham, UK: Ashgate, 2009, 205–07.

Leo VI, *Ekphrasis of the Church in the Monastery of Kauleas*. Greek edition in Hieromonachos Akakios. *Leontos tou Sophou Panygeērikoi Logoi*. Athens: Ek toōn piesteērioōn Nikolaou Rousopoulou, 1868. English translation in Cyril A. Mango. *The Art of the Byzantine Empire, 312–1453: Sources and Documents*. Englewood Cliffs, NJ: Prentice-Hall, 1972.

Libanius. *Progymnasmata*. English translation by Craig A. Gibson. *Libanius's Progymnasmata: Model Exercises in Greek Prose Composition and Rhetoric*. Atlanta, GA: Society of Biblical Literature, 2008.

Liber pontificalis. Edited by Raymond David. *The Book of the Popes (Liber Pontificalis): The Ancient Biographies of the First Ninety Roman Bishops to AD 715*. Translated Texts for Historians 6. Rev. ed. Liverpool: Liverpool University Press, 2000.

Longinus. *On the Sublime*. English translation and edited by Donald A. Russell. *"Longinos" on the Sublime*. Oxford: Clarendon Press, 1965.

Manuel Chrysaphes. *The Treatise of Manuel Chrysaphes*. English translation by Dimitri Conomos. *The Treatise of Manuel Chrysaphes: On the Theory of the Art of Chanting and on Certain Erroneous Views That Some Hold about It*. Vienna: Verlag der Österreichischen Akademie der Wissenschaften, 1985.

Maximus the Confessor. *Mystagogy*. Greek edition in Christian Boudignon. *Maximi Confessoris Mystagogia: Una cum latina interpretatione Anastasii Bibliothecarii*. Turnhout: Brepols, 2011. English translation by George C. Berthold. *Maximus Confessor: Selected Writings*. New York: Paulist Press, 1985. English translation by Julian Stead. *The Church, the Liturgy, and the Soul of Man: The Mystagogia of St. Maximus the Confessor*. Still River, MA: St. Bede's Publications, 1982.

Michael the Deacon. *Ekphrasis of Hagia Sophia*. English translation by Cyril Mango and John Parker. "A Twelfth-Century Description of St. Sophia." *Dumbarton Oaks Papers* 14 (1960): 23–45.

Missale Romanum: Edition Princeps (1570). Edited by Manlio Sodi and Achille Maria Triacca. Monumenta Liturgica Concilii Tridentini 2. Vatican City: Libreria Editrice Vaticana, 1998.

Nicholas and Theodore of Andida. *Protheōria Kephalaiōdēs peri tōn en tē Theia Leitourgia Ginomenōn Symbolōn kai Mystēriōn*.

Patrologia cursus completus: Series graeca. Edited by J.-P. Migne. Paris: Migne, 1857–66, 140:417–68.

Nikolaos the Sophist. *Progymnasmata*. Edited by J. Felten. *Nicolai Progymnasmata*. Leipzig: Teubner, 1913.

Nikolaos Mesarites. *Description of the Church of the Holy Apostles at Constantinople*. English translation by Glanville Downey, "Nikolaos Mesarites: Description of the Church of the Holy Apostles at Constantinople." *Transactions of the American Philosophical Society* 47 (1957): 855–924.

Paul of Monemvasia. *Spiritually Edifying Tales*. French translation and edited by John Wortley. *Les récits édifiants de Paul, évêque de Monembasie, et d'autres auteurs*. Paris: Éditions du Centre National de la Recherche Scientifique, 1987. English translation by John Wortley. *The Spiritually Beneficial Tales of Paul, Bishop of Monembasia and of Other Authors*. Kalamazoo, MI: Cistercian Publications, 1996.

Paul the Silentiary. *Description of Hagia Sophia*. French translation by Pierre Chauvin and Marie-Christine Fayant. *Paul le Silentiaire: Description de Sainte-Sophie de Constantinople*. Paris: Éditions A. Die, 1997.

Photius. *Homily 10*. Edited by B. Laourdas. Thessaloniki: Hetaireia Makedonikon Spoudon, 1959.

Procopius of Caesarea. *Buildings*. English translation by H. B. Dewing, *Procopius*, vol. 7, *On Buildings*. Cambridge, MA: Harvard University Press, 1940. English translation by Cyril A. Mango. *The Art of the Byzantine Empire, 312–1453: Sources and Documents*. Englewood Cliffs, NJ: Prentice-Hall, 1972, rpt. Toronto: University of Toronto Press, 1986.

Quintilian. *Institutes of Oratory*. English translation and edited by Donald A. Russell. *The Orator's Education*. 5 vols. Cambridge, MA: Harvard University Press, 2002.

Romanus the Melodist. *Hymns*. Edited by Paul Maas and C. A. Trypanis. *Sancti Romani melodi cantica: Cantica genuina*. Oxford: Clarendon Press, 1963. English translation by Marjorie Carpenter. *Kontakia of Romanos, Byzantine Melodist*. Columbia: University of Missouri Press, 1970.

[Sebeos]. *The Armenian History Attributed to Sebeos*. English translation by Robert W. Thomson, James Howard-Johnston, and Timothy Greenwood. *The Armenian History Attributed to Sebeos*. Liverpool: Liverpool University Press, 1999.

The Septuagint with Apocrypha. Greek edition by Sir Lancelot Charles Lee Brenton. *The Septuagint Version of the Old Testament: With an English Translation, and with Various Readings and Critical Notes*. Grand Rapids, MI: Zondervan Pub. House, 1971.

The Shepherd of Hermas. English translation in *The Apostolic Fathers*. Edited by Michael W. Holmes. Grand Rapids, MI: Baker Academic, 2007, 442–685.

Symeon of Thessaloniki. *Explanation of the Divine Temple and On the Sacred Liturgy*. Greek edition and English translation by Steven Hawkes-Teeples. *St. Symeon of Thessalonika: The Liturgical Commentaries*. Toronto: Pontifical Institute of Mediaeval Studies, 2011.

Symeon of Thessaloniki. *On the Sacred Temple*. *Patrologia cursus completus: Series graeca*. Edited by J.-P. Migne. Paris: Migne, 1857–66, 155:305–62.

Testament of Levy. English translation by H. C. Kee. "The Testament of Levi." In *The Old Testament Pseudepigrapha I: Apocalyptic Literature and Testaments*, edited by James H. Charlesworth, 788–95. Garden City, NY: Doubleday, 1983.

The Testaments of the Twelve Patriarchs. Greek edition in Marinus de Jonge. *The Testaments of the Twelve Patriarchs: A Critical Edition of the Greek Text*. Leiden: Brill, 1978. English translation by James H. Charlesworth. *The Old Testament Pseudepigrapha*, vol. 1, *Apocalyptic Literature and Testaments*. Garden City, NY: Doubleday, 1983.

Theodore of Mopsuestia. *Catechetical Homilies*. Syriac text and French translation by Raymond Tonneau. *Les homélies catéchétiques*. Vatican City: Biblioteca Apostolica Vaticana, 1949. English translation in Alphonse Mingana. *Commentary of Theodore of Mopsuestia on the Nicene Creed*. Woodbrooke Studies 5. Cambridge: W. Heffer, 1932.

Alphonse Mingana. *Commentary of Theodore of Mopsuestia on the Lord's Prayer and on the Sacraments of Baptism and the Eucharist*. Woodbrooke Studies 6. Cambridge: W. Heffer, 1933.

Alphonse Mingana. *Commentary of Theodore of Mopsuestia on the Nicene Creed*. Piscataway, NJ: Gorgias Press, 2009. Facsimile edition in Raymond Tonneau and Robert Devreese. *Les homélies catéchétiques: Reproduction phototypique du ms. Mingana Syr. 561*. (Selly Oak Colleges' Library, Birmingham). Vatican City: Biblioteca Apostolica Vaticana, 1949.

Theodore the Stoudite, *Epistulae*. Edited by Fatouros Georgios. *Theodori Studitae epistulae*. Berlin: Walter de Gruyter, 1992.

Triodion. *Triōdion katanyktikon: Periechon apasan tēn anēkousan autō akolouthian tēs hagias kai megalēs tessarakostēs*. Rome, 1879. English translation by Mother Mary and Kallistos Ware. *The Lenten Triodion*. London: Faber & Faber, 1984.

The Typikon of the Great Church. Translation and edited by Juan Mateos. *Le Typicon de la Grande Église*, 2 vols. Orientalia Christiana Analecta 165–66. Rome: Pont. Institutum Orientalium Studiorum, 1962–63.

Vitruvius. *The Ten Books on Architecture*. English translation by Morris Morgan. *The Ten Books on Architecture*. Cambridge, MA: Harvard University Press, 1914. Reprint, New York: Dover Publications, 1960.

Secondary Sources

Abel, Jonathan S., Nicholas J. Bryan, Patty P. Huang, Miriam A. Kolar, and Bissera V. Pentcheva. "Estimating Room Impulse Responses from Recorded Balloon Pops." In *Proceedings of the 129th Convention of the Audio Engineering Society (AES)*, San Francisco, CA, November 4–7, 2010. New York: Curran, 2011, www.aes.org/e-lib/browse.cfm?elib=15594.

Abel, Jonathan S., John W. Rick, Patty P. Huang, Miriam A. Kolar, Julius O. Smith, and John M. Chowning. "On the Acoustics of the Underground Galleries of Ancient Chavín de Huántar, Peru." In *Proceedings of Acoustics '08*, Paris, June 29–July 4, 2008. https://ccrma.stanford.edu/groups/chavin/publications/Acoustics08.pdf.

Abel, Jonathan S., Wieslaw Woszczyk, Doyuen Ko, Scott Levine, Jonathan Hong, Travis Skare, Michael J. Wilson, Sean Coffin, and Fernando Lopez-Lezcano. "Recreation of the Acoustics of Hagia Sophia in Stanford's Bing Concert Hall for the Concert Performance and Recording of Cappella Romana." In *Proceedings of the International Symposium on Room Acoustics (ISRA)*, Toronto, June 9–11, 2013. Essex: Multi-Science Publishing, 2014, www.caa-aca.ca/conferences/isra2013/en/index_en.html.

Alexandru, Maria. Studie über die "Grossen Zeichen" der byzantinischen musikalischen Notation unter Besonderer Berücksichtigung der Periode vom Ende des 12. bis Anfang des 19. Jahrhunderts. PhD diss., University of Copenhagen, 2000.

Almac, Umut, Karl Schweizerhof, Gunther Blankenhorn, Ch. Duppel, and F. Wenzel. "Structural Behaviour of Hagia Sophia under Dynamic Loads." In *Vienna Congress on Recent Advances in Earthquake Engineering and Structural Dynamics 2013*, Vienna, August 28–30, 2013, edited by C. Adam, R. Heuer, W. Lenhardt, and C. Schranz, paper no. 475. http://veesd2013.conf.tuwien.ac.at.

Alturo, Jesús. "Deux nouveaux fragments de l''Itinerarium Egeriae' du IXe–Xe siècle." *Revue Bénédictine* 115 (2005): 241–50.

Antonova, Clemena. *Space, Time, and Presence in the Icon: Seeing the World with the Eyes of God*. Farnham, UK: Ashgate, 2009.

Arranz, Miguel, Ed. *Le typicon du Monastère du Saint-Sauveur à Messine, Codex Messinensia gr 115, A.D. 1131*. Orientalia Christiana Analecta 185. Rome: Pontificium Institutum Orientalium Studiorum, 1969.

Auxentios, Bishop of Photiki. *The Paschal Fire in Jerusalem: A Study of the Rite of the Holy Fire in the Church of the Holy Sepulchre*. 3rd ed. Berkeley, CA: Saint John Chrysostom Press, 1999.

Bakker, Ron, and S. Gillian. "The History of Active Acoustic Enhancement Systems." In *Proceedings of the Institute of Acoustics: Reproduced Sound Conference* (Birmingham, UK), October 14–15, 2014, 36, no. 2 (2014): 56–65.

Baldovin, John F. *The Urban Character of Christian Worship: The Origins, Development, and Meaning of Stational Liturgy*. Rome: Pontificium Institutum Studiorum Orientalium, 1987.

Barbagallo, Salvatore. *Iconografia liturgica del Pantokrator*. Rome: Studia Anselmiana, 1996.

Barber, Charles. "From Transformation to Desire: Art and Worship after Byzantine Iconoclasm." *Art Bulletin* 75, no. 1 (1993): 7–16.

——. "Mimesis and Memory in the Narthex Mosaics at the Nea Moni, Chios." *Art History* 24 (2001): 323–37, 462.

——. "The Monastic Typikon for Art Historians." In *The Theotokos Evergetis and Eleventh-Century Monasticism: Papers of the Third Belfast Byzantine Colloquium*, edited by Margaret Mullett and Anthony Kirby, 198–214. Belfast: Belfast Byzantine Enterprises, 1994.

Belting, Hans. *Likeness and Presence: A History of the Image before the Era of Art*. Chicago, IL: University of Chicago Press, 1994.

Betancourt, Roland. "A Byzantine Liturgical Commentary in Verse: Introduction and Translation." *Orientalia Christiana Periodica* 81 (2015): 433–72.

Bilsen, Frans A. "Repetition Pitch Glide from the Step Pyramid at Chichen Itza (L)." *Journal of the Acoustical Society of America* 120 (2006): 594–96.

Blesser, Barry, and Linda-Ruth Salter. *Spaces Speak, Are You Listening? Experiencing Aural Architecture*. Cambridge, MA: MIT Press, 2007.

von Bogyay, Thomas. "Zur Geschichte der Hetoimasie." In *Akten des XI. internationalen Byzantinistenkongresses München 1958*. Munich: Beck, 1960. 58–61.

Bolens, Guillemette. *The Style of Gestures: Embodiment and Cognition in Literary Narrative*. Baltimore, MD: Johns Hopkins University Press, 2012.

Bolman, Elizabeth. "The White Monastery Federation and the Angelic Life." In *Byzantium and Islam: Age of Transition, 7th–9th Century*, edited by Helen C. Evans, 75–77. New Haven, CT: Yale University Press, 2015.

——, Ed. *Monastic Visions: Wall Paintings in the Monastery of St. Antony at the Red Sea*. New Haven, CT: Yale University Press, 2002.

Borgehammar, Stephan. "Heraclius Learns Humility: Two Early Latin Accounts Composed for the Celebration of Exaltatio Crucis." *Millennium: Jahrbuch zu Kultur und Geschichte des Ersten Jahrtausends n. Chr.* 6 (2009): 145–202.

Bornert, René. *Les commentaires byzantins de la Divine Liturgie du VIIe au XVe siècle*. Paris: Institut Français d'Études Byzantines, 1966.

Bouras, Charalampos. *Nea Moni on Chios: History and Architecture*, edited by David A. Hardy. Athens: Commercial Bank of Greece, 1982.

Boyd, Susan A., and Marlia Mundell Mango, Eds. *Ecclesiastical Silver Plate in Sixth-Century Byzantium*. Washington, DC: Dumbarton Oaks, 1992.

Bradshaw, Paul F. *Essays on Early Eastern Eucharistic Prayers*. Collegeville, MN: Liturgical Press, 1997.

Bradshaw, Paul F., and E. Johnson Maxwell. *The Eucharistic Liturgies: Their Evolution and Interpretation*. Collegeville, MN: Liturgical Press, 2012.

Braga, Carlo. *L'espace liturgique: Ses éléments constitutifs et leur sens; Conférences Saint-Serge, LIIe Semaine d'Études Liturgiques: Paris, 27–30 Juin 2005*. Rome: CLV–Edizioni Liturgiche, 2006.

Brandt, Ros, Michelle Duffy, and Dolly MacKinnon, Eds. *Hearing Places: Sound, Place, Time and Culture*. Newcastle upon Tyne: Cambridge Scholars Publishing, 2007.

Brereton, Jude, Damian T. Murphy, and David M. Howard. "Evaluating the Auralization of Performance Spaces and Its Effect on Singing Performance." In *Proceedings of the 130th Convention of the Audio Engineering Society (AES)*, London, May 13–16, 2011. New York: Curran, 2011, www.aes.org/e-lib/browse.cfm?elib=15847.

Brooks-Leonard, John K. "Easter Vespers in Early Medieval Rome: A Critical Edition and Study." PhD diss., University of Notre Dame, 1988.

Brown, Peter. *The Body and Society: Men, Women and Sexual Renunciation in Early Christianity*. New York: Columbia University Press, 1988.

Brubaker, Leslie, and John Haldon. *Byzantium in the Iconoclast Era, c. 680–850: A History*. Cambridge: Cambridge University Press, 2011.

Bryan, Nicholas J., and Jonathan S. Abel. "Methods for Extending Room Impulse Responses beyond Their Noise Floor." In *Proceedings of the 129th Convention of the Audio Engineering Society (AES)*, San Francisco, November 4–7, 2010. New York: Curran, 2011, www.aes.org/e-lib/browse.cfm?elib=15590.

Bucca, Donatella. *Catalogo dei manoscritti musicali greci del SS. Salvatore di Messina*. Rome: Comitato Nazionale per le Celebrazioni del Millenario della Fondazione dell'Abbazia di S. Nilo a Grottaferrata, 2011.

Buckton, David. "Byzantine Enamel and the West." *Byzantinische Forschungen* 13 (1988): 235–44.

Bull, Michael, and Les Back, Eds. *The Auditory Culture Reader*. Oxford: Berg, 2003.

Burrus, Virginia. "'In the Theater of This Life': The Performance of Orthodoxy in Late Antiquity." In *The Limits of Ancient Christianity: Essays on Late Antique Thought and Culture in Honor of R. A. Markus*, edited by William Klingshirn and Mark Vessey, 80–96. Ann Arbor, MI: University of Michigan Press, 1999.

Caillois, Roger. "Mimicry and Legendary Psychasthenia." *October* 31 (Winter 1984): 16 32.

Cameron, Alan. "Isidore of Miletus and Hypathia: On the Editing of Mathematical Texts." *Greek, Roman, and Byzantine Studies* 31 (1990): 103–27.

Carr, Annemarie Weyl. "Hell in the Sweet Land: Hell's Place in the Last Judgments of Byzantine and Medieval Cyprus." In *The Damned in Hell in the Frescoes of Venetian-Dominated Crete (13th–17th Centuries)*, edited by Angeliki Lymberopoulou and Vasiliki Tsamakda. Forthcoming.

——. "Images: Expressions of Faith and Power." In *Byzantium: Faith and Power (1261–1557)*, edited by Helen C. Evans, 143–52. New Haven, CT: Yale University Press, 2004.

Carr, Annemarie Weyl, and Laurence J. Morrocco. *A Byzantine Masterpiece Recovered, the Thirteenth-Century Murals of Lysi, Cyprus*. Austin, TX: University of Texas Press, 1991.

Carr, Annemarie Weyl, and Andreas Nicolaïdès. *Asinou across Time: Studies in the Architecture and Murals of the Panagia Phorbiotissa, Cyprus*. Washington, DC: Dumbarton Oaks Research Library and Collection, 2012.

Carruthers, Mary J. *The Book of Memory: A Study of Memory in Medieval Culture*. Cambridge: Cambridge University Press, 1990.

Chavasse, Antoine. *Les lectionnaires romains de la messe au VIIe et au VIIIe siècle: Sources et dérivés*. Spicilegii Friburgensis Subsidia 22. 2 vols. Fribourg: Éditions Universitaires, 1993.

Ch'ng, Eugene. "Experiential Archaeology: Is Virtual Time Travel Possible?" *Journal of Cultural Heritage* 10 (2009): 458–70.

Christoforaki, Ionna. "Cyprus between Byzantium and the Levant: Eclecticism and Interchange in the Cycle of the Life of the Virgin in the Church of the Holy Cross at Pelendri." *Epeteris Kentrou Epistimonikon Erevnon* 22 (1996): 215–45.

Chronz, Tinatin, and Alexandra Nikiforova. "Beobachtungen zum ältesten bekannten Tropologion-Codex Sinaiticus graecus ΜΓ 56+5 des 8.–9. Jhs. mit Erstedition ausgewählter Abschnitte." In *Σύναξις καθολική: Beiträge zu Gottesdienst und Geschichte der fünf altkirchlichen Patriarchate für Heinzgerd Brakmann zum 70. Geburtstag*, edited by Diliana Atanassova and Tinatin Chronz, 147–74. Orientalia–Patristica–Oecumenica. 6 vols. Berlin: LIT Verlag, 2014.

Cirillo, Ettore, and Francesco Martellotta. *Worship, Acoustics, and Architecture*. Brentwood, UK: Multi-Science Publishing, 2006.

Clay, Jennifer Strauss. *Homer's Trojan Theater: Space, Vision, and Memory in the Iliad*. Cambridge: Cambridge University Press, 2011.

Collecchia, Regina E., Miriam A. Kolar, and Jonathan S. Abel. "A Computational Acoustic Model of the Coupled Interior Architecture of Ancient Chavín." In *Proceedings of the 133rd Convention of the Audio Engineering Society (AES)*, San Francisco, October 26–29, 2012. New York: Curran, 2013, www.aes.org/e-lib/browse.cfm?elib=16439.

Conomos, Dimitri E. *Byzantine Trisagia and Cheroubika of the 14th and 15th Centuries: A Study of Late Byzantine Liturgical Chant*. Thessaloniki: Patriarchal Institute for Patristic Studies, 1974.

——. *The Late Byzantine and Slavonic Communion Cycle: Liturgy and Music*. Washington, DC: Dumbarton Oaks Research Library and Collection, 1985.

Conybeare, Frederick Cornwallis, and A. J. Maclean. *Rituale Armenorum: Being the Administration of the Sacraments and the Breviary Rites of the Armenian Church*. Oxford: Clarendon Press, 1905.

Cormack, Robin. *Byzantine Art*. Oxford: Oxford University Press, 2000.

Corrigan, Kathleen. *Visual Polemics in the Ninth-Century Byzantine Psalters*. Cambridge: Cambridge University Press, 1992.

Cosgrove, Charles H. *An Ancient Christian Hymn with Musical Notation: Papyrus Oxyrhynchus 1786; Text and Commentary*. Tübingen: Mohr Siebeck, 2011.

Cremer, Lothar, and Helmut A. Müller. *Principles and Applications of Room Acoustics*. 2 vols. London: Applied Science Publishers, 1982.

Damasio, Antonio. *The Self Comes to Mind: Constructing the Conscious Brain*. 2nd ed. London: Vintage, 2012.

Day, Peter D. *The Liturgical Dictionary of Eastern Christianity*. Collegeville, MN: Liturgical Press, 1993.

De Benedictis, Elaine. "The 'Schola Cantorum' in Rome during the High Middle Ages." PhD diss., Bryn Mawr College, 1983.

Declercq, Nico F., Joris Degrieck, Rudy Briers, and Oswald Leroy. "A Theoretical Study of Special Acoustic Effects Caused by the Staircase of the El Castillo Pyramid at the Maya Ruins of Chichen-Itza in Mexico." *Journal of the Acoustical Society of America* 116, no. 6 (December 2004): 3328–35.

De Jong, Irene, and René Nünlist. "From Bird's Eye View to Close-up: The Standpoint of the Narrator in Homeric Epics." In *Antike Literature in neuer Deutung*, edited by Anton Bierl, Arbogast Schmitt, and Andreas Willi, 63–83. Munich: K. G. Saur, 2004.

Demus, Otto. *Byzantine Mosaic Decoration: Aspects of Monumental Art in Byzantium*. Boston, MA: Boston Book & Art Shop, 1955.

Denysenko, Nicholas. "*The Life in Christ* by Nicholas Cabasilas: A Mystagogical Work." *Studia Liturgica* 38 (2009): 242–60.

Depuydt, Leo. *Catalogue of Coptic Manuscripts in the Pierpont Morgan Library*. Louvain: Uitgeverij Peeters, 1993.

Dix, Gregory. *The Shape of the Liturgy*. London: Dacre Press, 1945.

Donabédian, Patrick. *L'âge d'or de l'architecture arménienne*. Marseilles: Parenthèsis, 2008.

——. "Les thèmes bibliques dans la sculpture arménienne préarabe." *Revue des Études Arméniennes* 22 (1990–91): 253–314.

Doneda, Annalisa. "I manoscritti liturgico-musicali byzantine: Tipologie e organisazione." In *El palimpsesto grecolatino como fenómeno librario y textual*, edited by Àngel Escobar, 103–10. Zaragoza: Institución Fernando el Católico, 2006.

Dournovo, Lydia. "Stennaja Živopis' v Aruc." *Lraber hasarakakan gitut'yunneri*, 1952, 49–66.

Downey, George. "Byzantine Architects and Their Methods." *Byzantion* 18 (1948): 99–118.

Doyle, Peter. *Echo and Reverb: Fabricating Space in Popular Music Recording, 1900–1960*. Middletown, CT: Wesleyan University Press, 2005.

——. "From 'My Blue Heaven' to 'Race with the Devil': Echo, Reverb and (Dis)ordered Space in Early Popular Music Recording." *Popular Music* 23, no. 1 (2004): 31–49.

Durand, Jannic, Ioanna Rapti, and Dorota Giovannoni. *Armenia sacra: Mémoire chrétienne des Arméniens (IVe–XVIIIe siècles)*. Paris: Musée du Louvre, 2007.

Elsner, Jaś. "The Viewer and the Vision: The Case of the Sinai Apse." *Art History* 17, no. 1 (March 1994): 81–102.

Elsner, Jaś, and Gerhard Wolf. "The Transfigured Mountain: Icons and Transformations of Pilgrimage at the Monastery of St Catherine at Mount Sinai." In *Approaching the Holy Mountain: Art and Liturgy at St Catherine's Monastery in the Sinai*, edited by Robert S. Nelson and Sharon E. J. Gerstel, 37–71. Turnhout: Brepols, 2010.

Empereur, Jean-Yves. *Alexandria Rediscovered*. Translated by Margaret Maehler. New York: George Braziller, 1998.

Epstein, Ann Wharton. "Formulas for Salvation: A Comparison of Two Byzantine Monasteries and Their Founders." *Church History* 50 (1981): 385–400.

Ergin, Nina. "The Soundscape of Sixteenth-Century Istanbul Mosques: Architecture and Qu'ran Recital." *Journal of the Society of Architectural Historians* 67 (2008): 204–21.

Evangelatou, Maria. "Liturgy and the Illustration of the Ninth-Century Marginal Psalters." *Dumbarton Oaks Papers* 63 (2009): 59–116.

Evans, Helen C., Ed. *Byzantium: Faith and Power (1261–1557)*. New York: Metropolitan Museum of Art, 2004.

Farina, Angelo, and Regev Ayalon. "Recording Concert Hall Acoustics for Posterity." In *Proceedings of the Audio Engineering Society (AES) 24th International Conference on Multichannel Audio*, Banff, Canada, June 26–28, 2003. New York: Audio Engineering Society, 2003.

Farina, Angelo, Paolo Martingnon, Andrea Azzali, and Andrea Capra. "Listening Tests Performed Inside a Virtual Room Acoustic Simulator." In *Proceedings of I seminario Música Ciênca e Tecnologia "Acústica Musical,"* São Paolo, November 3–5, 2004. http://gsd.ime.usp.br/acmus/publi/seminario2004.htm.

Fausti, Patrizio, Roberto Pompoli, and Nicola Prodi. "Comparing the Acoustics of Mosques and Byzantine Churches." In *Proceedings of the 19th International Symposium of the International Committee for Documentation of Cultural Heritage (CIPA)*, Antalya, Turkey, September 30–October 4, 2003. Istanbul: CIPA 2003 Organizing Committee, 2003.

Fazenda, Bruno, and Ian Drumm. "Recreating the Sound of Stonehenge." *ACTA Acustica United with Acustica* 99 (2013): 110–17.

Feldman, Louis H. "The Case of the Blasphemer (Lev. 24:10–16) According to Philo and Josephus." In *Heavenly Tablets: Interpretation, Identity and Tradition in Ancient Judaism*, edited by Lynn LiDonnici and Andrea Lieber, 213–26. Leiden: Brill, 2007.

Fenwick, John. *The Missing Oblation: The Contents of the Early Antiochene Anaphora*. Bramcote, UK: Grove Books, 1989.

Findikyan, Michael Daniel. "Armenian Hymns of the Church and the Cross." *St. Nersess Theological Review* 11 (2006): 63–106.

——. "Armenian Hymns on the Holy Cross and the Jerusalem Encaenia." *Revue des Études Arméniennes* 32 (2010): 25–58.

——. "The Armenian Liturgy of Dedicating a Church: A Textual and Comparative Analysis of Three Early Sources." *Orientalia Christiana Periodica* 64, no. 1 (1998): 75–121.

Finney, Paul Corby. "Margherita Guarducci, La più antica icone di Maria: Un prodigioso vincolo fra oriente e occidente." *Journal of Early Christian Studies* 1 (1993): 104–05.

Fischinger, Timo, Klaus Frieler, and Jukka Louhivuori. "Influence of Virtual Room Acoustics on Choir Singing." *Psychomusicology: Music, Mind, and Brain* 25, no. 3 (2015): 208–18.

Floros, Constantin. *The Origins of Russian Music: Introduction to the Kondakarian Notation*. Frankfurt: Peter Lang, 2009.

——. *Universale Neumenkunde*. Kassel-Wilhelmshöhe: Bärenreiter, 1970.

Florovsky, Georges. *Christianity and Culture*. Belmont, MS: Nordland, 1974.

Follieri, Enrica. *Initia hymnorum ecclesiae graecae*. Vatican City: Biblioteca Apostolica Vaticana, 1960.
Fortescue, Adrian, and J. B. O'Connell. *The Ceremonies of the Roman Rite Described*. 5th ed., rev. London: Burns, Oates, & Washbourne, 1934.
Foster, Susan Leigh. *Choreographing Empathy: Kinesthesia in Performance*. London: Routledge, 2011.
Franses, Rico. "When All That Is Gold Does Not Glitter: On the Strange History of Looking at Byzantine Art." In *Icon and Word: The Power of Images in Byzantium: Studies Presented in Honor of Robin Cormack*, edited by Anthony Eastmond and Liz James, 13–24. Aldershot: Ashgate, 2003.
Frøyshov, Stig Simeon R. "The Early Development of the Liturgical Eight-Mode System in Jerusalem." *St Vladimir's Theological Quarterly* 51 (2007): 139–78.
——. "The Georgian Witness to the Jerusalem Liturgy: New Sources and Studies." In *Inquiries into Eastern Christian Worship: Selected Papers of the Second International Congress of the Society of Oriental Liturgy, Rome, 17–21 September 2008*, edited by Bert Groen, Steven Hawkes-Teeples, and Stefanos Alexopoulos, 227–67. Louvain: Peeters, 2012.
Gade, A. C. "Investigations of Musicians' Room Acoustic Conditions in Concert Halls, Part I: Methods and Laboratory Experiments." *Acustica* 69 (1989): 193–203.
Gamble, Harry Y. *Books and Readers in the Early Church: A History of Early Christian Texts*. New Haven, CT: Yale University Press, 1995.
Gampel, Alan. "The Origins of Musical Notation." In *Age of Transition: Byzantine Culture in the Islamic World*, edited by Helen C. Evans, 144–56. New Haven, CT: Yale University Press, 2015.
Garibian de Vartavan, Nazénie. *La Jérusalem nouvelle et les premiers sanctuaires chrétiens de l'Arménie*. Yerevan, Armenia: Isis Pharia, 2009.
Gell, Alfred. *Art and Agency: An Anthropological Theory*. Oxford: Clarendon Press, 1998.
Gerstel, Sharon., Ed. *Beholding the Sacred Mysteries. Programs of the Byzantine Sanctuary*. Seattle, WA: Published by College Art Association in association with University of Washington Press, 1999.
Gerzon, Michael A. "Ambisonics in Multichannel Broadcast and Video." *Journal of the Audio Engineering Society* 33, no. 11 (November 1985): 859–71.
——. "Recording Concert Hall Acoustics for Posterity." *Journal of the Audio Engineering Society* 23, no. 7 (1975): 569–71.
Getcha, Archimandrite Job. *The Typikon Decoded: An Explanation of Byzantine Liturgical Practice*. Translated by Paul Meyendorff. Yonkers, NY: St. Vladimir's Seminary Press, 2012.
Glancy, Jennifer. *Corporal Knowledge: Early Christian Bodies*. Oxford: Oxford University Press, 2010.
Gleason, Maud. *Making Men: Sophists and Self-Presentation in Ancient Rome*. Princeton, NJ: Princeton University Press, 1995.
Goddio, Franck, and Andre Bernard. *Alexandria: The Submerged Royal Quarters*. London: Periplus, 1998.
Golitzen, Alexander. *Mystagogy: A Monastic Reading of Dionysius Aeropagita*. Collegeville, MN: Liturgical Press, 2013.
Greenwood, Timothy. "A Corpus of Early Medieval Armenian Inscriptions." *Dumbarton Oaks Papers* 58 (2004): 27–91.
Greer, Rowan A. *Theodore of Mopsuestia: Exegete and Theologian*. London: Faith Press, 1961.
Grenfell, Bernard P., and Arthur S. Hunt. *New Classical Fragments and Other Greek and Latin Papyri*. Oxford: Clarendon Press, 1897.
Grisar, Hartmann. *Das Missale im Lichte römischer Stadtgeschichte: Stationen, Perikopen, Gebräuche*. Freiburg im Breisgau: Herder, 1925.
Guarducci, Margherita. *La più antica icona di Maria: Un prodigioso vincolo fra Oriente e Occidente*. Rome: Istituto Poligrafico e Zecca dello Stato, 1989.

Guidobaldi, Federico, Claudia Barsanti, and Alessandra Guiglia Guidobaldi. *San Clemente: La scultura del VI secolo*. San Clemente Miscellany 4. 2 vols. Rome: Collegio San Clemente, 1992.

Hadjisolomos, Solomon (Solon) J. *Byzantine Music: The Modal Structure of the 11 Eothina Anastassima Ascribed to the Emperor Leo (+912): A Musicological Study*. Nicosia, Cyprus: Holy Monastery of Kykko, 1986.

Haldon, John. "The Reign of Heraclius: A Context for Change?" In *The Reign of Heraclius: Crisis and Confrontation*, edited by Gerrit J. Reinink and Bernard H. Stolte, 1–16. Louvain: Peeters, 2002.

Hamasaki, Kimio, Toshiyuki Nishiguchi, Reiko Okumura, Yasushige Nakayama, and Akio Ando. "22.2 Multichannel Sound System for Ultra High-Definition TV." *Society of Motion Picture and Television Engineers Motion Imaging Journal* 117 (2008): 40–49.

Hamilakis, Yannis. *Archaeology and the Senses: Human Experience, Sense and Affect*. Cambridge: Cambridge University Press, 2013.

Hanggi, Anton. *Prex Eucharistica: Textus e variis liturgiis antiquioribus selecti*. Fribourg: Éditions Universitaires, 1968.

Hapgood, Isabel Florence. *Service Book of the Holy Orthodox–Catholic Apostolic Church*. Rev. ed. New York: Association Press, 1922; reprint, New York, then Englewood, NJ: Antiochian Orthodox Christian Archdiocese of North America, 1956, 1965, 1975, 1983, 1996.

Hardiman, Alan. "Electronic Acoustic Enhancement Systems: Part One." *Lighting & Sound America* (March 2009): 88–96.

——. "Electronic Acoustic Enhancement Systems: Part Two." *Lighting & Sound America* (April 2009): 74–79.

Harris, Simon, ed. *The Communion Chants of the Thirteenth-Century Byzantine Asmatikon*. Amsterdam: Harwood Academic Publishers, 1999.

Hawkins, Ernst, and Arthur Megaw. "The Church of the Holy Apostles at Perachorio and Its Frescoes." *Dumbarton Oaks Papers* 16 (1962): 277–350.

Haydn, Franz Joseph. *The Virtual Haydn: Complete Works for Solo Keyboard*. Performer, Tom Beghin. Producer, Martha de Francisco. Virtual Acoustics Architect, Wieslaw Woszczyk. Naxos, 2011. 4 Blu-ray discs.

Heath, Malcolm. "Longinus On Sublimity." *Proceedings of the Cambridge Philological Society* 45 (1999): 43–74.

Hewsen, Robert H. *Armenia: A Historical Atlas*. Chicago, IL: University of Chicago Press, 2001.

Heyser, Richard C. "More on 'Recording Concert Hall Acoustics for Posterity.'" *Journal of the Audio Engineering Society* 23, no. 7 (1975): 571–72.

Hiley, David. *Western Plainchant: A Handbook*. Oxford: Clarendon Press, 1993.

Hintze, Gisa. *Das byzantinische Prokeimena-Repertoire*. Vol. 9. Hamburg: Verlag der Musikalienhandlung Wagner, 1973.

Howard, Deborah, and Laura Moretti. *Sound and Space in Renaissance Venice: Architecture, Music, Acoustics*. New Haven, CT: Yale University Press, 2010.

Husson, Geneviève. "P. Strasb. inv. 1185: Hymne pour la fête de l'Hypapantè (2 février)." In *Atti del XXII Congresso internazionale di Papirologia, Firenze 1998*, edited by Isabella Andorlini, 681–87. Florence: Istituto Papirologico G. Vitelli, 2001.

Huxley, George. *Anthemius of Tralles: A Study of Later Greek Geometry*. Greek, Roman, and Byzantine Monographs 1. Cambridge, MA: Eaton Press, 1959.

Irving, John. "Digital Approaches to Haydn's Solo Keyboard Music." *Early Music* 40 (2012): 535–38.

Isar, Nicoletta. "'Χορός of Light': The Vision of the Sacred in Paulus the Silentiary's Poem *Descriptio S. Sophiae*." *Byzantinische Forschungen* 28 (2004): 215–42.

Jahn, Robert G., Paul Devereux, and Michael Ibison. "Acoustical Resonances of Assorted Ancient Structures." *Journal of the Acoustical Society of America* 99 (February 1996): 649–58.

James, Liz. "Art and Lies: Text, Image and Imagination in the Medieval World." In *Icon and Word: The Power of Images in Byzantium; Studies Presented to Robin Cormack*, edited by Antony Eastmond and Liz James, 59–72. Aldershot: Ashgate, 2003.

Jeffery, Peter. "The Earliest Christian Chant Repertory Recovered: The Georgian Witnesses to Jerusalem Chant." *Journal of the American Musicological Society* 47 (1994): 1–39.

——. "The Earliest Oktōēchoi: The Role of Jerusalem and Palestine in the Beginnings of Modal Ordering." In *The Study of Medieval Chant, Paths and Bridges, East and West: In Honor of Kenneth Levy*, edited by Peter Jeffery, 144–206. Woodbridge: Boydell Press, 2000.

——. "The Lost Chant Tradition of Early Christian Jerusalem: Some Possible Melodic Survivals in the Byzantine and Latin Chant Repertories." *Early Music History* 11 (1992): 151–90.

——. "Oktōēchos." In *The New Grove Dictionary of Music and Musicians*, 18:370–73. 2nd ed. London: Macmillan, 2001.

——. "Philo's Impact on Christian Psalmody." In *Psalms in Community: Jewish and Christian Textual, Liturgical, and Artistic Traditions*, edited by Harold W. Attridge and Margot E. Fassler, 147–87. Society of Biblical Literature Symposium Series 25. Leiden: E. J. Brill; Atlanta: Society of Biblical Literature, 2003.

——. "The Roman Liturgical Year and the Early Liturgy of St. Peter's." In *Old Saint Peter's, Rome*, edited by Rosamond McKitterick, John Osborne, Carol M. Richardson, and Joanna Story. British School at Rome Studies, 157–76. Cambridge: Cambridge University Press, 2013.

——. *The Secret Gospel of Mark Unveiled: Imagined Rituals of Sex, Death, and Madness in a Biblical Forgery*. New Haven, CT: Yale University Press, 2006.

——. "The Sunday Office of Seventh-Century Jerusalem in the Georgian Chantbook (Iadgari): A Preliminary Report." *Studia Liturgica* 21 (1991): 52–75.

Jordan, Robert H., Ed. *The Synaxarion of the Monastery of the Theotokos Evergetis*. Belfast Byzantine Texts and Translations. 3 vols. Belfast: Belfast Byzantine Enterprises, the Institute of Byzantine Studies, the Queen's University of Belfast, 2000.

Karabiber, Zerhan. "A New Approach to an Ancient Subject: CAHRISMA Project." In *Proceedings of the Seventh International Congress on Sound and Vibration, ICSV–7: Garmisch-Partenkirchen, Germany, July 4–7, 2000*. Vol. 4, 1661–68. Garmisch-Partenkirchen: International Institute of Acoustics and Vibration, 2000.

Karapetyan, Samvel. "Mrenə ev Nra Hušarjannerə" [Mren and its monuments]. *Vardsk'/Duty of Soul* 7 (2012): 31–63.

Kartsonis, Anna. *Anastasis: The Making of the Image*. Princeton, NJ: Princeton University Press, 1986.

Kaufman, Kay S., Peter Jeffery, and Ingrid T. Monson. "Oral and Written Transmission in Ethiopian Christian Chant." *Early Music History* 12 (1993): 55–117.

Kazaryan, Armen. "The Chancel and Liturgical Space in the Church of Zuart'noc'." In *Ikonostas: Proiskhozhdenie, Razvitie, Simvolika*, edited by Alexei M. Lidov, 85–117. Moscow: Progress-Traditsiia, 2000.

Kleiner, Mendel, Bengt-Inge Dalenbäck, and Peter Svensson. "Auralization—An Overview." *Journal of the Audio Engineering Society* 41, no. 11 (November 1993): 861–75.

Klentos, John Eugene. "Byzantine Liturgy in Twelfth-Century Constantinople: An Analysis of the Synaxarion of the Monastery of the Theotokos Evergetis (codex *Athens Ethnike Bibliotheke 788*)." PhD diss., University of Notre Dame, 1995.

Knight, David. "The Archaeoacoustics of San Vitale, Ravenna." MA thesis, University of Southampton, 2010.

——. "The Arcaeoacoustics of a Sixth-Century Christian Structure: San Vitale, Ravenna." In *Music & Ritual: Bridging Material and Living Cultures*, edited by Raquel Jiménez, Rupert Till, and Mark Howell, 133–46. Berlin: ēkhō Verlag, 2013.

Ko, Doyuen, Wieslaw Woszczyk, and Song Hui Chon, "Evaluation of a New Active Acoustics System in Performances of Five String Quartets." In *Proceedings of the 132nd Convention of the Audio Engineering Society, Budapest, Hungary, April 26–29, 2012*. www.aes.org/e-lib/browse.cfm?elib=16241.

Ko, Doyuen, Wieslaw Woszczyk, and Jonathan Hong. "Augmented Stage Support in Ensemble Performance Using Virtual Acoustics Technology." In *Proceedings of the 21st International Congress on Acoustics, Montreal, Canada, June 2–7, 2013*. https://ccrma.stanford.edu/~shchon/pubs/aes2012.pdf.

Kolar, Miriam. "Archaeological Psychoacoustics at Chavin de Huantar, Peru." PhD diss., Stanford University, 2013.

——. "Tuned to the Senses: An Archaeoacoustic Perspective on Ancient Chavín." *Appendix* 1 (2013), http://theappendix.net/issues/2013/7/tuned-to-the-senses-an-archaeoacoustic-perspective-on-ancient-chavin.

Kosso, Cynthia, and Anne Scott, Eds. *Fear and Its Representations in the Middle Ages and the Renaissance*. Turnhout: Brepols, 2002.

Krautheimer, Richard, and Slobodan Ćurčić. *Early Christian and Byzantine Architecture*. Pelican History of Art. 4th ed. New Haven, CT: Yale University Press, 1986.

Krautheimer, Richard, et al. *Corpus Basilicarum Christianarum Romae*. 5 vols. Vatican City: Pontificio Istituto di Archeologia Cristiana, 1937–77.

Krueger, Derek. *Liturgical Subjects: Christian Ritual, Biblical Narrative and the Formation of the Self in Byzantium*. Philadelphia, PA: University of Pennsylvania Press, 2014.

——. "Romanos the Melodist and the Formation of the Christian Self in Early Byzantium." In *Proceedings of the 21st International Congress of Byzantine Studies, London, 2006*. Vol. 1, *Plenary Papers*, edited by Elizabeth M. Jeffreys, Fiona K. Haarer, and Judith Gilliland, 247–66. Aldershot: Ashgate, 2006.

Kuttruff, Heinrich. *Room Acoustics*. Oxon: Spoon Press, 2009.

Laffan, Michael, and Max Weiss, Eds. *Facing Fear: The History of an Emotion in Global Perspective*. Princeton, NJ: Princeton University Press, 2012.

Lampe, Geoffrey W. H., and Henry G. Liddell. *A Patristic Greek Lexicon*. Oxford: Clarendon, 1961.

Larchet, Jean-Claude. "La symbolique spirituelle de l'église selon la *Mystagogie* de Saint Maxime le Confesseur." In *L'espace liturgique: Ses éléments constitutifs et leur sens*, edited by Carlo Braga, 49–63. Rome: Edizione Liturgiche, 2006.

Largier, Niklaus. *In Praise of the Whip: A Cultural History of Arousal*. New York: Zone Books, 2007.

La Riche, William. *Alexandria: The Sunken City*. London: Weidenfeld & Nicolson, 1996.

Larsen, Sven. "A Forerunner of Hagia Sophia." *American Journal of Archeology* 41, no. 1 (January–March 1937): 1–5.

Lehmann, Karl. "The Dome of Heaven." *Art Bulletin* 27, no. 1 (1945): 1–27.

Leibeschuetz, J. H. W. G. *Antioch: City and Imperial Administration in the Later Roman Empire*. Oxford: Clarendon Press, 1972.

Levy, Kenneth. "The Byzantine Sanctus and Its Modal Tradition in East and West." *Annales Musicologiques* 6 (1958–63): 7–67.

Liddell, Henry G., Robert Scott, Henry S. Jones, and Roderick McKenzie. *A Greek-English Lexicon*. Oxford: Clarendon Press, 1996.

Lingas, Alexander. "Festal Cathedral Vespers in Late Byzantium." *Orientalia Christiana Periodica* 63 (1997): 421–59.

——. "From Earth to Heaven: The Changing Musical Soundscape of Byzantine Liturgy." In *Experiencing Byzantium*, edited by C. Nesbitt and M. Jackson, 311–58. Farnham: Ashgate, 2013.

——. "Performance Practice and the Politics of Transcribing Byzantine Chant." *Acta Musicae Byzantinae* 6 (2003): 56–76.

———. "Sunday Matins in the Byzantine Cathedral Rite: Music and Liturgy." PhD diss., University of British Columbia, 1996.

Lopez-Lezcano, Fernando, Travis Skare, Michael J. Wilson, and Jonathan S. Abel. "Byzantium in Bing: Live Virtual Acoustics Employing Free Software." In *Proceedings of the Linux Audio Conference (LAC), Graz, Austria, May 9–12, 2013*. http://lac.linuxaudio.org/2013/papers/52.pdf.

Lot-Borodine, Myrrha. *Un maître de la spiritualité byzantine au XIVe siècle: Nicolas Cabasilas*. Paris: Éditions de l'Orante, 1958.

Lucey, Stephen J. "Art and Socio-Cultural Identity in Early Medieval Rome: The Patrons of Santa Maria Antiqua." In *Roma Felix: Formation and Reflections of Medieval Rome*, edited by Éamonn Ó Carragáin and Carol Neuman de Vegvar, 139–58. Aldershot: Ashgate, 2007.

Macrides, Ruth, and Paul Magdalino. "The Architecture of Ekphrasis: Construction and Context of Paul the Silentiary's Poem on Hagia Sophia." *Byzantine and Modern Greek Studies* 12 (1988): 47–82.

McKenzie, Judith. *The Architecture of Alexandria and Egypt, c. 300 B.C.–A.D. 700*. New Haven, CT: Yale University Press, 2007.

Maguire, Henry. *Art and Eloquence in Byzantium*. Princeton, NJ: Princeton University Press, 1981.

——. "From the Evil Eye to the Eye of Justice." In *Law and Society in Byzantium: Ninth–Twelfth Centuries*, edited by Angeliki E. Laiou and Dieter Simon, 217–40. Cambridge, MA: Harvard University Press, 1994.

——. *The Icons of Their Bodies: Saints and Their Images in Byzantium*. Princeton, NJ: Princeton University Press, 1996.

——. "The Mosaics of Nea Moni: An Imperial Reading." *Dumbarton Oaks Papers* 46 (1992): 205–14.

Mainstone, Rowland. *Hagia Sophia: Architecture, Structure and Liturgy of Justinian's Great Church*. London: Thames & Hudson, 1988.

Majeska, George P. "Notes on the Archeology of St. Sophia at Constantinople: The Green Marble Bands on the Floor." *Dumbarton Oaks Papers* 32 (1978): 299–308.

Mango, Cyril. "Byzantine Writers on the Fabric of Hagia Sophia." In *Hagia Sophia from the Age of Justinian to the Present*, edited Robert Mark and Ahmet Ş. Çakmak, 45–50. Cambridge: Cambridge University Press, 1992.

Manovich, Lev. "The Poetics of Augmented Space." *Visual Communication* 5, no. 2 (2006): 219–40.

Maranci, Christina. "The Humble Heraclius: Revisiting the North Portal at Mren." *Revue des Études Arméniennes* 31 (2008–09): 167–80.

——. "Performance and Church Exterior in Early Medieval Armenia." In *Visualizing Medieval Performance: Perspectives, Histories, Contexts*, edited by Elina Gertsman, 17–32. London: Ashgate, 2008.

Marinis, Vasileios. *Architecture and Ritual in the Churches of Constantinople, Ninth to Fifteenth Centuries*. Cambridge: Cambridge University Press, 2013.

Mark, Robert, and Ahmet Ş. Çakmak, Eds. *Hagia Sophia: From the Age of Justinian to the Present*. Cambridge: Cambridge University Press, 1992.

Mark, Robert, Ahmet Ş. Çakmak, K. Hill, and R. Davidson. "Structural Analysis of Hagia Sophia: A Historical Perspective." *WIT Transactions on the Built Environment* 4 (1993): 867–80.

Marshall, A. Harold, D. Gottlob, and Herbert Alrutz. "Acoustical Conditions Preferred for Ensemble." *Journal of the Acoustical Society of America* 64 (1978): 1437–42.

Markouris, Ioannis. "Apprenticeships in Greek Orthodox Chanting and Greek Language Learning in Venetian Crete (14th–15th Century)." In *I greci durante la Venetocrazia: Uomini, Spazio, Idee (XIII–XVIII); Atti del Convegno internazionale di studi, Venezia, 3–7 dicembre 2007*, edited by Chryssa Maltezou et al., 233–49. Venice: Istituto Ellenico di Studi Bizantini 2009.

Mateos, Juan. *La célébration de la parole dans la liturgie byzantine: Étude historique*. Rome: Pontificium Institutum Orientalium Studiorum, 1971.

Mathews, Thomas F. "Cracks in Lehmann's 'Dome of Heaven.'" *Source* (1982): 12–16.

——. "The Transformation Symbolism in Byzantine Architecture and the Meaning of the Pantokrator in the Dome." In *Church and People in Byzantium: Society for the Promotion of Byzantine Studies, Twentieth Spring Symposium of Byzantine Studies, Manchester 1986*, edited by Rosemary Morris, 200–20. Birmingham: Centre for Byzantine, Ottoman and Modern Greek Studies, University of Birmingham, 1990.

Mazza, Enrico. *Mystagogy: A Theology of Liturgy in the Patristic Age*. Translated by Matthew J. O'Connell. New York: Pueblo Publishing, 1989.

Metzger, Marcel, ed. *Les constitutions apostoliques*. Sources Chrétiennes 320, 329, 336. 3 vols. Paris: Cerf, 1985–87.

Meyendorff, John. *A Study of Gregory Palamas*. Translated by George Lawrence. London: Faith Press, 1964. 2nd ed. Crestwood, NY: St. Vladimir's Seminary Press, 1998.

Mnac'akanyan, Step'an. *Zuart'noc 'ə ev nuynatip hušarjannerə*. Erevan: Haykakan SSH Gitut'yunneri Akademiayi Hratarakč'ut'yun, 1971.

Montefusco, Lucia Calboli. "*Enargeia* et *energeia*: L'évidence d'une demonstration qui signifie les choses en acte." In *Demonstrare: Voir et faire voir; Forme de la démonstration à Rome. Actes du Colloque international de Toulouse, 18–20 novembre 2004*, edited by Mireille Armisen-Marchetti, 43–58. Toulouse: Presses Universitaires du Mirail, 2005.

Moran, Neil K. "The Choir of the Hagia Sophia," *Oriens Christianus* 89 (2005): 1–7.

——. "The Musical 'Gestaltung' of the Great Entrance Ceremony in the 12th Century in Accordance with the Rite of Hagia Sophia." *Jahrbuch der Österreichischen Byzantinistik* 28 (1979): 178–81.

——. *The Ordinary Chants of the Byzantine Mass*. Hamburg: Wagner 1975.

——. *Singers in Late Byzantine and Slavonic Painting*. Leiden: E. J. Brill, 1986.

——. "The Skeuophylakion of the Hagia Sophia." *Cahiers Archéologiques* 34 (1986): 29–32.

Morin, Germain. "Le plus ancien comes ou lectionnaire de l'Église romaine." *Revue Bénédictine* 27 (1910): 41–74.

Mouriki, Doula. *The Mosaics of Nea Moni on Chios*. Translated by Richard Burji. 2 vols. Athens: Commercial Bank of Greece, 1985.

Müller, Swen, and Paulo Massarani. "Transfer-Function Measurement with Sweeps." *Journal of the Audio Engineering Society* 49 (2001): 443–71.

Murphy, Damian T. "Archaeological Acoustic Space Measurement for Convolution Reverberation and Auralization Applications." In *Proceedings of the 9th International Conference on Digital Audio Effects (DAFx-06)*, 221–26. Montreal: McGill University, Schulich School of Music, Music Technology Area, 2006, www.dafx.ca/proceedings/papers/p_221.pdf or http://ant-s4.unibw-hamburg.de/paper-archive/2006/papers/p_221.pdf.

Nakayama, I. "Preferred Time Delay of a Single Reflection for Performers." *Acustica* 54 (1984): 217–21.

Navarro, Jaime, Juan J. Sendra, and Salvador Muñoz. "The Western Latin Church as a Place for Music and Preaching: An Acoustic Assessment." *Applied Acoustics* 70 (2009): 781–89.

Nelson, Robert S. "To Say and to See: Ekphrasis and Vision in Byzantium." In *Visuality before and beyond the Renaissance: Seeing as Others Saw*, edited by Robert S. Nelson, 143–68. Cambridge: Cambridge University Press, 2000.

Odahl, Charles Matson. *Constantine and the Christian Empire*. 2nd ed. London: Routledge, 2010.

Ong, Walter J. *Orality and Literacy: The Technologizing of the Word*. New York: Routledge, 1982.

Osborne, John, J. Rasmus Brandt, and Giuseppe Morganti, eds. *Santa Maria Antiqua al Foro Romano cento anni dopo: Atti del colloquio internazionale, Roma, 5–6 maggio 2000*. Rome: Campisano, 2004.

Ousterhout, Robert. "The Holy Space: Architecture and Liturgy." In *Heaven on Earth: Art and the Church in Byzantium*, edited by Linda Safran, 81–120. University Park, PA: Pennsylvania State University Press, 1998.

Papageorghiou, Athanasios. *Christian Art in the Turkish-Occupied Part of Cyprus*. Nicosia: Holy Archbishopric of Cyprus, 2010.

——. "The Paintings in the Dome of the Church of the Panagia Chryseleousa, Strovolos." In *Medieval Cyprus: Studies in Art, Architecture and History in Memory of Doula Mouriki*, edited by Christopher Moss and Nancy Patterson Ševcčenko, 147–61. Princeton, NJ: Princeton University Press, 1999.

Parenti, Stefano. “The Cathedral Rite of Constantinople: Evolution of a Local Tradition.” *Orientalia Christiana Periodica* 77 (2011): 449–69.

Patrich, Joseph. “The Transfer of Gifts in the Early Christian Churches of Palestine: Archaeological and Literary Evidence for the Evolution of the ‘Great Entrance.’” In *Pèlerinages et lieux saints dans l’Antiquité et le Moyen Âge: Mélanges offerts à Pierre Maraval*, edited by Béatrice Caseau-Chevallier, Jean-Claude Cheynet, and V. Déroche, 347–50. Centre de Recherche d’Histoire et Civilisation de Byzance Monographies 23. Paris: Association des Amis du Centre d’Histoire et Civilisation de Byzance, 2006.

Pätynen, Jukka, and Tapio Lokki. “Evaluation of Concert Hall Auralization with Virtual Symphony Orchestra.” In *Proceedings of the International Symposium on Room Acoustics (ISRA)*, Melbourne, Australia, August 29–31, 2010. Melbourne: Acoustical Society, 2010, www.odeon.dk/pdf/isra2010_patynen-lokki.pdf.

Pennington, Anne Elizabeth. “Seven Akolouthiai from Putna.” In *Studies in Eastern Chant*, edited by Egon Wellesz and Miloš Velimirović, 4:112–334. Oxford: Oxford University Press 1979.

Pentcheva, Bissera V. “The Aesthetics of Landscape and Icon at Sinai.” *RES: Anthropology and Aesthetics* 65–66 (2015): 192–209.

——. “Epigrams on Icons.” In *Art and Text in Byzantine Culture*, edited by Liz James, 120–38. Cambridge: Cambridge University Press, 2007.

——. “Hagia Sophia and Multisensory Aesthetics.” *Gesta* 50, no. 2 (2011): 93–111.

——. *Hagia Sophia: Sound, Space and Spirit in Byzantium*. University Park, PA: Pennsylvania State University Press, forthcoming (2017).

———. *Icons and Power: The Mother of God in Byzantium*. University Park, PA: Pennsylvania State University Press, 2006.

———. “In-Spiriting in the Byzantine Consecration (*Kathierosis*) Rite.” *Codex Aquilarensis* 30 (2014): 37–65.

——. “Mirror, Inspiration, and the Making of Art in Byzantium,” *Convivium* 1, no. 2 (2014): 10–39.

——. “Performing the Sacred in Byzantium: Image, Breath, and Sound.” *Performance Research International* 19, no. 3 (2014): 120–28.

——. *The Sensual Icon: Space, Ritual, and the Senses in Byzantium*. University Park, PA: Pennsylvania State University Press, 2010.

Piguet-Panayotova, Dora. “Recherches sur les tetraconques à déambulatoire et leur décor en Transcaucasie au VIIe siècle.” *Oriens Christianus* 73 (1989): 166–212.

Pintaudi, Rosario. *Papyri Graecae Wessely Pragenses (PPrag. II)*. Florence: Gonnelli, 1995.

Plamper, Jan, and Benjamin Lazier. *Fear across the Disciplines*. Pittsburgh, PA: University of Pittsburgh Press, 2012.

Plested, Marcus. *Orthodox Readings of Aquinas: Changing Paradigms in Historical and Systematic Theology*. Oxford: Oxford University Press, 2012.

Plontke-Lüning, Annegret. *Frühchristliche Architektur in Kaukasien*. Vienna: Verlag der Österreichischen Akademie der Wissenschaften, 2007.

Polemis, Ioannis. “Nikolaos Kabasilas’s De Vito in Christo and Its Context.” *Byzantinische Zeitschrift* 106 (2013): 101–32.

Pott, Thomas. *La réforme liturgique byzantine: Étude du phénomène de évolution non-spontanée de la liturgie byzantine*. Rome: CLV–Edizioni Liturgiche, 2000. Translated by Paul Metendorff as *Byzantine Liturgical Reform: A Study of Liturgical Change in the Byzantine Tradition*. Crestwood, NY: St. Vladimir’s Orthodox Seminary Press, 2010.

Pym, Anne. “Writing Oral Traditions: Clash between Worlds.” *Review of Communication* 7 (2007): 303–10.

Raasted, Jørgen. *Intonation Formulas and Modal Signatures in Byzantine Musical Manuscripts*. Copenhagen: Munksgaard, 1966.

——. “Koukouzeles’ Revision of the Sticherarion and Sinai gr. 1230.” In *Laborare Fratres in Unum: Festschrift László Dobszay zum 60. Geburtstag*, edited by Janka Szandrei et al. Hildesheim: Weidmann, 1995.

——. "Koukouzeles' Sticherarion." In *Byzantine Chant: Tradition and Reform; Acts of a Meeting Held at the Danish Institute in Athens, 1993*, edited by Christian Troelsgård, 9–21. Athens: Danish Institute at Athens, 1997.

Raczka, Gary Philippe. "The Lectionary at the Time of Saint John Chrysostom." PhD diss., University of Notre Dame, 2015.

Renoux, Athanase (Charles). "Le Codex Armenien Jerusalem 121." In *Patrologia Orientalis* 168, 36, no. 2. Turnhout: Brepols, 1971.

——. "De Jérusalem en Arménie: L'héritage liturgique de l'Église arménienne." In *Patterns of the Past, Prospects for the Future: The Christian Heritage in the Holy Land*, edited by Thomas Hummel, Kevork Hintlian, and Ulf Carmesund, 114–23. London: Melisende, 1999.

Rettinger, Michael. "Note on Reverberation Chambers." *Journal of the Audio Engineering Society* 5 (1957): 108.

Reznikoff, Igor. "The Evidence of the Use of Sound Resonance from Paleolithic to Medieval Times." In *Archaeoacoustics*, edited by Chris Scarre and Graeme Lawson, 77–84. Cambridge: McDonald Institute for Archaeological Research, 2006.

Rhoby, Andreas. *Byzantinische Epigramme auf Fresken und Mosaiken*. Vienna: Verlag der Österreichischen Akademie der Wissenschaften, 2009.

Riegl, Alois. *The Group Portraiture of Holland*. Translated by Evelyn M. Kain and David Britt. Los Angeles: Getty Research Center for the History of Art and the Humanities, 1999.

Rindel, Jens. "Echo Problems in Ancient Theatres and a Comment on 'Sounding Vessels' Described by Vitruvius." In *The Acoustics of Ancient Theatres Conference*, University of Patras, Patras, September, 18–21, 2011. www.odeon.dk/pdf/AA2011_9_Rindel.pdf.

Rindel, Jens Holger, Claus Lynge Christensen, and George Koutsouris. "Simulations, Measurements and Auralisations in Architectural Acoustics." In *Acoustics 2013 New Delhi*, New Delhi, India, November 10–15, 2013. www.odeon.dk/pdf/Acoustics%202013%20Rindel.pdf.

Roberts, Michael. *The Jeweled Style: Poetry and Poetics in Late Antiquity*. Ithaca, NY: Cornell University Press, 1989.

Safran, Linda. *Heaven on Earth: Art and the Church in Byzantium*. University Park, PA: Pennsylvania State University Press, 1998.

Sargsyan, Minas. "Mreni Tačarə Himnaderneri Patkerak'andaknerə" [The images of founders on the church of Mren]. *Patma–Banasirakan Handes* 35, no. 4 (1966): 241–50.

Scarre, Chris, "Sound, Place and Space: Towards an Archaeology of Acoustics." In *Archaeoacoustics*, edited by Chris Scarre and Graeme Lawson, 1–10. Cambridge: McDonald Institute for Archaeological Research, 2006.

Scarre, Chris, and Graeme Lawson, eds. *Archaeoacoustics*. Cambridge: McDonald Institute for Archaeological Research, 2006.

Schafer, R. Murray. *The Soundscape: Our Sonic Environment and the Tuning of the World*. Rochester, VT: Destiny Books, 1994.

Schibille, Nadine. *Hagia Sophia and the Byzantine Aesthetic Experience*. Farnham: Ashgate, 2014.

——. "The Profession of the Architect in Late Antique Byzantium." *Byzantion* 79 (2009): 27–46.

Schneider, Wolfgang Christian, and Rudolf H. W. Stichel. "Der 'Cherubinische Einzug' in der Hagia Sophia Justinians." In *Performativität und Ereignis*, edited by Erika Fischer-Lichte, 377–94. Tübingen: A. Francke, 2003.

Schulz, Hans-Joachim. *Die Byzantinische Liturgie: Vom Werden Ihrer Symbolgestalt*. Freiburg im Breisgau: Lambertus-Verlag, 1964. 2nd ed. Trier: Paulinus, 1980. 3rd ed. Trier: Paulinus, 2000. Translated as *The Byzantine Liturgy: Symbolic Structure and Faith Expression*. New York: Pueblo Publishing, 1986.

Serinus, Jason Victor. "Cappella Romana: Time-Travel to Constantinople." *San Francisco Classical Voice*, February 1, 2013.

Skarlakidis, Haris. *Holy Fire: The Miracle of the Light of the Resurrection at the Tomb of Jesus: Seventy Historical Accounts (4th–16th c.)*. Athens: Elea Publishing, 2011.

Small, Jocelyn Penny. *Wax Tablets of the Mind: Cognitive Studies of Memory and Literacy in Classical Antiquity*. London: Routledge, 1997.
Smith, Mark M., Ed. *Hearing History: A Reader*. Athens, GA: University of Georgia Press, 2004.
Spinks, Bryan D. *Do This in Remembrance of Me: The Eucharist from the Early Church to the Present Day*. Norwich: SCM Press, 2013.
——. *The Sanctus in the Eucharistic Prayer*. Cambridge: Cambridge University Press, 1991.
Spyrakou, Evangelia. *Οι χοροί ψαλτών κατά την Βυζαντινή π αράδοση* [The choirs of singers according to the Byzantine tradition]. Athens: Institute for Byzantine Musicology, 2008.
Stathes, Gregorios. *Heē exeēgeēsis teēs palaias vyzantineēs seēmeiographias: kai ekdosis anoōnymou syngrapheēs tou koōdikos Xeēropotamou 357 hoōs kai epilogeēs teēs Mousikeēs techneēs tou Apostolou Koōnsta Chiou ek tou koōdikos Docheiariou 389* [Interpretation of Paleo-Byzantine Notation and the Edition of the Anonymous Composition of the Codex Xeropotamou 357 and a Selection of the Music of Apostolos Kosta of Chios from the Codex Docheiariou 389]. Athens: Institute for Byzantine Musicology, 1978.
Sterne, Jonathan. "Sonic Imaginations." In *The Sound Studies Reader*, edited by Jonathan Sterne, 1–17. Abingdon, UK: Routledge University Press, 2012.
——. "Space within Space: Artificial Reverb and the Detachable Echo." *Grey Room* 60 (2015): 110–31.
——, Ed. *The Sound Studies Reader*. Abingdon, UK: Routledge University Press, 2012.
Stichel, Rainer. "Die musizierenden Hirten von Bethlehem." In *Beiträge zum Symposion zur Byzantinischen Lexikographie*, edited by Wolfgang Hörandner et al., 249–82. Vienna: Verlag der Österreichische Akademie der Wissenschaften, 1989.
Stichel, Rudolph, and Helge Svenshon. *Einblicke in den virtuellen Himmel: Neue und alte Bilder vom Inneren der Hagia Sophia in Istanbul*. Tübingen: Ernst Wasmuth Verlag, 2008.
Strunk, Oliver. "The Antiphons of the Oktoechos." In *Essays on Music in the Byzantine World*, edited by Oliver Strunk, 165–90. New York: W. W. Norton, 1977.
——. "The Byzantine Office at Hagia Sophia." *Dumbarton Oaks Papers* 9–10 (1956): 175–202.
——. "Melody Construction in Byzantine Chant." In *Essays on Music in the Byzantine World*, edited by Oliver Strunk, 191–201. New York: W. W. Norton, 1977.
——. "P. Lorenzo Tardo and His 'Ottoeco nei mss. Melurgici.'" In *Essays on Music in the Byzantine World*, edited by Oliver Strunk, 255–67. New York: W. W. Norton, 1977.
——. "S. Salvatore di Messina and the Musical Tradition of Magna Graecia" and "The Byzantine Office at Hagia Sophia." In *Essays on Music in the Byzantine World*, edited by Oliver Strunk, 45–54 and 112–50. New York: W. W. Norton, 1977.
——. *Specimina notationum antiquiorum*. Monumenta Musicae Byzantinae, Subsidia, vol. 7. Copenhagen: Ejnar Munksgaard, 1966. Pars suppletoria, 1–7.
Strzygowski, Josef. *Die Baukunst der Armenier und Europa*. Vienna: Anton Schroll, 1918.
Stylianou, Andreas, and Judith Stylianou. *The Painted Churches of Cyprus: Treasures of Byzantine Art*. London: Trigraph for the A. G. Leventis Foundation, 1985.
Suárez, Rafael, Alicia Alonso, and Juan J. Sendra. "Intangible Cultural Heritage: The Sound of the Romanesque Cathedral of Santiago de Compostela." *Journal of Cultural Heritage* 16 (2015): 239–43.
Svenshon, Helge Olaf. "Heron of Alexandria and the Dome of Hagia Sophia in Istanbul." *Proceedings of the Third International Congress on Construction History: Brandenburg University of Technology Cottbus, Germany, 20th–24th May 2009*, edited by Karl-Eugen Kurrer, Werner Lorenz, and Volker Wetzk, 1387–94. Cottbus: Brandenburgische Technische Universität, 2009.
Taft, Robert F. *The Byzantine Rite: A Short History*. Collegeville, MN: Liturgical Press, 1992.
——. "Christian Liturgical Psalmody: Origins, Development and Decomposition, Collapse." In *Psalms in Community: Jewish and Christian Textual, Liturgical, and Artistic Traditions*, edited by Harold Attridge and Margot Fassler, 7–32. Society of Biblical Literature Symposium Series 25. Leiden: E. J. Brill; Atlanta, GA: Society of Biblical Literature, 2003.

——. *The Great Entrance: A History of the Transfer of Gifts and Other Preanaphoral Rites of the Liturgy of St. John Chrysostom*. Rome: Pontificium Institutum Orientalium Studiorum, 1975.

——. *A History of the Liturgy of St. John Chrysostom: The Precommunion Rites*. Rome: Pontificio Istituto Orientale, 2000.

——. "Is the Liturgy Described in the Mystagogia of Maximus Confessor Byzantine, Palestinian, or Neither?" *Bollettino della Badia Greca di Grottaferrata* 8 (2011): 233–70.

——. "The Liturgy of the Great Church: An Initial Synthesis of Structure and Interpretation on the Eve of Iconoclasm." *Dumbarton Oaks Papers* (1980–81): 45–75.

——. "St. John Chrysostom and the Byzantine Anaphora That Bears His Name." In *Essays in Early Eastern Eucharistic Prayers*, edited by Paul Bradshaw, 195–226. Collegeville, MN: Liturgical Press, 1997.

——. "The *Synaxarion* of Evergetis in the History of Byzantine Liturgy." In *The Theotokos Evergetis and Eleventh-Century Monasticism: Papers of the Third Belfast Byzantine Colloquium*, edited by Margaret Mullett and Anthony Kirby, 274–93. Belfast: Belfast Byzantine Enterprises, 1994.

Teteriatnikov, Natalia. "Private Salvation Programs and Their Effect on Byzantine Church Decoration." *Arte Medievale* 7 (1993): 47–63.

Thierry, Jean-Michel, and Nicole Thierry. "La cathédrale de Mren et sa decoration." *Cahiers Archéologiques* 21 (1971): 43–77.

Thierry, Nicole. "Héraclius et la vraie croix en Arménie." In *From Byzantium to Iran: Armenian Studies in Honour of Nina G. Garsoïan*, edited by Jean-Pierre Mahé and Robert W. Thomson, 165–86. Atlanta, GA: Scholars Press, 1997

Thodberg, Christian. *Der byzantinische Alleluiarionzyklus: Studien im kurzen Psaltikonstil.* Copenhagen: Monumenta Musicae Byzantinae Subsidia, 1966.

Thodberg, Christian, and Holger Hamann. *Der byzantinische Alleluiarionzyklus: Studien im kurzen Psaltikonstil*. Copenhagen: E. Munksgaard, 1966.

Thompson, Emily Ann. *The Soundscape of Modernity: Architectural Acoustics and the Culture of Listening in America, 1900–1933*. Cambridge, MA: MIT Press, 2002.

Thomson, Robert W. "Jerusalem and Armenia." In *Papers of the 1983 Oxford Patristic Conference, Kalamazoo*, edited by Elizabeth A. Livingstone, 77–91. Leuven: Peeters, 1985.

Thunberg, Lars. *Man and the Cosmos: The Vision of St. Maximus the Confessor*. Crestwood, NY: St. Vladimir's Seminary Press, 1985.

Thunø, Erik. *Image and Relic: Mediating the Sacred in Early Medieval Rome. Analecta Romana Studi Danici 32*. Rome: L'Erma di Bretschneider, 2002.

Thurn, Hans, Ed. *Comes Romanus Wirziburgensis: Faksimileausgabe des Codex M. p. th. f. 62 der Universitätsbibliothek Würzburg*. Graz: Akademische Druck, 1968.

Timken-Matthews, Jane, "Pantocrator: Title and Image." PhD diss., New York University, 1980.

Touliatos, Diane. "Nonsense Syllables in the Music of the Ancient Greek and Byzantine Traditions." *Journal of Musicology* 7, no. 2 (1989): 231–43.

Treu, Kurt, and Johannes M. Diethart. *Griechische Literarische Papyri Christlichen Inhaltes II*. Vienna: Österreichische Nationalbibliothek, 1993.

Troelsgård, Christian. *Byzantine Neumes: A New Introduction to the Middle Byzantine Musical Notation*. Copenhagen: Museum Tusculanum Press, 2011.

——. "The Exaposteilaria Anastasima with Round Notation in MS Athos, Iberon 953." In *Studi di musica bizantina in onore di Giovanni Marzi*, edited by Alberto Doda, 15–28. Studi e Testi Musicali, n.s., 6. Lucca: Libreria Musicale Italiana, 1995.

——. "A List of Sticheron Call-Numbers of the Standard Abridged Version of the Sticherarion." *Cahiers de l'Institut du Moyen-Âge Grec et Latin* 74 (2003): 3–20.

——. "Melodic Variation in the 'Marginal' Repertories of Byzantine Musical MSS, Exemplified by Apolytikia/Kontakia and Exaposteilaria Anatasima." In *Cantus Planus: Papers Read at the 7th Meeting, Sopron, Hungary, 1995/International Musicological Society Study Group*, edited by Laszlo Dobszay, 601–09. Budapest: Hungarian Academy of Sciences Institute for Musicology, 1998.

——. "A New Source for the Early Octoechos? Papyrus Vindobonensis G. 19.934 and Its Musical Implications." In *Byzantine Musical Culture: Proceedings of the 1st International Conference of the American Society of Byzantine Music and Hymnology*, 2008, 668–79.

——. "The Prokeimena in Byzantine Rite: Performance and Tradition." In *Cantus Planus: Papers Read at the 6th Meeting, Eger, Hungary*, 65–77. Budapest: Hungarian Academy of Sciences, 1995.

Tronchin, Lamberto, and Angelo Farina. "Acoustics of the Former Teatro 'La Fenice' in Venice." *Journal of the Audio Engineering Society* 45, no. 12 (December 1997): 1051–62.

Tronzo, William. "Mimesis in Byzantium: Notes toward a History of the Function of the Image." *RES: Anthropology and Aesthetics* 25 (Spring 1994): 61–76.

Tsirpanlis, Constantine N. *The Liturgical and Mystical Theology of Nicolas Cabasilas*. New York: Tsirpanlis, 1979.

Tversky, Barbara. "Spatial Cognition." In *The Cambridge Handbook of Situated Cognition*, edited by Philip Robbins and Murat Aydede, 201–16. Cambridge: Cambridge University Press, 2012.

Ueno, Kanako, Kosuke Kato, and Keiji Kawai. "Effect of Room Acoustics on Musicians' Performance. Part I: Experimental Investigation with a Conceptual Model." *ACTA Acustica United with Acustica* 96 (2010): 505–15.

Väänänen, Veikko. *Le journal-épître d'Egérie (Itinerarium Egeriae): Étude linguistique*. Helsinki: Suomalainen Tiedeakatemia, 1987.

Välimäki, Vesa, Julian D. Parker, Lauri Savioja, Julius O. Smith, and Jonathan S. Abel. "Fifty Years of Artificial Reverberation." *IEEE Transactions on Audio, Speech, and Language Processing* 20, no. 50 (July 2012): 1421–48.

van de Paverd, Frans. *St John Chrysostom, the Homilies on the Statues: An Introduction*. Orientalia Christiana Analecta 239. Rome: Pontificium Institutum Orientalium Studiorum, 1991.

——— *Zur Geschichte der Messliturgie in Antiocheia und Konstantinopel gegen Ende des vierten Jahrhunderts: Analyse der Quellen bei Johannes Chrysostomos*. Rome: Pontificium Institutum Orientalium Studiorum, 1970.

Van Dijk, S. J. P. "The Medieval Easter Vespers of the Roman Clergy." *Sacris Erudiri* 19 (1969–70): 261–363.

van Haelst, Joseph. "Cinq textes provenant de Khirbet Mird." *Ancient Society* 22 (1991): 297–317.

Velimirović, Milos. "The Byzantine Heirmos and Heirmologion." In *Gattungen der Musik in Einzeldarstellungen: Gedenkschrift für Leo Schrade*, edited by Wulf Arlt et al., 192–244. Bern: Francke, 1973.

Vörlander, Michael. *Auralization: Fundamentals of Acoustics, Modelling, Simulation, Algorithms and Acoustic Virtual Reality*. Berlin: Springer-Verlag, 2008.

Wanek, Nina-Maria. *Sticheraria in spät und postbyzantinischer Zeit: Untersuchungen Anhand der Stichera für August*. Vienna: Praesens Verlag, 2013.

Webb, Matilda. *The Churches and Catacombs of Early Christian Rome: A Comprehensive Guide*. Brighton: Sussex Academic Press, 2001.

Webb, Ruth. "The Aesthetics of Sacred Space: Narrative, Metaphor and Motion in Ekphraseis of Church Buildings." *Dumbarton Oaks Papers* 53 (1999): 59–74.

——. *Ekphrasis, Imagination and Persuasion in Ancient Rhetorical Theory and Practice*. Farnham, UK: Ashgate, 2009.

Weitze, Christoffer Andres, Jens Holger Rindel, Claus Lynge Christensen, and Anders Christian Gade. "Computer Simulation of the Acoustics of Mosques and Byzantine Churches." In *Proceedings of the 17th International Congress on Acoustics*, Rome, Italy, September 2–7, 2001.

——. "The Acoustical History of Hagia Sophia Revived through Computer Simulation." In *Proceedings of Forum Acusticum, Seville, Spain, 2002*. Stuttgart: Hirzel, 2002.

Weitzmann, Kurt, and John Galey. *The Monastery of Saint Catherine at Mount Sinai, the Icons*. Princeton, NJ: Princeton University Press, 1976.

Wellesz, Egon. *A History of Byzantine Music and Hymnography*. Oxford: Clarendon Press, 1961.

——. "Words and Music in Byzantine Liturgy." *Musical Quarterly* 33 (1947): 297–310.

Wescoat, Bona, and Robert Ousterhout, Eds. *Architecture of the Sacred: Space, Ritual, and Experience*. Cambridge: Cambridge University Press, 2012.

Whitby, Mary. "From Moschus to Nonnus: The Evolution of Nonnian Style." In *Studies in the Dionysiaca of Nonnus*, edited by Neil Hopkinson, 99–155. Cambridge: Cambridge Philological Society, 1994.

Williams, Edward V. "John Koukouzeles' Reform of Byzantine Chanting for Great Vespers in the Fourteenth Century." PhD diss., Yale University, 1968.

Winfield, David. "Hagios Chrysostomos, Trikomo, Asinou: Byzantine Painters at Work." In *Praktika tou protou diethnous Kyprologikou synedriou, Leukosia, 1969*, 285–91. Nicosia: Hetaireia Kypriakōn Spoudon, 1972.

Winter, Irene J. "Agency Marked, Agency Ascribed: The Affective Object in Ancient Mesopotamia." In *Art's Agency and Art History*, edited by Robin Osborne and Jeremy Tanner, 42–69. Oxford: Blackwell Publishing, 2007.

Woszczyk, Wieslaw, and William L. Martens. "Evaluation of Virtual Acoustic Stage Support for Musical Performance." In *Proceedings of Acoustics '08*, Paris, France, June 29–July 4, 2008. www.conforg.fr/acoustics2008/cdrom/data/articles/000968.pdf.

Wybrew, Hugh. *The Orthodox Liturgy: The Development of the Eucharistic Liturgy in the Byzantine Rite*. Crestwood, NY: St. Vladimir's Seminary Press, 1990.

Yarnold, Edward. *The Awe-Inspiring Rites of Initiation: Baptismal Homilies of the Fourth Century*. Slough: St. Paul Publications, 1972. Rev. ed. Edinburgh: T&T Clark, 1994.

Young, Frances M. *Biblical Exegesis and the Formation of Christian Culture*. Cambridge: Cambridge University Press, 1997.

——. *The Use of Sacrificial Ideas in Greek Christian Writers from the New Testament to John Chrysostom*. Cambridge, MA: Philadelphia Patristic Foundation, 1979. 2nd ed. Eugene, OR: Wipf & Stock, 2004.

Index

Bold indicates glossary entries and italics indicate figures

For Product Safety Concerns and Information please contact our EU
representative GPSR@taylorandfrancis.com
Taylor & Francis Verlag GmbH, Kaufingerstraße 24, 80331 München, Germany

www.ingramcontent.com/pod-product-compliance
Ingram Content Group UK Ltd.
Pitfield, Milton Keynes, MK11 3LW, UK
UKHW050754180926
475701UK00007B/1051

* 9 7 8 0 3 6 7 2 3 1 8 4 2 *